Intuitive Biostatistics

Intuitive Biostatistics

A Nonmathematical Guide to Statistical Thinking

HARVEY MOTULSKY, M.D.

GraphPad Software, Inc.

THIRD EDITION

New York   Oxford

OXFORD UNIVERSITY PRESS

Oxford University Press is a department of the University of Oxford.
It furthers the University's objective of excellence in research, scholarship,
and education by publishing worldwide.

Oxford   New York
Auckland   Cape Town   Dar es Salaam   Hong Kong   Karachi
Kuala Lumpur   Madrid   Melbourne   Mexico City   Nairobi
New Delhi   Shanghai   Taipei   Toronto

With offices in
Argentina   Austria   Brazil   Chile   Czech Republic   France   Greece
Guatemala   Hungary   Italy   Japan   Poland   Portugal   Singapore
South Korea   Switzerland   Thailand   Turkey   Ukraine   Vietnam

Copyright © 2014 by Oxford University Press.

Published by Oxford University Press.
198 Madison Avenue, New York, NY 10016
www.oup.com

Oxford is a registered trademark of Oxford University Press.

Library of Congress Cataloging-in-Publication Data

ISBN 978-0-19-994664-8

Printing number: 9 8 7 6 5 4 3 2 1

Printed in the United States of America
on acid-free paper

I dedicate this book to my wife Lisa, to my kids (Wendy, Nat, Naomi and Ruby), to readers who encouraged me to continue with a third edition, and to future scientists who I hope will avoid common mistakes in biostatistics.

PRAISE FOR INTUITIVE BIOSTATISTICS

Intuitive Biostatistics is a beautiful book that has much to teach experimental biologists of all stripes. Unlike other statistics texts I have seen, it includes extensive and carefully crafted discussions of the perils of multiple comparisons, warnings about common and avoidable mistakes in data analysis, a review of the assumptions that apply to various tests, an emphasis on confidence intervals rather than P values, explanations as to why the concept of statistical significance is rarely needed in scientific work, and a clear explanation of nonlinear regression (commonly used in labs; rarely explained in statistics books).

In fact, I am so pleased with *Intuitive Biostatistics* that I decided to make it the reference of choice for my postdoctoral associates and graduate students, all of whom depend on statistics, and most of whom need a closer awareness of precisely why. Motulsky has written thoughtfully, with compelling logic and wit. He teaches by example what one may expect of statistical methods and perhaps just as importantly, what one may not expect of them. He is to be congratulated for this work, which will surely be valuable and perhaps even transformative for many of the scientists who read it.

—Bruce Beutler, 2011 Nobel Laureate, Physiology or Medicine
Director, Center for the Genetics of Host Defense,
UT Southwestern Medical Center

GREAT FOR SCIENTISTS

This splendid book meets a major need in public health, medicine, and biomedical research training — a user-friendly biostatistics text for non-mathematicians that clearly explains how to make sense of statistical results, how to avoid common mistakes in data analysis, and how to avoid being confused by statistical nonsense. You may enjoy statistics for the first time!

—Gilbert S. Omenn, Professor of Medicine,
Genetics, Public Health, and Computational Medicine &
Bioinformatics, University of Michigan

I am entranced by the book. Statistics is a topic that is often difficult for many scientists to fully appreciate. The writing style and explanations of *Intuitive Biostatistics* makes the concepts accessible. I recommend this text to all researchers. Thank you for writing it.

—Tim Bushnell, Director of Shared Resource Laboratories,
Univ. Rochester Med. Center

GREAT FOR STUDENTS

After struggling with books that weren't right for my class, I was delighted to find *Intuitive Biostatistics*. It is the best starting point for undergraduate students seeking to learn the fundamental principles of statistics because of its unique presentation of the important concepts behind statistics. Lots of books give you the "recipe" approach, but only *Intuitive Biostatistics* explains what it all means. It meticulously goes through common mistakes and shows how to correctly choose, perform, and interpret the proper statistical test. It is accessible to new learners without being condescending.

—Beth Dawson,
The University of Texas at Austin

This textbook emphasizes the thinking needed to interpret statistical analysis in published research over knowledge of the mathematical underpinnings. The basics of choosing tests and doing simpler analyses are covered very clearly and simply. The language is easy to understand yet accurate. It brings in the higher level of intuitive understanding that we hope students will have at the end of an honors undergraduate or MSc. program, skipping over the mathematical details that are now handled by software anyway. It is the prefect approach and level for undergraduates beginning research.

—Janet E. Kübler, Biology Dept.,
California State University at Northridge

GREAT FOR EVERYONE

I've read several statistics books, but found that some concepts I was interested in were not mentioned and other concepts were hard to understand. You can ignore the "bio" in *Intuitive Biostatistics,* as it is the best applied statistics books I have come across, period. Its clear, straightforward explanations have allowed me to better understand research papers and select appropriate statistical tests. Highly recommended.

—Ariel H. Collis, Economist,
Georgetown Economic Services

BRIEF CONTENTS

CONTENTS

PART E   **Challenges in Statistics**

PART F   **Statistical Tests**

PREFACE

My approach in this book is informal and brisk (at least I hope
it is), not ceremonious and plodding (at least I hope it isn't).

JOHN ALLEN PAULOS (2008)

Intuitive Biostatistics provides a comprehensive overview of statistics without getting bogged down in the mathematical details. I've been gratified to learn that many people have found my approach refreshing and useful. Some scientists have told me that statistics had always been baffling until they read *Intuitive Biostatistics*. This enthusiasm encouraged me to write this third edition.

Who is this book for?
I wrote *Intuitive Biostatistics* for three main audiences:

- Medical (and other) professionals who want to understand the statistical portions of journals they read. These readers don't need to analyze any data, but they do need to understand analyses published by others. I've tried to explain the big picture without getting bogged down in too many details.
- Undergraduate and graduate students, postdocs, and researchers who analyze data. This book explains general principles of data analysis, but it won't teach you how to do statistical calculations or how to use any particular statistical program. It makes a great companion to the more traditional statistics texts and to the documentation of statistical software.
- Scientists who consult with statisticians. Statistics often seems like a foreign language, and this text can serve as a phrase book to bridge the gap between scientists and statisticians. Sprinkled throughout the book are "Lingo" sections that explain statistical terminology and point out when ordinary words are given very specialized meanings (the source of much confusion).

I wrote *Intuitive Biostatistics* to be a guidebook, not a cookbook. The focus is on how to interpret statistical results, rather than how to analyze data. This book presents few details of statistical methods and only a few tables required to complete the calculations.

Nonmathematical?

Statistics is a branch of mathematics, so you can't truly understand statistics without studying many equations. In fact, some aspects of statistics cannot be understood unless you first master calculus and matrix algebra. Don't despair. This book is for the many students and scientists who may find math confusing and prefer verbal explanations. I only include equations in a few places as a way to understand an idea or to explain how to do some simple calculations by hand. Compared with most statistics books, this one uses very little math.

Statistical methods are tools, and scientists use lots of tools they don't always fully understand. For example, one can effectively use a radioactively labeled compound in an experiment without really understanding the nuclear physics of radioactive decay. You can do experiments with chemical reagents that you don't know how to synthesize. You can use scientific instruments without understanding exactly how they work. Similarly, you can interpret the results calculated by statistical software without understanding how those tests were derived and without having studied mathematical proofs that they work. But you do have to know the assumptions the tests are based on, and be alert to the many conceptual traps scientists commonly fall into.

Statistical lingo

In Lewis Carroll's *Through the Looking Glass*, Humpty Dumpty says, "When I use a word, it means exactly what I say it means—neither more nor less" (Carroll, 1871). Lewis Carroll (the pseudonym for Charles Dodgson) was a mathematician, and it seems as though he was thinking of statisticians when he wrote that line. Well it would, except for the inconvenient fact that little statistical terminology had been invented by 1871!

Statistics books can get especially confusing when they use words and phrases that have both ordinary meanings and technical meanings. The problem is that you may think the author is using the ordinary meaning of a word or phrase, when in fact the author is using that word or phrase as a technical term with a very different meaning. I try hard to point out these potential ambiguities when I use potentially confusing terms such as:

- Significant
- Error
- Hypothesis
- Model
- Power
- Variance
- Residual
- Normal
- Independent
- Sample
- Population

- Fit
- Confidence
- Distribution
- Control

What makes the book unique?
I include many topics often omitted from short introductory texts, including:

- How common sense can mislead. Chapter 1 is a fun chapter that explains how common sense can lead you astray and why we therefore need to understand statistical principles.
- Multiple comparisons. It is simply impossible to understand statistical results without a deep understanding of how to think about multiple comparisons. This isn't just a practical issue, but almost a philosophical issue in analyzing data. Chapters 22, 23, and 40 are devoted to this topic. I explain several approaches used to deal with multiple comparisons, including the false discovery rate (FDR).
- Nonlinear regression. In many fields of science, nonlinear regression is used more often than linear regression, but most introductory statistics books ignore nonlinear regression completely. This book gives them equal weight. Chapters 34 and 35 set the stage by explaining the concept of fitting models to data and comparing alternative models. Chapter 36 then discusses nonlinear regression.
- Bayesian logic. Bayesian thinking is briefly mentioned in Chapter 2 and is then explored in Chapter 18 as a way to interpret a finding that a comparison is statistically significant. This topic returns in Chapter 42, which compares interpreting statistical significance to interpreting the results of clinical laboratory tests. These are only brief introductions to Bayesian thinking. This book is about conventional (Frequentist) statistics, and only briefly introduces Bayesian approaches to data analysis.
- Lognormal distributions. These are commonly found in scientific data, but not in statistics books. They are explained in Chapter 11 and are touched upon again in several examples that appear in later chapters. Logarithms and antilogarithms are reviewed in Appendix E.
- Testing for equivalence. Sometimes the goal is not to prove that two groups differ, but rather to prove that they are the same. This requires a different mindset, as explained in Chapter 21.
- Normality tests. Many statistical tests assume data are sampled from a Gaussian (also called normal) distribution, and normality tests are used to test this assumption. Chapter 24 explains why these tests are less useful than many hope.
- Outliers. Values far from the other values in a set are called outliers. Chapter 25 explains how to think about outliers.

- Comparing the fit of alternative models. Statistical hypothesis testing is usually viewed as a way to test a null hypothesis. Chapter 35 explains an alternative way to view statistical hypothesis testing as a way to compare the fits of alternative models.
- Meta-analysis as a way to reach conclusions by combining data from several studies. This topic is the subject of new chapter (Chapter 43).
- Detailed review of assumptions. All analyses are based on a set of assumptions, and many chapters discuss these assumptions in depth.
- Lengthy discussion of common mistakes in data analysis. Most chapters include lists (with explanations) of common mistakes and misunderstandings.

To make space for these topics, I have left out many topics that are traditionally included in introductory texts:

- Probability. I assume that you have at least a vague familiarity with the ideas of probability, and this book does not explain these principles in much depth. I have added a new chapter (Chapter 2) to this edition that explains why probability can seem confusing. But you can still understand the rest of the book even if you skip this chapter.
- Equations needed to compute statistical tests. I assume that you will be either interpreting data analyzed by others or using statistical software to run statistical tests. In only a few places do I give enough details to compute the tests by hand.
- Statistical tables. If you aren't going to be analyzing data by hand, there is very little need for statistical tables. I include only a few tables in places where it might be useful to do simple calculations by hand.
- Statistical distributions. You can choose statistical tests and interpret the results without knowing much about z, t, and F distributions. This book mentions them but goes into very little depth.

A unique organization

The organization of the book is unique.

Part A has three introductory chapters. Chapter 1 explains how "common sense" can mislead us when thinking about probability and statistics. Chapter 2 briefly explains some of the complexities of dealing with probability. Chapter 3 explains the basic idea of statistics—to make general conclusions from limited data, to extrapolate from sample to population.

Part B explains confidence intervals in three contexts. Chapter 4 introduces the concept of a confidence interval (CI) in the context of confidence intervals of a proportion. I think this is the simplest example of a confidence interval, since it requires no background information. Most books would start with the confidence interval of the mean, but this would require first explaining the Gaussian distribution and the standard deviation. Chapters 5 and 6 are short chapters that explain confidence intervals of survival data and Poisson (counted) data. Many instructors will choose to skip these two chapters. I think they drive home the point that different kinds of data require different kinds of analyses, as well as what it means to sample from a population and to estimate a population parameter from that sample.

Part C finally gets to continuous data, the concept with which most statistics books start. The first three chapters are fairly conventional, explaining how to graph continuous data, how to quantify variability, and the Gaussian distribution. Chapter 11 is about lognormal distributions, which are common in biology but rare in statistics texts. Chapter 12 explains the confidence interval of the mean, and Chapter 13 is an optional chapter that gives a taste of the theory behind confidence intervals. Then comes an important chapter (14) which explains the different kinds of error bars, emphasizing the difference between the standard deviation and the standard error of the mean (which are frequently confused).

Part D is unconventional, as it explains the ideas of P values, statistical hypothesis testing and statistical power without explaining any statistical tests. I think it is easier to learn the concepts of a P value and statistical significance apart from the details of a particular test. This section also includes a chapter (unusual for introductory books) on testing for equivalence.

Part E explains challenges in statistics. The first two chapters of this section explain the problem of multiple comparisons. This is a huge challenge in data analysis but a topic that is not covered by most introductory statistics books. The next two chapters briefly explain the principles of normality and outlier tests, topics that most statistics texts omit. Finally, Chapter 26 is an overview of determining necessary sample size.

Part F explains the basic statistical tests, including those that compare survival curves (an issue omitted from many introductory texts).

Part G is about fitting models to data. It begins, of course, with linear regression. Later chapters in the section explain the ideas of creating models and briefly explain the ideas of nonlinear regression (a method used commonly in biological research but omitted from most introductory texts), multiple regression, logistic regression, and proportional hazards regression.

Part H contains miscellaneous chapters briefly introducing analysis of variance (ANOVA), nonparametric methods, and sensitivity and specificity. It ends with a chapter (43, new to this edition) on meta-analysis, a topic covered by few introductory texts.

Part I tries to put it all together. Chapter 44 is a very brief summary of the key ideas of statistics. Chapter 45 is a much longer chapter explaining common traps in data analysis, a reality check missing from most statistics texts. Chapter 46 goes through one example in detail as a review. Chapter 47 is a list of problems, which are answered in Chapter 48.

If you don't like the order of the chapters, read (or teach) them in a different order. It is not essential that you read the chapters in order. Some teachers have told me that they have successfully presented the chapters in a very different order than I present them.

What's new in the third edition?

The second edition (published in 2010, 15 years after the first edition) was a complete rewrite with new chapters, expanded coverage of some topics that were only touched upon in the first edition, and a complete reorganization. If you own the first edition (published in 1995) and are wondering if you should buy the third, the answer is yes. The improvements are enormous.

This third edition is not a complete rewrite, but I have substantially edited nearly every chapter to clarify concepts and add examples. Some chapters were essentially rewritten. Additionally, the third edition has these improvements:

- New chapter 2, addressing the complexities of probability. Statistical thinking is based on understanding the basic concepts of probability, so this chapter explains the basics.
- New chapter 43, addressing meta-analysis. Combining the results of many studies, or meta-analysis, is becoming more and more common. This chapter helps you understand a published meta-analysis.
- New chapter 45, reviewing statistical traps to avoid. Chapter 45 of the previous edition, which explained how to choose a statistical test, is now Appendix F.
- More examples. New examples have been added to seven chapters in order to better illustrate the specific, real-world applications of statistical topics.
- Reorganized chapters. Six chapters of the previous edition have been extensively rewritten. Chapter 23 has been reorganized and expanded to emphasize the many ways in which multiple comparisons occur. Chapter 26, about choosing sample size, is expanded and reorganized from the former chapter 43. Chapter 27 has been modified to include material from the previous edition's chapter 26. Chapters 37 (multiple regression) and 38 (logistic regression) have been rethought and expanded. Chapter 44 has been pared down to a concise summary of the important points of statistics.
- Emphasis on how to avoid common mistakes. About half the chapters have sections on common mistakes.
- Summary of terminology. Each chapter now ends with a list of terms introduced in that chapter. Much of the challenge in learning statistics is learning the terminology, and these lists will make that process easier.
- End of chapter summaries. Each chapter now ends with a list of the most important points you should remember.
- More Q&A sections. The question and answer sections of the second edition were popular, so I've now included one at the end of almost every chapter.
- More figures. The new edition now includes fifteen new figures, ten of which appear in the new chapter on statistical traps to avoid.
- More topics. New to this edition are discussions of pseudoreplicates, genome-wide association studies, primary versus secondary outcomes in clinical trials, researcher degrees of freedom, P-hacking, multiplicity-adjusted P values, five-sigma cutoff, reproducibility of P values, the relationship between sample size and P value, five-number summary, Type S errors, and more.

Which chapters are essential?

If you don't have time to read this entire book, read these 16 chapters to learn the essential concepts of statistics:

1. Statistics and Probability Are Not Intuitive
3. From Sample to Population
4. Confidence Interval of a Proportion
9. Quantifying Scatter
10. The Gaussian Distribution
12. Confidence Interval of a Mean
14. Error Bars
15. Introducing P Values
16. Statistical Significance and Hypothesis Testing
18. Interpreting a Result That Is Statistically Significant
19. Interpreting a Result That Is Not Statistically Significant
22. Multiple Comparisons Concepts
23. The Ubiquity of Multiple Comparisons
33. Simple Linear Regression
44. The Key Concepts of Statistics
45. Statistical Traps to Avoid

Who helped?

A huge thanks to the many people listed below who reviewed draft chapters of this edition. Their contributions were huge, and immensely improved this book:

Raid Amin, University of West Florida
Timothy Bell, Chicago State University
Patrick Breheny, University of Kentucky
William M. Cook, Saint Cloud State University
Jesse Dallery, University of Florida
Beth Dawson, The University of Texas at Austin
Vincent A. DeBari, Seton Hall University
Lisa A. Gardner, Drake University
William Greco, University at Buffalo, The State University of New York
Heather J. Hoffman, George Washington University
Dennis A. Johnston, Baylor University
Martin Jones, College of Charleston
Janet E. Kübler, California State University, Northridge
Lee Limbird, Fisk University
Leonard C. Onyiah, Saint Cloud State University
Nancy Ostiguy, Pennsylvania State University
Carol Paronis, Northeastern University
Evelyn Schlenker, University of South Dakota
Manfred Stommel, Michigan State University
Liansheng Tang, George Mason University

William C. Wimley, Tulane University School of Medicine
Gary Yellen, Harvard Medical School

Thanks also to the reviewers of past editions, as many of their ideas and corrections survived into this edition: Jan Agosti, David Airey, William (Matt) Briggs, Peter Chen, Cynthia J Coffman, Jacek Dmochowski, Jim Ebersole, Gregory Fant, Joe Felsenstein, Harry Frank, Joshua French, Phillip Ganter, Cedric Garland, Steven Grambow, John Hayes, Ed Jackson, Laurence Kamin, Eliot Krause, James Leeper, Yulan Liang, Longjian Liu, Lloyd Mancl, Sheniz Moonie, Arno Motulsky, Lawrence "Doc" Muhlbaier, Pamela Ohman-Strickland, Lynn Price, Jeanette Ruby, Soma Roychowdhury, Andrew Schaffner, Paige Searle, Christopher Sempos, Arti Shankar, Patricia A Shewokis, Jennifer Shook, Sumihiro Suzuki, Jimmy Walker, Paul Weiss, and Dustin White.

I would also like to express appreciation to everyone at Oxford University Press: Jason Noe, Senior Editor; Andrew Heaton and Caitlin Kleinschmidt, Editorial Assistants; Patrick Lynch, Editorial Director; John Challice, Publisher and Vice President; Frank Mortimer, Director of Marketing; Lisa Grzan, Manager In-House Production; Shelby Peak, Senior Production Editor; Michele Laseau, Art Director; Bonni Leon-Berman, Designer; Sarah Vogelsong, copy editor; and Pamela Hanley, Production Editor.

Who am I?
After graduating from medical school and completing an internship in internal medicine, I switched to research in receptor pharmacology (and published over 50 peer-reviewed articles). While I was on the faculty of the Department of Pharmacology at the University of California, San Diego, I was given the job of teaching statistics to first-year medical students and graduate students. The syllabus for those courses grew into the first edition of this book.

I hated creating graphs by hand, so I created some programs to do it for me! I also created some simple statistics programs after realizing that the existing statistical software, while great for statisticians, was overkill for most scientists. These efforts constituted the beginnings of GraphPad Software, Inc., which has been my full-time endeavor for many years (see Appendix A). In this role, I exchange emails with students and scientists almost daily, which makes me acutely aware of the many ways that statistical concepts can be confusing or misunderstood.

I have organized this book in a unique way, and have chosen an unusual set of topics to include in an introductory text. However none of the ideas are particularly original. All of the statistical concepts are standard and have been discussed in many texts. I include references for some concepts that are not widely known, but I don't provide citations for methods that are in common usage.

Please email me with your comments, corrections, and suggestions for the next edition. I'll post errata at www.intuitivebiostatistics.com.

Harvey Motulsky
hmotulsky@graphpad.com
July 2013

ABBREVIATIONS

Abbreviation	Definition	Chapter where defined
α (alpha)	Significance level	16
ANOVA	Analysis of variance	39
CI	Confidence interval	4
CV	Coefficient of variation	9
df	Degrees of freedom	9
FDR	False discovery rate	18
n	Sample size	4
OR	Odds ratio	28
SD or s	Standard deviation	9
SE	Standard error	14
SEM	Standard error of the mean	14
p (lower case)	Proportion	4
P (upper case)	P value	15
r	Correlation coefficient	32
ROC	Receiver-operating characteristic curve	42
RR	Relative risk	27
W	Margin of error	4

Introducing Statistics

CHAPTER 1

❖❖❖

Statistics and Probability Are Not Intuitive

If something has a 50% chance of happening, then 9 times out of 10 it will.

YOGI BERRA

The word *intuitive* has two meanings. One is "easy to use and understand," which is my goal for this book—hence its title. The other meaning is "instinctive, or acting on what one feels to be true even without reason." This fun (really!) chapter shows how our instincts often lead us astray when dealing with probabilities.

WE TEND TO JUMP TO CONCLUSIONS

A three-year-old girl told her male buddy, "You can't become a doctor; only girls can become doctors." To her, this statement made sense, because the three doctors she knew were all women.

When my oldest daughter was four, she "understood" that she was adopted from China, whereas her brother "came from Mommy's tummy." When we read her a book about a woman becoming pregnant and giving birth to a baby girl, her reaction was, "That's silly. Girls don't come from Mommy's tummy. Girls come from China." With only one person in each group, she made a general conclusion. Like many scientists, when new data contradicted her conclusion, she questioned the accuracy of the new data rather than the validity of the conclusion.

The ability to generalize from a sample to a population is hardwired into our brains and has even been observed in eight-month-old babies (Xu & Garcia, 2008). To avoid our natural inclination to make overly strong conclusions from limited data, scientists need to use statistics.

WE TEND TO BE OVERCONFIDENT

How good are people at judging how confident they are? You can test your own ability to quantify uncertainty using a test devised by Russo and Schoemaker (1989). Answer each of these questions with a range. Pick a range that you think

3

has a 90% chance of containing the correct answer. Don't use Google to find the answer. Don't give up and say you don't know. Of course you won't know the answers precisely! The goal is not to provide precise answers, but rather to correctly quantify your uncertainty and come up with ranges of values that you think are 90% likely to include the true answer. If you have no idea, answer with a superwide interval. For example, if you truly have no idea at all about the answer to the first question, answer with the range 0 to 120 years old, which you can be 100% sure includes the true answer. But try to narrow your responses to each of these questions to a range that you are 90% sure contains the right answer:

- Martin Luther King Jr.'s age at death
- Length of the Nile River, in miles or kilometers
- Number of countries in OPEC
- Number of books in the Old Testament
- Diameter of the moon, in miles or kilometers
- Weight of an empty Boeing 747, in pounds or kilograms
- Year Mozart was born
- Gestation period of an Asian elephant, in days
- Distance from London to Tokyo, in miles or kilometers
- Deepest known point in the ocean, in miles or kilometers

Compare your answers with the correct answers listed at the end of this chapter. If you were 90% sure of each answer, you would expect about nine intervals to include the correct answer and one to exclude it.

Russo and Schoemaker (1989) tested more than 1,000 people and reported that 99% of them were overconfident. The goal was to create ranges that included the correct answer 90% of the time, but most people created narrow ranges that included only 30 to 60% of the correct answers. Similar studies have been done with experts estimating facts in their areas of expertise, and the results have been similar.

Since we tend to be too sure of ourselves, scientists must use statistical methods to quantify confidence properly.

WE SEE PATTERNS IN RANDOM DATA

Most basketball fans believe in "hot hands"—that players occasionally have streaks of successful shots. People think that once a player has successfully made a shot, he or she is more likely to make the next shot, and that clusters of successful shots will happen more often than is predicted by chance.

Gilovich (1985) analyzed data from the Philadelphia 76ers during the 1980–1981 basketball season. The data are clear: "hot hands" don't exist. The chance that a player will put a ball through the basket is the same after making a shot as it is after missing one. Additionally, streaks of four, five, or six successful shots in a row happened no more often than was predicted by chance. The sequence of successes and failures was entirely random, yet almost everyone (players and fans) perceived patterns.

–	–	X	–	X	–	X	X	X	–	–	–	–	X	X	X	–	X	X	–	X	X	–	–	–	–	X	X	–	–	–	X X
X	–	–	X	–	X	X	–	–	X	X	–	–	X	–	X	–	X	–	–	–	X	X	X	X	–	–	X	X	–	–	– –
X	X	X	X	–	X	X	–	X	–	X	–	X	X	X	–	–	–	–	–	–	X	–	X	–	X	X	X	–	–	–	– X
–	X	–	X	–	–	X	X	–	X	X	–	X	X	–	–	X	X	X	X	–	–	–	–	–	X	X	–	X	–	X	– –
–	X	–	X	–	X	X	–	–	–	X	X	–	–	–	–	–	–	X	–	X	–	X	–	–	X	–	–	X	–	X	X
–	–	X	X	X	–	X	–	X	–	–	–	X	X	X	X	–	X	X	X	X	–	–	–	–	–	–	X	X	–	X	X X
X	–	–	X	X	–	–	X	X	X	X	–	X	X	X	–	–	X	–	–	X	X	X	X	X	–	X	X	X	–	–	–
X	–	X	–	–	–	X	X	X	X	X	–	X	–	X	–	X	X	–	X	X	–	X	X	X	–	X	X	–	X	–	X
X	X	X	–	–	X	X	X	X	X	–	X	–	X	–	X	X	–	X	–	X	X	X	X	–	X	X	–	X	X	X	X
–	–	–	X	X	X	–	–	X	X	X	–	X	X	X	–	–	X	–	–	X	–	X	X	X	X	X	–	–	–	X	–

Table 1.1. Random patterns don't seem random.
Table 1.1 represents 10 basketball players (1 per row) shooting 32 baskets each. An "X" represents a successful shot, and a "–" represents a miss. Is this pattern random? Or does it show signs of nonrandom streaks? Most people tend to see patterns, but in fact the arrangement is entirely random. Each spot in the table had a 50% chance of having an "X."

Table 1.1 demonstrates the problem. It presents simulated data from 10 basketball players (1 per row) shooting 30 baskets each. An "X" represents a successful shot and a "–" represents a miss. Is this pattern random? Or does it show signs of nonrandom streaks? Look at Table 1.1 before continuing.

Most people see patterns. Yet Table 1.1 was generated randomly. Each spot had a 50% chance of being "X" (a successful shot) and a 50% chance of being "–" (an unsuccessful shot), without taking into consideration previous shots. We see clusters perhaps because our brains have evolved to find patterns and do so very well. This ability may have served our ancestors well to avoid predators and poisonous plants, but it is important that we recognize this built-in mental bias. Statistical rigor is essential to avoid being fooled by seeing apparent patterns among random data.

WE DON'T REALIZE THAT COINCIDENCES ARE COMMON

In November 2008, I attended a dinner for the group Conservation International. The actor Harrison Ford is on their board, and I happened to notice that he wore an ear stud. The next day, I watched an episode of the TV show *Private Practice*, and one character pointed out that another character had an ear stud that looked just like Harrison Ford's. The day after that, I happened to read (in a book on serendipity!) that the Nobel Prize–winning scientist Baruch Blumberg looks like Indiana Jones, a movie character played by Harrison Ford (Meyers, 2007).

What is the chance that this particular set of events would happen? Tiny, but that doesn't mean much. While it is highly unlikely that any particular coincidence will occur, it is almost certain that some seemingly astonishing set of unspecified events will happen often, since we notice so many things each day. Remarkable coincidences are always noted in hindsight and never predicted with foresight.

WE HAVE INCORRECT INTUITIVE
FEELINGS ABOUT PROBABILITY

Imagine that you can choose between two bowls of jelly beans. The small bowl has nine white and one red jelly bean. The large bowl has 93 white beans and 7 red beans. Both bowls are well mixed, and you can't see the beans. Your job is to pick one bean. You win a prize if your bean is red. Should you pick from the small bowl or the large one?

When you choose from the small bowl, you have a 10% chance of picking a red jelly bean. When you pick from the large bowl, the chance of picking a red one is only 7%. So your chances of winning are higher if you choose from the small bowl. Yet about two-thirds of people prefer to pick from the larger bowl (Denes-Raj & Epstein, 1994). Many of these people do the math and know that the chance of winning is higher with the small bowl, but they feel better about choosing from the large bowl because it has more red beans and offers more chances to win. Of course, the large bowl also has more white beans and more chances to lose. Our brains have simply not evolved to deal sensibly with probability, and most people make the illogical choice.

WE FIND IT HARD TO COMBINE PROBABILITIES

Here is a classic brainteaser called the Monty Hall problem, named after the host of the game show *Let's Make a Deal*: You are a contestant on a game show and are presented with three doors. Behind one is a fancy new car. You must choose one door, and you get to keep whatever is behind it. You pick a door. At this point, the host chooses one of the other two doors to open and shows you that there is no car behind it. He now offers you the chance to change your mind and choose the other door (the one he has not opened).

Should you switch?

Before reading on, you should think about the problem and decide whether you should switch. There are no tricks or traps. Exactly one door has the prize, all doors appear identical, and the host—who knows which door leads to the new car—has a perfect poker face and gives you no clues. There is never a car behind the door the moderator chooses to open. Don't cheat. Think it through before continuing.

When you first choose, there are three doors and each is equally likely to have the car behind it. So your chance of picking the winning door is one-third. Let's separately think through the two cases: originally picking a winning door or originally picking a losing door.

If you originally picked the winning door, then neither of the other doors has a car behind it. The host opens one of these two doors. If you now switch doors, you will have switched to the other losing door.

What happens if you originally picked a losing door? In this case, one of the remaining doors has a car behind it and the other one doesn't. The host knows which is which. He opens the door without the car. If you now switch, you will win the car.

Let's recap. If you originally chose the correct door (an event that has a one-third chance of occurring), then switching will make you lose. If you originally picked either of the two losing doors (an event that has a two-thirds chance of occurring), then switching will definitely make you win. Switching from one losing door to the other losing door is impossible, because the host will have opened the other losing door.

Your best choice is to switch! Of course, you can't be absolutely sure that switching doors will help. One-third of the time you will be switching away from the prize. But the other two-thirds of the time you will be switching to the prize. If you repeat the game many times, you will win twice as often by switching doors every time. If you only get to play once, you have twice the chance of winning by switching doors.

Almost everyone (including mathematicians and statisticians) intuitively reaches the wrong conclusion and thinks that switching won't be helpful (Vos Savant, 1997).

WE DON'T DO BAYESIAN CALCULATIONS INTUITIVELY

Imagine this scenario: You are screening blood donors for the presence of human immunodeficiency virus (HIV). Only a tiny fraction (0.1%) of the blood donors has HIV. The antibody test is highly accurate but not quite perfect. It correctly identifies 99% of infected blood samples but also incorrectly concludes that 1% of noninfected samples have HIV. When this test identifies a blood sample as having HIV present, what is the chance that the donor does, in fact, have HIV, and what is the chance the test result is an error (a false positive)?

Try to come up with the answer before reading on.

Let's imagine that 100,000 people are tested. Of these, 100 (0.1%) will have HIV, and the test will be positive for 99 (99%) of them. The other 99,900 people will not have HIV, but the test will incorrectly return a positive result in 1% of cases. So there will be 999 false positive tests. Altogether, there will be 99 + 999 = 1,098 positive tests, of which only 99/1,098 = 9% will be true positives. The other 91% of the positive tests will be false positives. So if a test is positive, there is only a 9% chance that there is actually HIV in that sample.

Most people, including most physicians, intuitively think that a positive test almost certainly means that HIV is present. Our brains are not adept at combining what we already know (the prevalence of HIV) with new knowledge (the test is positive).

Now imagine that the same test is used in a population of intravenous drug users, of which 10% have HIV. Again, let's imagine that 100,000 people are tested. Of these, 10,000 (10%) will have HIV, and the test will be positive for 9,900 (99%) of them. The other 90,000 people will not have HIV, but the test will incorrectly return a positive result in 1% of cases. So there will be 900 false positive tests. Altogether, there will be 9,900 + 900 = 10,800 positive tests, of which 9,900/10,800 = 92% will be true positives. The other 8% of the positive tests will be false positives. So if a test is positive, there is a 92% chance that there is HIV in that sample.

The interpretation of the test result depends greatly on what fraction of the population has the disease. This example gives you a taste of what is called Bayesian logic (a subject that will be discussed again in Chapters 2 and 18).

WE ARE FOOLED BY MULTIPLE COMPARISONS

Austin and colleagues (2006) sifted through a database of health statistics of 10 million residents of Ontario, Canada. They examined 223 different reasons for hospital admission and recorded the astrological sign of each patient (computed from his or her birth date). They then asked if people with certain astrological signs are more likely to be admitted to the hospital for certain conditions.

The results seem impressive. Seventy-two diseases occurred more frequently in people with one astrological sign than in people with all the other astrological signs put together, with the difference being statistically significant. Essentially, a result that is "statistically significant" would occur by chance less than 5% of the time (you'll learn more about what "statistically significant" means in Chapter 16).

Sounds impressive, doesn't it? Indeed, the data make you think that there is a convincing relationship between astrology and health. But there is a problem. It is misleading to focus on the strong associations between one disease and one astrological sign without considering all the other combinations. Austin et al. (2006) tested the association of 223 different reasons for hospital admissions with 12 astrological signs and so tested 2,676 distinct hypotheses (223 × 12 = 2,676). Therefore, they would expect to find 134 statistically significant associations just by chance (5% of 2,676 = 134) but in fact only found 72.

Note that this study wasn't really done to ask about the association between astrological sign and disease. It was done to demonstrate the difficulty of interpreting statistical results when many comparisons are performed.

Chapters 22 and 23 explore multiple comparisons in more depth.

WE TEND TO IGNORE ALTERNATIVE EXPLANATIONS

Imagine you are doing a study on the use of acupuncture in treating osteoarthritis. Patients who come in with severe arthritis pain are treated with acupuncture. They are asked to rate their arthritis pain before and after the treatment. The pain decreases in most patients after treatment, and statistical calculations show that such consistent findings are exceedingly unlikely to happen by chance. Therefore, the acupuncture must have worked. Right?

Not necessarily. The decrease in recorded pain may not be caused by the acupuncture. Here are five alternative explanations (adapted from Bausell, 2007):

- If the patients believe in the therapist and treatment, that belief may reduce the pain considerably. The pain relief may be a placebo effect and have nothing to do with the acupuncture itself.

- The patients want to be polite and may tell the experimenter what he or she wants to hear (that the pain decreased). Thus, the decrease in reported pain may be because the patients are not accurately reporting pain after therapy.
- Before, during, and after the acupuncture treatment, the therapist talks with the patients. Perhaps he or she recommends a change in aspirin dose, a change in exercise, or the use of nutritional supplements. The decrease in reported pain might be due to these aspects of the treatment, rather than the acupuncture.
- The experimenter may have altered the data. For instance, what if three patients experience worse pain with acupuncture, whereas the others get better? The experimenter carefully reviews the records of those three patients and decides to remove them from the study because one of those people actually has a different kind of arthritis than the others, and two had to climb stairs to get to the appointment because the elevator didn't work that day. The data, then, are biased or skewed because of the omission of these three participants.
- The pain from osteoarthritis varies significantly from day to day. People tend to seek therapy when pain is at its worst. If you start keeping track of pain on the day when it is the worst, it is quite likely to get better, even with no treatment. The next section explores this idea of *regression to the mean*.

WE ARE FOOLED BY REGRESSION TO THE MEAN

Figure 1.1 illustrates simulated blood pressures before and after a treatment. Figure 1.1A includes 24 pairs of values. The "before" and "after" groups are about the same. In some cases, the value goes up after treatment, and in others it goes down. If these were real data, you'd conclude that there is no evidence at all that the treatment had any effect on the outcome (blood pressure).

Now imagine the study was designed differently. You've made the before measurements and want to test a treatment for high blood pressure. There is no point in treating individuals whose blood pressure is not high, so you select the people with the highest pressures to study. Figure 1.1B illustrates data for only those 12 individuals with the highest before values. In every case but one, the after values are lower. If you performed a statistical test (e.g., a paired t test; see Chapter 31), the results would convince you that the treatment decreased blood pressure. Figure 1.1C illustrates a similar phenomenon with the other 12 pairs, those with low before values. In all but two cases, these values go up after treatment. This evidence would convince you that the treatment increases blood pressure.

But these are random data! The before and after values came from the same distribution. What happened?

This is an example of *regression to the mean*: the more extreme a variable is upon its first measurement, the more likely it is to be closer to the average the second time it is measured. People who are especially lucky at picking stocks one year are likely to be less lucky the next year. People who get extremely high

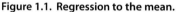

Figure 1.1. Regression to the mean.
All data in (A) were drawn from random distributions (Gaussian; mean = 120, SD = 15) without regard to the designations "before" and "after" and without regard to any pairing. (A) shows 48 random values, divided arbitrarily into 24 before–after pairs (which overlap enough that you can't count them all). (B) shows only the 12 pairs with the highest before values. In all but one case, the after values are lower than the before values. (C) shows the pairs with the lowest before measurements. In 10 of the 12 pairs, the after value is higher than the before value. If you only saw the graph in (B) or (C), you'd probably conclude that whatever treatment came between the before and after measurements had a large impact on blood pressure. In fact, these graphs simply show random values, with no systematic change between before and after. The apparent change is called *regression to the mean*.

scores on one exam are likely to get lower scores on a repeat exam. An athlete who does extremely well in one season is likely to perform more poorly the next season. This probably explains much of the *Sports Illustrated* cover jinx—many believe that appearing on the cover of *Sports Illustrated* will bring an athlete bad luck (Wolff, 2002).

WE CRAVE CERTAINTY, BUT STATISTICS OFFERS PROBABILITIES

Many people expect statistical calculations to yield definite conclusions. But in fact, every statistical conclusion is stated in terms of probability. Statistics can be very difficult to learn if you keep looking for definitive conclusions. As statistician Myles Hollander reportedly said, "Statistics means never having to say you're certain!" (quoted in Samaniego, 2008).

CHAPTER SUMMARY

- Our brains do a really bad job of interpreting data. We see patterns in random data, tend to be overconfident in our conclusions, and mangle interpretations that involve combining probabilities.
- Our intuitions tend to lead us astray when interpreting probabilities, and when interpreting multiple comparisons.
- Statistical (and scientific) rigor is needed to avoid reaching invalid conclusions.

TERM INTRODUCED IN THIS CHAPTER

- Regression to the mean (p. 9)

Answers to the 10 questions in the "We Tend to Be Overconfident" section.
Martin Luther King Jr.'s age at death: 39
Length of the Nile River: 4,187 miles or 6,738 kilometers
Number of countries in OPEC: 13
Number of books in the Old Testament: 39
Diameter of the moon: 2,160 miles or 3,476 kilometers
Weight of an empty Boeing 747: 390,000 pounds or 176,901 kilograms
Year Mozart was born: 1756
Gestation period of an Asian elephant: 645 days
Distance from London to Tokyo: 5,989 miles or 9,638 kilometers
Deepest known point in the ocean: 6.9 miles or 11.0 kilometers

CHAPTER 2

❖❖❖

The Complexities of Probability

Statistical thinking will one day be as necessary for efficient citizenship as the ability to read and write.

H. G. WELLS

Entire books have been written on probability. It is a topic that seems simple at first but becomes more complicated as you delve into the details. This chapter provides a very broad overview that will help you understand why probability is confusing, make sense of statements about probability, and provide a solid background if you decide to learn more about the subject.

BASICS OF PROBABILITY

Probabilities range from 0.0 to 1.0 (or 100%) and are used to quantify a prediction about future events or the certainty of a belief. A probability of 0.0 means either that an event can't happen or that someone is absolutely sure that a statement is wrong. A probability of 1.0 (or 100%) means that an event is certain to happen or that someone is absolutely certain a statement is correct. A probability of 0.50 (or 50%) means that an event is equally likely to happen or not happen, or that someone believes that a statement is equally likely to be true or false.

This chapter uses the terminology of Kruschke (2011) to distinguish two uses of probability:

- Probability that is "out there," or outside your head. This is *probability as long-term frequency*. The probability that a certain event will happen has a definite value, but we rarely have enough information to know that value with certainty.
- Probability that is inside your head. This is probability as strength of subjective beliefs, so it may vary among people and even among different assessments by the same person.

PROBABILITY AS LONG-TERM FREQUENCY

Probabilities as predictions from a model

This chapter will focus on one simple example: a woman plans to get pregnant and wants to know the chance that her baby will be a boy. One way to think about probabilities is as the predictions of future events that are derived by using a model. A *model* is a simplified description of a mechanism. For this example, we can create the following simple model:

- Each ovum has an X chromosome, and none have a Y chromosome.
- Half the sperm have an X chromosome (but no Y) and half have a Y chromosome (and no X).
- Only one sperm will fertilize the ovum.
- Each sperm has an equal chance of fertilizing the ovum.
- If the winning sperm has a Y chromosome, the fetus will have both an X and a Y chromosome and so will be male. If the winning sperm has an X chromosome, the fetus will have two X chromosomes and so will be female.
- Any miscarriage or abortion is equally likely to happen to male and female fetuses.

If you assume that this model is true, then the predictions are easy to figure out. Since all sperm have the same chance of being the one to fertilize the ovum, and since the sperm are equally divided between those with an X chromosome and those with a Y chromosome, the chance that the fetus will have a Y chromosome is 50%. Thus, our model predicts that the chance that the fetus will be male is 0.50, or 50%. In any particular group of babies, the fraction of boys might be more or less than 50%. But in the long run, you'd expect 50% of the babies to be boys.

You can make predictions about the occurrence of future events from any model, even if the model doesn't reflect reality. Some models will prove to be useful, and some won't. The model described here is pretty close to being correct, so its predictions are pretty useful but not perfect.

Probabilities based on data

Of all the babies born in the world in 2011, 51.7% were boys (Central Intelligence Agency [CIA], 2012). We don't need to know *why* more boys than girls were born. And we don't need to know why the CIA tabulates this information. It is simply a fact that this sex ratio has stayed fairly consistent over many years in many countries. Based on these data, we can answer the question, What is the chance that my baby will be a boy? The answer is 51.7%. If you had a group of 1,000 pregnant women, you'd expect about 517 to have male fetuses and 483 to have female fetuses. In any particular set of 1,000 pregnant women, the number of males might be higher or lower than 517, but that is what you'd expect in the long run.

PROBABILITY AS STRENGTH OF BELIEF

Subjective probabilities

You badly want a boy. You search the Internet and read about an interesting book:

> *How to Choose the Sex of Your Baby* explains the simple, at-home, noninvasive Shettles method and presents detailed steps to take to conceive a child of a specific gender. The properly applied Shettles method gives couples a 75 percent or better chance of having a child of the desired sex. (Shettles, 1996)

The reviews of this book are glowing and convince you that its premise is correct. Therefore, you plan to follow the book's recommendations to increase your chance of having a boy.

What is the chance that you'll have a boy? If you have complete faith that the method is correct, then you believe that the probability, as stated on the book jacket, is 75%. You don't have any data supporting this number, but you believe it strongly. It is your firmly held *subjective probability*.

But what if you have doubts? You think the method probably works, and you will quantify that by saying you think there is an 85% chance that the Shettles method works (that is, your chance of having a boy would be 75%). That leaves a 15% chance that the method is worthless and your chance of having a boy is 51.7%. What is your chance of having a boy? Calculate a weighted average of the two predictions, weighting on your subjective assessment of the chance that each theory is correct. So the chance is $(0.850 \times 0.750) + (0.150 \times 0.517) = 0.715 = 71.5\%$.

Of course, different people have different beliefs about the efficacy of the method. I didn't look into the matter carefully, but I did spend a few minutes searching and found no studies actually proving that the Shettles method works, and one (in a prestigious journal) saying it doesn't work (Wilcox et al., 1995). Given that limited amount of research, I will quantify my beliefs by saying that I believe there is only a 1% chance that the Shettles method works (and that your chance of having a boy is 75%) and thus a 99% chance that the method does not work (and so your chance of having a boy is 51.7%). The weighted average is thus $(0.010 \times 0.750) + (0.990 \times 0.517) = 0.519 = 51.9\%$.

Since you and I have different assessments of the probability of whether the Shettles method works, my answer for the chance of you having a boy (51.9%) doesn't match your answer (71.5%). The calculations depend upon a subjective assessment of the likelihood that the Shettles method works, so it isn't surprising that my answer and yours differ.

"Probabilities" used to quantify ignorance

Assume that you (or someone you know) are pregnant but haven't yet done any investigations (e.g., ultrasound, karyotype) to determine whether the fetus is male or female or undergone any interventions (see previous section) that purport to change the chances of having a boy or a girl. What is the probability that the fetus is male?

In one sense, the concept of probability is simply not involved. The random event was the race of many sperm to fertilize the egg. That already happened. One sperm won, and the fetus either is male or is female. Since the random event has already happened and the gender of the fetus is now a fact, you could argue that it makes no sense to ask about probability or chance or odds. Chance is no longer involved. The issue is ignorance, not randomness.

Another perspective: Before you became pregnant, there was a 51.7% chance that your fetus would be male. If many thousands of women were pregnant without knowing the sex of their fetus, you'd expect about 51.7% to have male fetuses.

Therefore, it seems sensible to say that there is about a 51.7% chance that your fetus is male and a 48.3% chance that it is female. That was your chance before you got pregnant, so it still is your chance now.

But notice the change in perspective. Now you are no longer talking about predicting the outcome of a random future event but rather are quantifying your ignorance of an event that has already happened.

Quantitative predictions of one-time events

As I began to write this chapter in October 2012, the U.S. presidential election was two weeks away. What was the chance that Obama would be reelected? Different people had different opinions and interpreted the polling data differently. Nate Silver (2012), a respected political analyst, stated on October 22, 2012, that there was a 67.6% chance that Obama would be reelected.

Was this probability subjective? Partly. The 67.6% probability was computed from lots of poll data. But there are many polls, and there were definitely some subjective choices in how Silver constructed a model to combine the polls.

Can you interpret the 67.6% value as a long-term frequency? No. It doesn't mean that Silver expected Obama to win 67.6% of all the presidential elections he will ever enter, or that 67.6% of all voters would vote for Obama. Based on analysis of polling data, Silver said that he was 67.6% sure that Obama would win. That was his belief on October 22. By November 4, two days before the election, Silver raised that value to 73.6%. The probabilities he published were quantitative statements about his informed beliefs. Other knowledgeable pundits at the same time offered very different assessments (much closer to 50%).

CALCULATIONS WITH PROBABILITIES CAN BE EASIER IF YOU SWITCH TO CALCULATING WITH WHOLE NUMBERS

Gigerenzer (2002) showed that rephrasing probability problems can greatly impact the chance that they will be answered correctly. His example is about interpreting the results of mammograms given to asymptomatic women between the ages of 40 and 50. He asked 48 physicians to interpret a positive (abnormal) mammogram. Half of the physicians were given the question in terms of probabilities:

> The probability that one of these women has breast cancer is 0.8 percent. If a woman has breast cancer, the probability is 90 percent that she will have a positive mammogram. If a woman does not have breast cancer, the probability is 7 percent that she will have a positive mammogram. Imagine a woman who has a positive mammogram. What is the probability that she actually has breast cancer?

Go ahead. Try to figure out the answer.

The other half of the physicians were given the same question but in natural frequencies (whole numbers) rather than percentages:

Eight out of every 1,000 women will have breast cancer. Of those 8 women with breast cancer, 7 will have a positive mammogram. Of the remaining 992 women who don't have breast cancer, about 70 will still have a positive mammogram. Imagine a woman who has a positive mammogram. What is the probability that she actually has breast cancer?

Now try to answer the question. Is it easier?

The physicians in both groups took the time to think the problem through. But those in the first group (percentages) did terribly. Only 2 out of the 24 answered the question correctly (the probability of breast cancer is about 9%). A few others were close. The majority of the physicians were very wrong, and the most common answer was 90%. The physicians in the second group (natural frequencies) did much better, with the majority answering correctly.

Why is the answer 9%? Of the 1,000 women tested, we will see 7 positive mammograms among those who actually have breast cancer and 70 among those who don't. In other words, given the setup of this problem, we expect 7 out of 77 people with positive mammograms to turn out to actually have breast cancer (about 9%).

If you get stuck thinking through problems with probabilities, try changing the problem to thinking about whole numbers. That trick often makes things much easier to think through.

COMMON MISTAKES: PROBABILITY

Mistake: Ignoring the assumptions

The examples above asked, "What is the chance that a fetus will be male?" However, that question is meaningless without context. For it to be meaningful, you must accept a list of assumptions, including:

- We are asking about human babies. Sex ratios may be different in other species.
- There is only one fetus. The question, Will the baby be a boy or girl? is ambiguous or needs elaboration if you allow for the possibility of twins or triplets.
- There is only a tiny probability (which we ignore) that the baby is neither completely female nor completely male.
- The sex ratio is the same for all countries and all ethnic groups.
- The sex ratio does not change from year to year, or between seasons.
- There will be (and have been) no sex-selective abortions or miscarriages, so the sex ratio at conception is the same as the sex ratio at birth.

Whenever you hear questions or statements about probability, remember that probabilities are *always* contingent on a set of assumptions. To think clearly about probability in any situation, you must know what those assumptions are.

Mistake: Trying to understand a probability without clearly defining both the numerator and the denominator

The example we've used for this chapter is the fraction of babies who are boys. The numerator is the number of boy babies. The denominator is the number of babies born in a certain place in a certain time period. In this case, the fraction is simple and unambiguous.

Sometimes, probabilities are confusing or ambiguous until you carefully define the numerator and denominator. Say you are told that a lab test is 98% accurate. What does this number mean? Without further definition, it doesn't mean anything. The 98% is a probability, but a probability of what? Without clearly defining both the numerator and denominator, the number is basically meaningless.

Table 2.1 defines four distinct ways to define the accuracy of a diagnostic test that will be explained in more detail in Chapter 42. For now, the key point is that these definitions are very different. If you determine all four values for a particular diagnostic test used in a defined population, each of the four values will probably be very different from the others. It is simply ambiguous to ask about the probability that a test will be "accurate" without further defining "accurate." You can't understand a probability without carefully defining both the numerator and denominator.

Another example of an area in which it is easy to misunderstand probabilities is DNA forensics. DNA testing can provide the *random match probability*. Imagine that DNA is collected at a crime scene and it matches DNA collected from the defendant. The random match probability is the probability that the DNA of a randomly selected person would also match the DNA collected at the crime scene. To understand this probability, imagine doing DNA testing on many people. The numerator of the random match probability is the number of those people whose DNA matches the DNA collected at the crime scene. The denominator is the number of people tested. Let's say the random match probability is 0.01%, or 1 in 10,000. If you tested 100,000 people, you'd expect 10 to match.

A prosecutor might try to imply to a jury that there is only a 1 in 10,000 (or 0.01%) chance that the DNA profile came from someone other than the defendant. This is called the *prosecutor's fallacy*, because it tends to convince

DEFINITION	NUMERATOR: WHICH RESULTS ARE LABELED "ACCURATE"?	DENOMINATOR: WHO IS TESTED?
Sensitivity	This test result (of unknown accuracy) is positive.	People known to have the disease (by other tests known to be superaccurate)
Specificity	This test result (of unknown accuracy) is negative.	People known to not have the disease (by other superaccurate tests)
Predictive value of a positive test	A superaccurate test (or the test of time) proves the patient has the disease being tested for.	All people whose test results (of unknown accuracy) are positive
Predictive value of a negative test	A superaccurate test (or the test of time) proves that the patient does not have the disease being tested for.	All people whose test results (of unknown accuracy) are negative

Table 2.1. Four definitions of the *accuracy* of a diagnostic test.

juries to convict. Thinking through the numerator and denominator helps you to keep things straight. To think about the probability to which the prosecutor refers, you have to imagine many similar crime scenes. The numerator of that probability is the number of times the DNA at the crime scene actually came from the defendant. The denominator is the total number of crime scenes processed. That ratio is very different from the random match probability defined in the prior paragraph. Its value would depend on the criteria used to identify and charge the defendants.

A defense attorney might point out that there are about 10 million people who could have committed the crime (based on location, age, and sex). Since 1 out of 10,000 of these people would have matching DNA, there are 1,000 people whose DNA matches the DNA found at the crime scene (10,000,000 people times 0.0001). Therefore, the attorney might argue, the chance that the defendant is actually guilty is only 1 in 1,000. This is called the *defense attorney's fallacy*, since it can be used to convince a jury that the defendant is not guilty. Of course, this is only correct if there is no other evidence that connects the defendant to the crime scene or victim.

Mistake: Reversing probability statements

With the examples we've used so far, there is no danger of accidentally reversing probability statements by mistake. The probability that a baby is a boy is obviously very different from the probability that a boy is a baby. But in many situations, it is easy to get things backwards.

- The probability that a heroin addict first used marijuana is not the same as the probability that a marijuana user will later become addicted to heroin.
- The probability that someone with abdominal pain has appendicitis is not the same as the probability that someone diagnosed with appendicitis will have had abdominal pain.
- The probability that a statistics books will be boring is not the same as the probability that a boring book is about statistics.
- The fraction of studies performed under the null hypothesis in which the P value is less than 0.05 is not the same as the fraction of studies with a P value less than 0.05 for which the null hypothesis is true. (This will make more sense once you read Chapter 15.)

Knowing how easy it is to mistakenly reverse a probability statement will help you avoid that error.

Mistake: Believing that probability has a memory

If a couple has four children, all boys, what is the chance that the next child will be a boy? A common mistake is to think that since this couple already has four boys, they are somehow due to have a girl, so the chance of having a girl is elevated. This is simply wrong.

This mistake is often made in gambling, especially when playing roulette or a lottery. Some people bet on a number that has not come up for a long while, believing that there is an elevated chance of that number coming up on the next spin. This is called the *gambler's fallacy*. Probability does not have a memory.

LINGO: PROBABILITY AND ODDS

So far, this chapter has quantified chance as probabilities. But it is also possible to express the same values as *odds*. Odds and probability are two alternatives for expressing precisely the same concept. Every probability can be expressed as odds. Every odds can be expressed as a probability. Some scientific fields tend to prefer using probabilities; other fields tend to favor odds. There is no consistent advantage to using one or the other.

If you search for demographic information on the fraction of babies who are boys, what you'll find is the *sex ratio*. This term, as used by demographers, is the ratio of males to females born. Worldwide, the sex ratio at birth in many countries is about 1.07. Another way to say this is that the odds of having a boy versus a girl are 1.07 to 1.00, or 107 to 100.

It is easy to convert the odds to a probability. If there are 107 boys born for every 100 girls born, the chance that any particular baby will be male is 107/(107 + 100) = 0.517, or 51.7%.

It is also easy to convert from a probability to odds. If the probability of having a boy is 51.7%, then you expect 517 boys to be born for every 1,000 births. Of these 1,000 births, you expect 517 boys and 483 girls (which is 1,000 − 517) to be born. So the odds of having a boy versus a girl are 517/483 = 1.07 to 1.00. The odds are defined as the probability that the event will occur divided by the probability that the event will not occur.

Odds can be any positive value or zero, but they cannot be negative. A probability must be between 0 and 1 if expressed as a fraction, or between 0 and 100 if expressed as a percentage.

A probability of 0.5 is the same as odds of 1.0. The probability of flipping a coin to heads is 50%. The odds are 50:50, which equals 1.0. As the probability goes from 0.5 to 1.0, the odds increase from 1.0 to approach infinity. For example, if the probability is 0.75, then the odds are 75:25, 3 to 1, or 3.0.

THE DISTINCTION BETWEEN PROBABILITY AND STATISTICS

The words *probability* and *statistics* are often linked together in course and book titles, but they are distinct.

This chapter has discussed probability calculations. While the details can get messy and make it easy to apply probability theory incorrectly, the concepts are pretty simple. Start with the general case, called the population or model, and predict what will happen in many samples of data. Probability calculations go from

PROBABILITY

General	→	Specific
Population	→	Sample
Model	→	Data

STATISTICS

General	←	Specific
Population	←	Sample
Model	←	Data

Table 2.2. The distinction between probability and statistics.
Probability theory goes from general to specific, from population to sample, from model to data. Statistics theory works in the opposite direction.

general to specific, from population to sample (as you'll see in Chapter 3), and from model to data (as you'll see in Chapter 34).

Statistical calculations work in the opposite direction (Table 2.2). You start with one set of data (the sample) and make inferences about the overall population or model. The logic goes from specific to general, from sample to population, and from data to model.

PROBABILITY IN STATISTICS

Earlier in this chapter, I pointed out that probability can be "out there" or "in your head." The rest of this book mostly explains confidence intervals and P values, which both use "out there" probabilities. This style of analyzing data is called *frequentist statistics*. Prior beliefs, or prior data, never enter frequentist calculations. Only the data from a current set are used as inputs when calculating P values (see Chapter 15) or confidence intervals (see Chapters 4 and 12). However, scientists often account for prior data and theory when interpreting these results, a topic to which we'll return in Chapters 18, 19, and 42.

Many statisticians prefer an alternative approach called *Bayesian statistics*, in which prior beliefs are quantified and used as part of the calculations. These prior probabilities can be subjective (based on informed opinion), objective (based on solid data or well-established theory), or uninformative (based on the belief that all possibilities are equally likely). Bayesian calculations combine these prior probabilities with the current data to compute probabilities and Bayesian confidence intervals called *credible intervals*. This book does not explain Bayesian statistics.

CHAPTER SUMMARY

- Probability is a confusing topic.
- There are two meanings of probability. One meaning is long-range chance that an event will happen. It is probability "out there." The other meaning is

that probability quantifies how sure one is about the truth of a proposition. This is probability "in your head."
- All probability statements are based on a set of assumptions.
- It is impossible to understand a probability until you have clearly defined both its numerator and its denominator.
- Probability calculations go from general to specific, from population to sample, and from model to data. Statistical calculations work in the opposite direction: from specific to general, from sample to population, and from data to model.
- Frequentist statistical methods calculate probabilities from the data (P values, confidence intervals). Your beliefs before collecting the data do not enter into the calculations, but they often are considered when interpreting the results.
- Bayesian statistical methods combine prior probabilities (which may be either subjective or based on solid data) and experimental evidence as part of the calculations. This book does not explain how these methods work.

TERMS INTRODUCED IN THIS CHAPTER

- Bayesian statistics (p. 20)
- Credible intervals (p. 20)
- Defense attorney's fallacy (p. 18)
- Frequentist statistics (p. 20)
- Gambler's fallacy (p. 19)
- Model (p. 13)
- Odds (p. 19)
- Probability (p. 12)
- Probability as long term frequency (p. 12)
- Prosecutor's fallacy (p. 17)
- Random match probability (p. 17)
- Sex ratio (p. 19)
- Statistics (p. 19)
- Subjective probability (p. 13)

CHAPTER 3

※※※

From Sample to Population

There is something fascinating about science. One gets such a wholesale return of conjecture out of a trifling investment of fact.

MARK TWAIN (*LIFE ON THE MISSISSIPPI*, 1850)

Before you can learn about statistics, you need to understand the big picture. This chapter concisely explains what statistical calculations are used for and what questions they can answer.

SAMPLING FROM A POPULATION

The distinction between a sample and a population is key to understanding much of statistics. Statisticians say that you extrapolate from a sample to make conclusions about a population. Note that the terms *sample* and *population* have specialized meanings in statistics that differ from the ordinary uses of those words. As you learn statistics, distinguish the specialized definitions of statistical terms from their ordinary meanings.

Here are four of the different contexts in which the terms are used:

- Quality control. A factory makes lots of items (the population) but randomly selects a few items to test (the sample). The results obtained from the sample are used to make inferences about the entire population.
- Political polls. A random sample of voters is used to make conclusions about the entire population of voters.
- Clinical studies. Data from the sample of patients included in a study are used to make conclusions useful to a larger population of future patients.
- Laboratory experiments. The data you collect are the sample. From the sample data, you want to make reliable inferences about the ideal, true underlying situation (the population).

In quality control and in political and marketing polls, the population is usually much larger than the sample, but it is finite and known (at least approximately). In biomedical research, we usually *assume* that the population is infinite or at least very large compared with our sample. Standard statistical methods assume that the population is much larger than the sample. If the population has a defined size and you have sampled a substantial fraction of the population (>10% or so), then you must use special statistical methods designed for use when a large fraction of the population is sampled.

SAMPLING ERROR AND BIAS

Much of statistics is based on the assumption that the data you are analyzing are randomly sampled from a larger population. Therefore, the values you compute from the sample (e.g., the mean, the proportion of cases that have a particular attribute, the best-fit slope of a linear regression line) are considered to be *estimates* of the true population values.

There are several reasons why a value computed from a sample differs from the true population value:

- Random sampling error. Just by chance, the sample you collected might have a higher (or lower) mean than that of the population. Just by chance, the proportion computed from the sample might be higher (or lower) than the true population value. Just by chance, the best-fit slope of a regression line (which you will learn about in Chapter 33) might be steeper (or shallower) than the true line that defines the population.
- *Selection bias*. The difference between the value computed from your sample and the true population value might be due to more than random sampling. It is possible that the way you collected the sample is not random, but rather preferentially samples subjects with particular values. For example, political polls are usually done by telephone and thus select for people who own a phone (and especially those with several phone lines) and for people who answer the phone and don't hang up on pollsters. These people may have opinions very different than those of a true random sample of voters.
- Other forms of *bias*. The experimental methodology may be imperfect and give results that are systematically either too high or too low. Bias is not random but rather produces errors in a consistent direction.

MODELS AND PARAMETERS

The concept of sampling from a population is an easy way to understand the idea of making generalizations from limited data, but it doesn't always perfectly describe what statistics and science are all about. Another way to think about statistics is in terms of models and parameters.

A *model* is a mathematical description of a simplified view of the world. A model consists of both a general description and *parameters* that have particular values. For example, the model might describe temperature values that follow a Gaussian bell-shaped distribution. The parameters of that distribution are its mean (average) and standard deviation (a measure of variation, see Chapter 9). Another example: After a person ingests a drug, its concentration in his or her blood may decrease according to an exponential decay with consistent half-life. The parameter is the half-life. In one half-life, the drug concentration goes down to 50% of the starting value; in two half-lives, it goes down to 25%; in three half-lives, it goes down to 12.5%; and so on.

One goal of statistics is to analyze your data to make inferences about the values of the parameters that define the model. Another goal is to compare alternative models to see which best explains the data. Chapters 34 and 35 discuss this idea in more depth.

MULTIPLE LEVELS OF SAMPLING

Roberts (2004) published 10 different self-experiments that he performed to test ideas about sleep, mood, health, and weight. Here is one example: When Roberts looked at faces on television in the morning, his mood was lower that evening (>10 hours later) but elevated the next day (>24 hours later). In another example from that same paper, Roberts found that when he drank unflavored fructose water between meals, he lost weight and maintained the lower weight for more than a year. Statistical analysis of the sample of data he collected enabled him to make conclusions that he found helpful for managing his life. But can the conclusions be extrapolated to others? These are intriguing ideas that provide a starting point for designing broader experiments, but you can't extrapolate from a sample of data drawn from one person to make firm conclusions about others.

A political poll might first randomly choose counties and then randomly select voters in those counties to survey. Here, there are two levels of extrapolation: from the voters surveyed in a county to all the voters in that county, and from the counties actually surveyed to all the counties in the state or country. Special methods are needed to deal with this kind of *hierarchical* or *multilevel sampling*.

WHAT IF YOUR SAMPLE *IS* THE ENTIRE POPULATION?

In some situations, your sample of data *is* the entire population, or so it seems. A political scientist may analyze all the election data for a certain election. Someone doing quality control may be able to analyze every single item being manufactured. A hospital epidemiologist will have data about every single infection that occurs in a hospital. How does statistics work in these situations? There are two alternative answers.

In some cases, you truly only care about the data you have collected and have no interest in making any general conclusions. In these cases, statistical inference won't be helpful, because there is no inference needed. All you need to do is describe and plot the data.

In many cases, you do want to make inferences beyond the data at hand. The political scientist probably wants to make conclusions about future elections. The quality control expert will want to make inferences to improve quality in the future. The epidemiologist will want to compare the current data with past data to make inferences to help future patients. In all these cases, the current set of data can sort of be considered a sample of a larger population that includes data to be collected in the future. Another perspective is that you might really care about the underlying probability model that generated your data, so the inferences aren't about the data at hand. Instead, you want to make inferences about the model that generated those data (Gelman, 2009).

CHAPTER SUMMARY

- The goal of data analysis is simple: to make the strongest possible conclusions from limited amounts of data.
- Statistics help you extrapolate from a particular set of data (your sample) to make a more general conclusion (about the population). Statistical calculations go from specific to general, from sample to population, from data to model.
- Bias occurs when the experimental design is faulty, so that on average the result is too high (or too low). Results are biased when the result computed from the sample is not, in fact, the best possible estimate of the value in the population.
- The results of statistical calculations are always expressed in terms of probability.
- If your sample *is* the entire population, you only need statistical calculations to extrapolate to an even larger population, other settings, or future times.

TERMS INTRODUCED IN THIS CHAPTER

- Bias (p. 23)
- Hierarchical/multilevel sampling (p. 24)
- Model (p. 23)
- Parameter (p. 23)
- Population (p. 22)
- Sample (p. 22)
- Selection bias (p. 23)

PART B

Introducing Confidence Intervals

$$\begin{array}{c} \diamond\!\diamond\!\diamond \\ \diamond\!\diamond\!\diamond \end{array}$$

Confidence Interval of a Proportion

> Confidence intervals can merely indicate how great our igno-
> rance might be, even if we had strictly random samples of an
> adequately defined population.
>
> D. Mainland

A fundamental concept in statistics is the use of a confidence interval, abbreviated CI, to make a general conclusion from limited data. This chapter defines confidence intervals in the context of results that have two possible outcomes summarized as a proportion or fraction. This chapter introduces fundamental concepts that apply to much of statistics, so it is worth reading even if you don't work with data expressed as proportions. Chapters 27 and 28 will explain how to compare two proportions.

DATA EXPRESSED AS PROPORTIONS

This chapter considers only results expressed as a proportion. Some examples of this kind of results include the proportion of patients who already suffered a myocardial infarction (heart attack) who then go on to have heart failure, the proportion of students who pass a particular course, and the proportion of coin flips that are heads. Later chapters discuss different kinds of data.

THE BINOMIAL DISTRIBUTION: FROM POPULATION TO SAMPLE

If you flip a coin fairly, there is a 50% probability that it will land on heads and a 50% probability that it will land on tails. That means that in the long run, it will land on heads as often as it lands on tails. But it any particular series of tosses, you may not see equal numbers of heads and tails, and you may even see only heads or only tails. The *binomial distribution* calculates the likelihood of observing any particular outcome when you know its proportion in the overall population.

The binomial distribution can answer questions like these:

- If you flip a coin fairly 10 times, what is the chance of observing exactly 7 heads?

- If, on average, 5% of the patients undergoing a particular operation get an infection, what is the chance that 10 of the next 30 patients will get an infection?
- If 40% of voters are Democrats (one of the two major political parties in the United States), what is the chance that a random sample of 600 voters will be 45% Democrat?

The *cumulative binomial distribution* answers questions like these:

- If you flip a coin fairly 10 times, what is the chance of observing 7 *or more* heads?
- If, on average, 5% of the patients undergoing a particular operation get infected, what is the chance that 10 *or more* of the next 30 patients will get infected?
- If 40% of voters are Democrats, what is the chance that a random sample of 600 voters will be *at least* 45% Democrat?

This book won't show you how to answer these questions, but it is easy to find the relevant equations or to compute the answers with Excel. A helpful Web calculator can be found at www.graphpad.com/quickcalcs/probability1/.

The binomial distribution is rarely useful in analyzing data because it goes the wrong way. It starts with a known probability for the population and computes the chances of observing various outcomes in *random samples*. When analyzing data, we usually need to work the opposite way. We know what the result is in a sample expressed as a single proportion and want to use that proportion to make an inference about the overall proportion in the population.

A CONFIDENCE INTERVAL: FROM SAMPLE TO PROPORTION

To better counsel the parents of premature babies, Allen, Donohue, and Dusman (1993) investigated the survival of premature infants. They retrospectively studied all premature babies born at 22 to 25 weeks of gestation at the Johns Hopkins Hospital during a three-year period. The investigators tabulated infant deaths by gestational age. Of 29 infants born at 22 weeks of gestation, none survived past six months. Of 39 infants born at 25 weeks of gestation, 31 (79.5%) survived for at least six months.

Is it worth doing any calculations to make more general conclusions? Only if you think it is reasonable to assume that these babies are somewhat representative of a larger population of premature babies born at that time in similar hospitals, and you wish to make inferences about that larger group of babies.

If you collected data on a much larger population of babies, the percentage of survival in the two groups would probably not be exactly 0% and 79.5%. How far away from those numbers are the percentages likely to be? If you want to create a range of values that you are absolutely certain will contain the overall percentage from the much larger hypothetical population, it would have to run all the way from 0 to 100%. That wouldn't be useful. So let's create ranges of percentages that we are 95% sure will contain the true population values for the two groups of babies. Such a range is called a *confidence interval*, abbreviated CI.

Before reading on, write down what you think the CIs are for the two examples (0/29 and 31/39. Calculate the answer using the equation on page 37, a computer program, or a web calculator www.graphpad.com/quickcalcs/confInterval1/). For the infants born at 25 weeks of gestation, the 95% CI of 31/39 ranges from 64.3 to 89.5% (calculated by a computer program using the modified Wald method, which will be explained later in this chapter). For the infants born at 22 weeks of gestation, the 95% CI of 0/29 extends from 0 to 13.9%. We can be 95% sure that this range includes the overall proportion of surviving infants.

These CIs only account for sampling variability, also called *sampling error*. When you extrapolate these results to another hospital, you also must account for the different populations served by different hospitals and the different methods used to care for premature infants. Also note that these data are a few decades old, and the care of premature infants has improved since then.

EXAMPLE: POLLING VOTERS

Say you polled 100 randomly selected voters just before an election and 33 said they would vote for your candidate. What can you say about the proportion of all voters in the population who will vote for your candidate?

Again, there are two issues to deal with. First, you must think about whether your sample is really representative of the population of voters *and* whether people tell pollsters the truth about how they will vote. Statistical calculations cannot help you grapple with those issues! We'll assume that the sample is perfectly representative of the population of voters and that all people will vote as they said they would when polled. Second, you must think about *sampling error*. Just by chance, your 100-voter sample will almost certainly contain either a smaller or a larger fraction of people voting for your candidate than does the overall population.

Because we only know the proportion of favorably inclined voters in one sample, there is no way to be sure about the proportion in the population. The best we can do is to calculate a range of values that bracket the true population proportion. How wide does this range of values have to be? In the overall population, the fraction of voters who will vote for your candidate could be almost anything. So, to create a 95% CI, we accept a 5% chance that the range will not include the true population value.

Before reading on, make your best guess for the 95% CI when your sample has 100 people and 33 of those people have said they would vote for your candidate (so the proportion in the sample is 0.33, or 33%).

Note that there is no uncertainty about what we observed in the voter sample. We are absolutely sure that 33.0% of the 100 people polled said they would vote for our candidate. If we weren't sure of that count, calculation of a CI could not overcome any mistakes that were made in tabulating those numbers or any ambiguity in defining "vote." What we don't know is the proportion of favorable voters in the entire population. Write down your guess before continuing.

One way to calculate the confidence interval is explained later in this chapter. You can also use free Web-based calculators to do the math. In this case, the 95% CI extends from 25 to 43% (you might get slightly different results depending on which program you use). We can be 95% confident that somewhere between 25 and 43% of the population you polled preferred your candidate on the day the poll was conducted. The phrase "95% confident" is explained in more detail below.

Don't get confused by the two different uses of percentages. The confidence interval is for 95% confidence. That quantifies how sure you want to be. You could ask for a 99% confidence interval if you wanted your range to have a higher chance of including the true value. Each confidence limit (25% and 43%) represents the percentage of voters who will vote for your candidate.

How good was your guess? Many people guess that the interval is narrower than it is.

ASSUMPTIONS: CONFIDENCE INTERVAL OF A PROPORTION

The whole idea of a confidence interval is to make a general conclusion from some specific data. This generalization is only useful if certain assumptions are true. A huge part of the task of learning statistics is to learn the assumptions upon which statistical inferences are based.

Assumption: Random (or representative) sample

All statistical conclusions are based on the assumption that the sample of data you are analyzing is drawn from a larger population of data about which you want to generalize. This assumption would be violated in the premature baby example if the infants included in the sample were somehow selected to be sicker (or less sick) than other premature infants of similar age. This assumption would be violated in the election example if the sample was not randomly selected from the population of voters. In fact, this mistake was made in the 1936 Roosevelt–Landon U.S. presidential election. To find potential voters, the pollsters relied heavily on phone books and automobile registration lists. In 1936, Republicans were far more likely than Democrats to own a phone or a car, and therefore, the poll selected too many Republicans. There also was a second problem. For some reason, supporters of Landon were much more likely to return replies to the poll than were supporters of Roosevelt (Squire, 1988). The poll predicted that Landon would win by a large margin, but in fact, Roosevelt won.

Assumption: Independent observations

The 95% CI is only valid when all subjects are sampled from the same population and each has been selected independently of the others.

In the premature infant example, the data would not be independent if some of the infants in the sample were twins, because they would share genetic or environmental factors that might influence survival, or if some of the deaths were caused by an infection or malfunctioning equipment that affected several infants in the study. The assumption would be violated in the election example if the pollsters questioned both the husband and the wife in each family, or if some voters were polled more than once.

Assumption: Accurate data

The 95% CI is only valid when the number of subjects in each category is tabulated correctly.

In the premature infant example, the outcome is death, which is pretty unambiguous. But even here, it is possible for the data to be distorted. If the treating physicians knew that six-month survival was being tracked, they may have used heroic measures to keep a five-month-old infant alive a few days longer until it was older than six months of age. Although this effort wouldn't ultimately help the infant, it would improve the six-month survival rate and distort the statistics. The data would not be accurate in the election example if the pollster deliberately recorded some of the opinions incorrectly or coerced or intimidated the respondents to answer a certain way.

Results are said to be *biased* if the experimental methods or analysis procedures ensure that the calculated results, on average, deviate from the true value.

WHAT DOES 95% CONFIDENCE REALLY MEAN?

The true population value either lies within the 95% CI you calculated or it doesn't. There is no way for you to know. If you were to calculate a 95% CI from many samples, you would expect it to include the population proportion in about 95% of the samples and to not include the population proportion in about 5% of the samples.

A simulation

To demonstrate this, let's switch to a *simulation* in which you know *exactly* the population from which the data were selected. Assume that you have a bowl of 100 balls, 25 of which are red and 75 of which are black. Mix the balls well and choose one randomly. Put it back in, mix again, and choose another one. Repeat this process 15 times, record the fraction of the sample that are red, and compute the 95% CI for that proportion. Figure 4.1 shows the simulation of 20 such samples. Each 95% CI is shown as a bar extending from the lower confidence limit to the upper confidence limit. The value of the observed proportion in each sample is shown as a line in the middle of each CI. The horizontal line shows the true population value (25% of the balls are red). In about half of the samples, the sample proportion is less than 25%, and in the other half, the sample proportion is higher than 25%. In one of the samples, the population value lies outside the 95% CI. In the long run, given the assumptions listed earlier, 1 of 20 (5%) of the

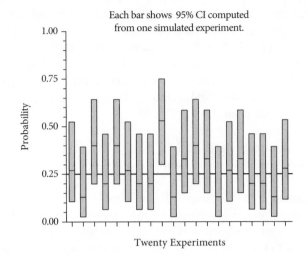

Figure 4.1. **What would happen if you collected many samples and computed a 95% CI for each?**

Each bar shows one simulated experiment, indicating the proportion of red balls chosen from a mixture of 25% red balls (the rest are black), with n = 15. The percentage success is shown as a line near the middle of each bar, which extends from one 95% confidence limit to the other. In all but one of the simulated samples, the CI includes the true population proportion of red balls (shown as a horizontal line). The 95% CI of Sample 9, however, does not include the true population value. You expect this to happen in 5% of samples. Because Figure 4.1 shows the results of simulations, we know when the CI doesn't include the true population value. When analyzing data, however, the population value is unknown, so you have no way of knowing whether the CI includes the true population value.

95% CIs will not include the population value, and that is exactly what we see in this set of 20 simulated experiments.

Figure 4.1 helps explain confidence intervals, but you cannot create such a figure when you analyze data. When you analyze data, you don't know the actual population value. You only have results from a single experiment. In the long run, 95% of such intervals will contain the population value and 5% will not. But there is no way you can know whether a particular 95% CI includes the population value or not, because you don't know that population value (except when running simulations).

95% chance of what?
A confidence interval has a 95% (or some other specified value) chance of containing the true population value. It is a mistake to flip things around and think that it specifies the chance that the population value lies within the calculated confidence interval. The two aren't quite the same.

Random chance affects which data you collect, which in turn affects the range of the CI. If you repeat the experiment, you'll get a different interval.

However, random chance does not affect the true population value, which is fixed (but unknowable). It doesn't matter how many times you repeat the experiment; the true population value won't change. It is considered incorrect to say: "There is a 95% chance that the population value is within this 95% CI." That statement implies that the population value is subject to random variation. Instead, it is proper to say: "There is a 95% chance that this 95% CI contains the population value." That statement correctly implies that the confidence interval is subject to random variation (if you were to repeat the experiment). Yes, this is a subtle distinction!

What is special about 95%?
Tradition!

CIs can be computed for any degree of confidence. By convention, 95% CIs are presented most commonly, although 90% and 99% CIs are sometimes published. With a 90% CI, there is a 10% chance that the interval does not include the population value. With a 99% CI, that chance is only 1%.

If you are willing to be less confident, then the interval will be *shorter*, so 90% CIs are narrower than 95% CIs. If you want to be more confident that your interval contains the population value, the interval must be *longer*. Thus, 99% CIs are wider than 95% CIs. If you want to be 100% confident that the interval contains the population value, the interval would need to be very wide so as to include every possible value, so this is not done.

ARE YOU QUANTIFYING THE EVENT YOU REALLY CARE ABOUT?

The 95% CI allows you to extrapolate from the sample to the population for the event that you tabulated. For the first example (premature babies), the tabulated event (death in the first month of life) is exactly what you care about. But sometimes you really care about an event that is different from the one you tabulated.

In the voting example, we assessed our sample's responses to a poll conducted on a particular date, so the 95% CI lets us generalize those results to predict how the population would respond to that poll on that date. We wish to extrapolate to election results in the future but can do so only by making an additional assumption: that people will vote as they said they would. In the polling prior to the 1948 Dewey versus Truman presidential election in the United States, polls of many thousands of voters indicated that Dewey would win by a large margin. Because the CI was so narrow, the pollsters were very confident. Newspapers were so sure of the results that they prematurely printed the headline "Dewey Beats Truman." In fact, Truman won. Why was the poll inaccurate? The polls were performed in September and early October, and the election was held in November. Many voters changed their minds in the interim period. The 95% CI computed from data collected in September was inappropriately used to predict voting results two months later.

LINGO

Confidence intervals versus confidence limits

The two ends of the CI are called the *confidence limits*. The confidence interval extends from one confidence limit to the other. The CI is a range, whereas each confidence limit is a value.

Many scientists use the terms *confidence interval* and *confidence limits* interchangeably. Fortunately, mixing up these terms does not get in the way of understanding statistical results.

Estimate

The sample proportion is said to be a *point estimate* of the true population proportion. The CI covers a range of values so is said to be an *interval estimate*.

Note that *estimate* has a special meaning in statistics. It does not mean an approximate calculation or an informed hunch, but rather is the result of a defined calculation. The term *estimate* is used because the value computed from your sample is only an estimate of the true value in the population (which you can't know).

Confidence level

A 95% CI has a confidence level of 95%. If you generate a 99% CI, the confidence level equals 99%. The term *confidence level* is used to describe the desired amount of confidence.

CALCULATING THE CI OF A PROPORTION

Several methods are commonly used

Several methods have been developed for computing the CI, and there is no consensus about which is best (Brown, Cai, & DasGupta, 2001). These methods all produce results that are pretty similar, especially with large samples. The three methods you are most likely to encounter are:

- The so-called *exact method* of Clopper and Pearson (1934) always has a confidence level of at least 95%, but the intervals are sometimes a bit wider than necessary. The Clopper method cannot be easily computed by hand.
- The *standard Wald method* (explained in the first edition of this book, and many others) is easy to compute by hand, but the modified Wald method (see next bullet point) is not much harder to compute and is much more accurate.
- The *modified Wald method,* developed by Agresti and Coull (1998) is quite accurate and is also easy to compute by hand (see next section). Simulations with many sets of data demonstrate that it works very well (Brown, Cai, & DasGupta, 2001; Ludbrook & Lew, 2009).

You can find a Web calculator that computes both the modified Wald and the Clopper/Pearson method at www.graphpad.com/quickcalcs/confInterval1/.

How to compute the modified Wald method by hand

If you're likely to only use a calculator or a program, you can skip this section. But if you want to compute a confidence interval by hand, it is easy to do so using the following steps.

1. Look up the value of z based on the confidence level you want. You'll learn more about z in Chapter 10. For 90% intervals, z = 1.645. For 99% intervals, z = 2.576. For 95% CI, z = 1.960, which is very close to 2, hence the approximations below.

2. Calculate a value called p' from the number of successes (S) and the number of trials (n). Its value will be between the observed proportion (S/n) and 0.5. This is the general equation, followed by the approximation for 95% confidence:

$$p' = \frac{S + 0.5z^2}{n + z^2} \approx \frac{S + 2}{n + 4}$$

3. Compute W, the margin of error (or half-width) of the CI:

$$W = z\sqrt{\frac{p'(1-p')}{n+z^2}} \approx 2\sqrt{\frac{p'(1-p')}{n+4}}$$

4. Compute the 95% CI.
   From (p' − W) to (p' + W)

Three notes:

- The variable p' used here is not the same as a P value (which we will discuss extensively in later chapters). This book uses an uppercase P for P values and a lowercase p for proportions, but not all books follow this convention.
- The margin of error is the direction the confidence interval extends in both directions. The length of the interval equals twice the margin of error.
- The confidence interval is symmetrical around p' but is not symmetrical around the observed proportion p. The lack of symmetry is noticeable with small samples and when p is far from 0.5.

For the survival data for the 25-week infants given at the beginning of this chapter, the sample proportion p is 31/39, or 79.5%; p' = 76.8%; W = 12.6%; and the 95% CI ranges from 64.2 to 89.5%.

Shortcut for proportions near 50%

When the proportion is near 50%, the margin of error of the CI is approximately equal to the square root of 1/n. This leads to the rules of thumb listed in Table 4.1. Many political polls use a sample size of about 1,000, and many polls have results very near 50:50. In this case, the margin of error is 3%, a value you'll often see reported in newspapers as the margin of error of a poll.

PROPORTION	MARGIN OF ERROR (%)	APPROXIMATE 95% CI
5/10	32	18 to 82
50/100	10	40 to 60
500/1,000	3	47 to 53
5,000/10,000	1	49 to 51

Table 4.1. Approximate CI of a proportion when the observed proportion is 50%.
This table's values should not be used for formal data analysis, because they are approximations. Remember these values when you wish to rapidly evaluate published data. The margin of error is approximately equal to the reciprocal of the square root of n.

Shortcut for proportions far from 50%
When the proportion is far from 50%, the confidence intervals will be narrower than if the proportion is close to 50%. Here is one rough rule of thumb for large n. If the proportion is 80% or 20%, then the confidence interval will be about 80% the length (so 20% shorter) than it would have been had the proportion been 50%.

Shortcut when the numerator is zero: The rule of three
Hanley and Lippman-Hand (1983) devised a simple shortcut equation, called the *rule of three*, for determining the 95% CI of a proportion when the numerator is zero. If you observe zero events in n trials, the 95% CI extends from 0.0 to approximately 3.0/n.

In the preceding example, 0 of 29 infants born at 22 weeks survived. Using the rule of three, the 95% CI for the survival proportion extends from 0.0 to 3/29, or 10.3%.

Another example: You have observed no adverse drug reactions in the first 250 patients treated with a new antibiotic. The CI for the true rate of drug reactions extends from 0.0 to about 3/250, or 1.20%.

Shortcut when the numerator is one or two: The rules of five and seven
Similar shortcut equations let you approximate the CI when the numerator is one or two (Montori et al., 2004).

When n is large, the lower confidence limit of 1/n will be close to zero and probably won't be of interest. The upper limit can be approximated as 5/n. For example, if you observe 1 success in 50 tries, the upper 95% confidence limit is very close to 5/50, or 0.10.

Similarly, if the numerator is two and n is large, the upper 95% confidence limit is close to 7/n. So if you observe 2 successes in 50 tries, the upper 95% confidence limit is very close to 7/50, or 0.14.

Table 4.2 summarizes the rules of three, five, and seven.

AMBIGUITY IF THE PROPORTION IS 0% OR 100%

In one of the previous examples, none of the 29 babies born at 22 weeks of gestation survived. It still makes sense to compute a CI, because the true proportion in the population may not be zero. Of course, the lower confidence limit must be 0%,

PROPORTION	UPPER 95% CONFIDENCE LIMIT (APPROXIMATE)
0/n	3/n
1/n	5/n
2/n	7/n

Table 4.2. The rules of three, five, and seven.

If the proportion is zero, one, or two over n (and n is fairly large), these rules will compute an approximation of the upper 95% confidence limit. The lower limit will be a small value (zero, when the numerator is zero) and usually is not of interest.

because there is no possibility of a proportion being negative. There are two ways to define the upper confidence limit in this situation.

As usually defined, the 95% CI allows for a 2.5% chance that the upper confidence limit doesn't go high enough to include the true population proportion and a 2.5% chance that the lower confidence limit doesn't go low enough to include the true population proportion. This leaves a 95% chance that the interval includes the true population proportion.

If you use this approach when the numerator is zero, there is a 2.5% chance that the upper interval doesn't go high enough. But there is no chance that the lower interval doesn't go low enough, because the lower limit is zero. Because the uncertainty only goes in one direction, this "95%" CI really gives you 97.5% confidence. Calculated in this way, the 95% CI for the percentage of surviving babies extends from 0 to 13.9%.

The alternative approach is to compute the upper confidence limit such that there is a 5% chance it doesn't go high enough to include the true population value. This approach creates a true 95% CI but is less consistent with other intervals. Calculated this way, the 95% CI for babies surviving extends from 0 to 10.2%.

If the numerator and denominator are equal, the sample proportion is 100%, which is also the upper limit of the CI. The same logic applies, and there are two ways to define the lower limit mirroring the approach used above to calculate the upper limit.

COMMON MISTAKES: CI OF A PROPORTION

Mistake: Using 100 as the denominator when the value is a percentage

The calculation of the confidence interval depends on the sample size. If the proportion is expressed as a percent, it is easy to mistakenly enter 100 as the sample size. If you do this, you'll get the wrong answer unless the sample size is 100 or close to it. Look at Table 4.1. In each case, the proportion is 0.5, so the percentage is 50%. But the confidence intervals vary enormously depending on sample size.

Mistake: Computing binomial confidence intervals from percentage change in a continuous variable

The methods explained in this chapter apply when there are two possible outcomes (i.e., binomial outcomes), so the result is expressed as a proportion of the time that one particular outcome occurs. This value is often expressed as a percentage. But beware of percentages, which can be used to express many kinds of data. Results such as percent change in weight or percent change in price, or percent change in survival time cannot be analyzed with the methods explained in this chapter. If you try to do so, the results won't make any sense.

Q & A

Which is wider, a 95% CI or a 99% CI?

To be more certain that an interval contains the true population value, you must generate a wider interval. A 99% CI is wider than a 95% CI. See Figure 4.2.

Is it possible to generate a 100% CI?

A 100% CI would have to include every possible value, so it would always extend from 0.0 to 100.0% and not be the least bit useful.

How do CIs change if you increase the sample size?

The width of the CI is approximately proportional to the reciprocal of the square root of the sample size. So if you increase the sample size by a factor of four, you can expect to cut the length of the CI in half. Figure 4.3 illustrates how the CI gets narrower if the sample size increases.

Can you compute a confidence interval of a proportion if you know the proportion but not the sample size?

No. The width of the confidence interval depends on the sample size. See Figure 4.3.

Why isn't the CI symmetrical around the observed proportion?

Because a proportion cannot go below 0.0 or above 1.0, the CI will be lopsided when the sample proportion is far from 0.50 or the sample size is small. See Figure 4.4.

You expect the population proportion to be outside your 95% CI in 5% of samples. Will you know when this happens?

No. You don't know the true value of the population proportion (except when doing simulations), so you won't know if it lies within your CI or not.

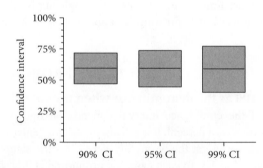

Figure 4.2. Effect of degree of confidence on the width of a CI.

Each bar represents a sample with a 60% success rate and n = 40. The graph illustrates the 90, 95, and 99% CIs for the percentage success in the population from which the data were drawn. When you choose to have less confidence, the CI is narrower.

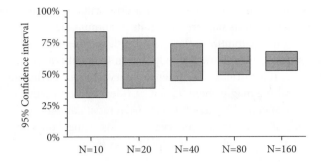

Figure 4.3. Effect of sample size on the width of a CI.

Each bar represents a sample with a 60% success rate, and the graph shows the 95% CI for the percentage success in the population from which the data were drawn. When the samples are larger, the CI is narrower.

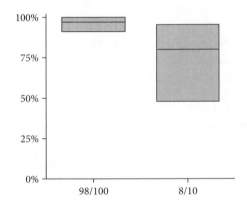

Figure 4.4. Asymmetrical CI.

If the proportion is far from 50%, the 95% CI of a proportion is noticeably asymmetrical, especially when the sample size is small.

CHAPTER SUMMARY

- This chapter discusses results that have two possible outcomes and are summarized as a proportion. The proportion is computed by dividing the number of times one outcome happened by the total number of trials.
- The binomial distribution predicts the distribution of results when you create many random samples and know the overall proportion of the two possible outcomes.
- A fundamental concept in statistics is the use of a confidence interval to analyze a single sample of data to make conclusions about the population from which the data were sampled.

- To compute a confidence interval of a proportion, you only need the two numbers that form its numerator and its denominator.
- Given a set of assumptions, 95% of confidence intervals computed from a proportion will include the true population value of that proportion. You'll never know if a particular confidence interval is part of that 95% or not.
- There is nothing magic about 95%. Confidence intervals can be created for any degree of confidence, but 95% is used most commonly.
- The width of a confidence interval depends in part on sample size. The interval is narrower when the sample size is larger.

TERMS INTRODUCED IN THIS CHAPTER

- Bias (p. 33)
- Binomial distribution (p. 29)
- Confidence interval (p. 31)
- Confidence level (p. 36)
- Confidence limit (p. 36)
- Cumulative binomial distribution (p. 30)
- Interval estimate (p. 36)
- Modified Wald method (p. 37)
- Point estimate (p. 36)
- Random sample (p. 30)
- Sampling error (p. 31)
- Simulation (p. 33)

CHAPTER 5

❖❖❖

Confidence Interval of Survival Data

In the long run, we are all dead.

JOHN MAYNARD KEYNES

Outcomes that can only happen once (e.g., death) are often displayed as graphs showing percentage survival as a function of time. This chapter explains how survival curves are created and how to interpret their confidence intervals. Even if you don't use survival data, skim this chapter to review the concept of confidence intervals with a different kind of data. This chapter explains how to analyze a single survival curve. Chapter 29 explains how to compare survival curves.

SURVIVAL DATA

The term *survival curve* is a bit limiting, because these kinds of graphs are used to plot time to any well-defined end point or event. The event is often death, but it could also be the time until occlusion of a vascular graft, first metastasis, or rejection of a transplanted kidney. The same principles are used to plot time until a lightbulb burns out, until a router needs to be rebooted, until a pipe leaks, and so on. In these cases, the term *failure time* is used instead of survival time. The event does not have to be dire. The event could be restoration of renal function, discharge from a hospital, resolution of a cold, or graduation. The event must be a one-time event. Special methods are needed when events can recur.

The methods described here (and in Chapter 29) apply when you know the survival time of each subject (or know when the data were censored, as explained in the next section). These methods are not appropriate for analyzing, for example, the survival of thousands or millions of cells, because you don't know the survival time of each individual cell. Instead, you would simply plot the percentage survival versus time and fit a curve to the data or connect the points with point-to-point lines.

CENSORED SURVIVAL DATA

If each subject's survival time were known, creating a survival curve would be trivial. But many survival times are not known.

Many studies enroll patients over a period of several years. The patients who enroll later are not followed for as many years as patients who enrolled early.

43

Imagine a study of a cancer treatment that enrolled patients between 1995 and 2000 and ended in 2008. If a patient enrolled in 2000 and was still alive at the end of the study, his survival time would be unknown but must exceed eight years. Although the study lasted 13 years, the fate of that patient after Year 8 is unknown.

During the study, imagine that some subjects dropped out—perhaps they moved to a different city or wanted to take a medication disallowed on the protocol. If a patient moved after two years in the study, his survival time would be unknown but must exceed two years. Even if you knew how long he lived, you couldn't use the data, because he would no longer be taking the experimental drug. However, the analysis must account for the fact that he lived at least two years on the protocol.

Information about these patients is said to be censored. The word censor has a negative connotation. It sounds like the subject has done something bad. Not so. It's the data that have been censored, not the subject! These censored observations should not be removed from the analyses—they must just be accounted for properly. The survival analysis must take into account how long each subject is known to have been alive and following the experimental protocol, and it must not use any information gathered after that time.

Table 5.1 presents data for a survival study that only includes seven patients. That would be a very small study, but the size makes the example easy to follow. The data for three patients are censored for three different reasons. One of the censored observations is for someone who is still alive at the end of the study. We don't know how long he will live after that. Another person moved away from the area and thus left the study protocol. Even if we knew how much longer he lived, we couldn't use the information, because he was no longer following the study protocol. One person died in a car crash. Different investigators handle this kind of situation differently. Some define a death to be a death, no matter what the cause (termed all-cause mortality). Some investigators present the data two ways, first using all subjects, called an intent-to-treat analysis, and then censoring subjects who did not adhere to the full protocol, called an according-to-protocol analysis. Here we will define a death from a clearly unrelated cause (such as a car crash) to be a censored observation. We know he lived 3.67 years on the treatment and don't know how much longer he would have lived, because his life was cut short.

Table 5.2 demonstrates how these data are entered into a computer program when using censoring. The codes for death (1) and censored (0) are commonly used but are not completely standard.

STARTING DATE	ENDING DATE	WHAT HAPPENED?
Feb 7, 1998	Mar 2, 2002	Died
May 19, 1998	Nov 30, 2004	Moved. No longer on protocol
Nov 14, 1998	Apr 3, 2000	Died
Mar 4, 1999	May 4, 2005	Study Ended
Jun 15, 1999	May 4, 2005	Died
Dec 1, 1999	Sep 4, 2004	Died
Dec 15, 1999	Aug 15, 2003	Died in car crash

Table 5.1. Sample survival data details for the data plotted in Figure 5.1.

YEARS	CODE
4.07	1
6.54	0
1.39	1
6.17	0
5.89	1
4.76	1
3.67	0

Table 5.2. How the data of Table 5.1 are entered into a computer program.

CALCULATING PERCENTAGE SURVIVAL
AT VARIOUS TIMES

There are two slightly different methods for calculating percentage survival over time. With the *actuarial method*, the X-axis is divided up into regular intervals, perhaps months or years, and survival is calculated for each interval. This method is used when the actual time of death is not known (Wormuth, 1999) or when the sample size is enormous. With the *Kaplan-Meier method*, survival time is recalculated with each patient's death. This method is preferred and is used more frequently, at least in clinical research. The term *life-table analysis* seems to be used inconsistently but usually includes both methods.

The Kaplan-Meier method is logically simple. To calculate the fraction of patients who survive on a particular day, simply divide the number alive at the end of the day by the number alive at the beginning of the day (excluding any who were censored on that day from both the numerator and the denominator). This gives you the fraction of patients who were alive at the beginning of a particular day and are still alive at the beginning of the next day. To calculate the fraction of patients who survive from Day 0 until a particular day, multiply the fraction of patients who survive Day 1 by the fraction of those patients who survive Day 2 and then by the fraction of those patients who survive Day 3, and so on, until you eventually multiply by the fraction who survive Day k. This method automatically accounts for censored patients, because both the numerator and the denominator are reduced on the day a patient is censored. Because we calculate the product of many survival fractions, this method is also called the *product-limit method*. This method is part of many statistical computer programs, so there is no need to learn the tedious details.

Time zero is not some specified calendar date; rather, it is the time that each patient entered the study. In many clinical studies, time zero spans several calendar years as patients are enrolled. At time zero, by definition, all patients are alive, so survival equals 100%. Whenever a patient dies, the percentage of surviving patients decreases. If the study (and thus the X-axis) were extended far enough, survival would eventually drop to zero.

Table 5.3 includes the computed survival percentages, with 95% CIs, at each time a patient dies.

NO. OF YEARS	LOWER LIMIT	% SURVIVAL	UPPER LIMIT
0.00	100.00	100.00	100.00
1.39	33.39	85.71	97.86
3.67	33.39	85.71	97.86
4.07	21.27	68.57	91.21
4.76	11.77	51.43	81.33
5.89	4.81	34.29	68.56
6.17	4.81	34.29	68.56
6.54	4.81	34.29	68.56

Table 5.3. Kaplan-Meier survival proportions with 95% confidence limits computed by computer (not using the approximate equations presented in this chapter).

GRAPHING SURVIVAL CURVES WITH CONFIDENCE BANDS

Figure 5.1 plots the example data four ways. Each time a patient dies, the curve drops down. The censored subjects are shown as circles in the upper- and lower-right panels and as ticks in the lower-left panel.

An essential part of data analysis includes interpreting confidence intervals. Survival curve programs compute a 95% CI at each time that a death occurs. These intervals can then be joined to plot 95% confidence bands, as shown in the lower-left panel of Figure 5.1. The stepped survival curve shows exactly what happened in the sample we studied. Given certain assumptions (explained later in the chapter), we can be 95% sure that 95% confidence bands at any time include the true percentage survival in the population.

Note that the confidence interval at any particular time is asymmetrical. This is because the percentage survival cannot possibly go below 0.0% or above 100%. The asymmetry of the CI will be especially noticeable when the sample survival percentage is far from 50 or the sample size is small. The interval is symmetrical when the survival percentage equals 50% and is always nearly symmetrical when the sample size is huge.

SUMMARIZING SURVIVAL CURVES

Median survival

It can be convenient to summarize an entire survival curve by one value, the *median survival*. The median is the middle value (the 50th percentile) of a set of numbers, so median survival is how long it takes until half the subjects have died.

It is easy to derive the median survival time from a survival curve: Draw a horizontal line at 50% survival and note where it crosses the survival curve. Then look down at the X-axis to read the median survival time. The lower-right panel of Figure 5.1 indicates that the median survival in this example is 5.89 years.

Calculations of median survival are ambiguous in two situations. Median survival is undefined when more than half the subjects are still alive at the end of the

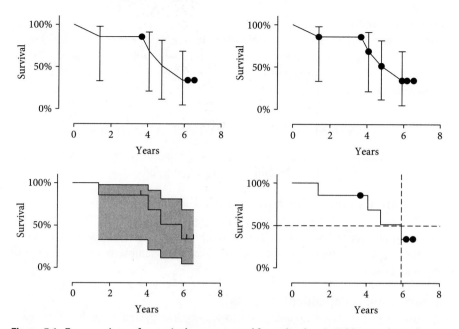

Figure 5.1. Four versions of a survival curve created from the data in Tables 5.1 through 5.3.
The X-axis plots the number of years after each subject was enrolled. Note that time zero does not have to be any particular day or year. The Y-axis plots percentage survival. The three censored patients are shown as symbols in three of the graphs, and as upward blips in the fourth. Four of the subjects died. You can see each death as a downward step in the curves. The 95% CIs for the populations are shown as error bars in the top two graphs, and as a shaded region (confidence bands) in the bottom-left graph. The bottom-right graph shows the median survival time, which is the time at which half of the subjects have died and half are still alive. Read across at 50% to determine median survival, which is 5.89 years in this example.

study. In this case, the survival curve never crosses the 50% survival line. The median survival is greater than the last duration of time plotted on the survival curve, but there is no way to know how much greater. The other ambiguous situation is when the survival curve is horizontal at 50% survival. One way to define median survival in this case is to average the first and last times when survival is 50%.

Mean survival?

Mean survival is rarely computed. This is because calculating a mean survival time requires that you know every single time of death. Thus, the mean survival cannot be computed if any observations are censored or if any subjects are still alive when the study ends.

In contrast, median survival can be computed when some observations are censored and when the study ends before all subjects have died. Once half of the subjects have died, the median survival is unambiguous, even without knowing how long the others will live.

Five-year survival

Survival with cancer is often quantified as *five-year survival*. A vertical line is drawn at X = 5 years, and the Y value that line intersects is the five-year percentage survival. Of course, there is nothing special about five years (rather than two or four or six years) except tradition.

ASSUMPTIONS: SURVIVAL ANALYSIS

When evaluating any statistical analysis, it is critical to review all assumptions. Analyses of survival curves depend on these assumptions.

Assumption: Random (or representative) sample

The whole idea of all statistical analyses is to generalize the data from your sample to a more general situation. If your sample is not randomly selected from a defined population, then you must *assume* that your sample is representative of that population.

Assumption: Independent subjects

The results of any statistical analyses can only be interpreted if the data from each subject offer independent information. If the study pools data from two hospitals, the subjects are not independent. It is possible that subjects from one hospital have different average survival times than subjects from another hospital, and you could alter the survival curve by choosing more subjects from one hospital and fewer from the other hospital. Because most diseases have a genetic component, another way to violate the assumption of independence would be to include two (or more) people from one family in one treatment group.

Assumption: Entry criteria are consistent

Typically, subjects are enrolled over a period of months or years. In these studies, it is important that the entry criteria do not change during the enrollment period.

Imagine a cancer survival curve starting from the date that the first metastasis was detected. What would happen if improved diagnostic technology detected metastases earlier? Patients would die at the same age they otherwise would, but now they would be diagnosed at a younger age and therefore would live longer with the diagnosis. Even with no change in therapy or in the natural history of the disease, survival time would apparently increase simply because the entry criteria changed.

Airlines use this trick to improve their on-time departure rates. Instead of planning to close the doors at the scheduled departure time, they now plan to close the airplane's door 10 minutes before the scheduled departure time. This means that the doors can close 10 minutes later than actually planned and the flight can still be defined as on time.

Assumption: End point defined consistently

If the curve is plotting duration of time until death, there can be ambiguity about which deaths to count. In a cancer trial, for example, what happens to subjects who die in car crashes? Some investigators count these as deaths; others count them as censored subjects. Both approaches can be justified, but the approach should be decided before the study begins or both approaches presented in the results. If there is any ambiguity about which deaths to count, the decision should be made by someone who doesn't know which patient received which treatment.

If the curve plots duration of time until an event other than death, it is crucial that the event be assessed consistently throughout the study.

Assumption: Starting time clearly defined

The starting point should be an objective date—perhaps the date of first diagnosis or first hospital admission. You may be tempted to use an earlier starting criterion instead, such as the time when a patient remembers first observing symptoms. Don't do it. Such data are invalid because a patient's recollection of early symptoms may be altered by later events.

What happens when subjects die before they receive the treatment they were supposed to get? It is tempting to remove these subjects from the study. But this can lead to biases, especially if one treatment (such as a medication) is started immediately, but another (surgery) requires preparation or scheduling. If you remove the patients who die early from the surgical group but not from the medication group, the two groups will have different survival times, even if the treatments are equivalent. To avoid such biases, most studies follow a policy of *intention to treat*. Each subject's survival is analyzed as if he or she received the assigned treatment, even if the treatment was not actually given. The justification for using the intention-to-treat criterion will be discussed in Chapter 29.

Assumption: Censoring is unrelated to survival

The survival analysis is only valid when the reasons for censoring are unrelated to survival. If data from a large fraction of subjects are censored, the validity of this assumption is critical to the integrity of the results.

Data for some patients are censored because they are alive at the end of the study. With these patients, there is no reason to doubt the assumption that censoring is unrelated to survival.

The data for other patients are censored because they drop out of the study. If the reason these patients dropped out could be related to survival, then the analysis won't be valid. Examples include patients who quit the study because they were too sick to come to the clinic, patients who felt well and stopped treatment, and patients who quit because they didn't think the treatment was working (or who were taken off the study by a physician who thought the treatment wasn't working). These reasons all relate to disease progression or response to therapy, and it is quite possible that the survival of these patients is related to their reason for dropping out. Including these subjects (and censoring their data) violates the assumption that censoring is unrelated to survival. But excluding these subjects entirely can also lead to biased results. The best plan is to analyze the data both ways. If the conclusions of the two analyses are similar, then the results are straightforward to interpret. If the conclusions of the two analyses differ substantially, then the study results are simply ambiguous.

Assumption: Average survival doesn't change during the study

Many survival studies enroll subjects over a period of several years. The analysis is only meaningful if you assume that the first few patients are sampled from the same population as the last few subjects.

If the nature of the disease changes during the time the study accrues patients, the results will be difficult to interpret. This is quite possible when studying rapidly evolving infectious diseases. It is also important that the treatment (including supportive care) does not change over the course of the study.

Q & A

Why is percentage survival graphed as staircases, rather than as point-to-point lines?
> The tradition is to plot the actual experience in the sample. So when a subject dies, the curve drops down in a staircase fashion. Connecting the points diagonally might do a better job of demonstrating your best estimate for survival in the overall population, but this is not standard.

What if subjects enter the study on different dates?
> That's OK and is common. The X-axis shows time from entry in the study, so it does not correspond to calendar dates.

Why isn't a CI plotted at time zero?
> At time zero, 100% survival is a given. Subjects who are not alive at time zero are not included in the study! So the survival at time zero is not a value subject to sampling error. It is 100% for sure.

My confidence band is too wide to be useful. How can I make it narrower?
> Collect more data, a lot more data. The width of the CI is approximately proportional to the reciprocal of the square root of the sample size. So if you increase the sample size by a factor of four, you can expect the CI to be half as wide.

In the example, three of seven subjects survived. Why doesn't the curve end at Y = 3/7, or 43%?
> Survival curve calculations properly account for censored data. After data are censored, that subject doesn't contribute to computing the percentage survival. Data for two subjects were censored prior to the last time point, so the simple computation (2/7 = 29% still survive) is not correct. In fact, the curve ends with 34.29% survival.

Can the data be plotted as percentage death rather than percentage survival?
> Yes! See Figure 5.2.

How does one compare two survival curves?
> See Chapter 29.

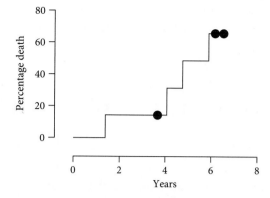

Figure 5.2. A graph of percentage death, rather than percentage survival, illustrates the same information.

When only a small percentage of subjects have died by the end of the study, this kind of graph can be more informative (because the Y-axis doesn't have to extend all the way to 100).

CHAPTER SUMMARY

- Survival analysis is used to analyze data for which the outcome is elapsed time until a one-time event (often death) occurs.
- The tricky part of survival analysis is dealing with censored observations. An observation is censored when the one-time event hasn't happened by the time the study ends, or when the patient stops following the study protocol.
- Survival curves can be plotted with confidence bands.
- A survival curve can be summarized by the median survival or the five-year survival.
- Survival curves, and their confidence bands, can only be interpreted if you accept a list of assumptions.

TERMS INTRODUCED IN THIS CHAPTER

- Actuarial method (p. 45)
- Censored data (p. 44)
- Failure time (p. 31)
- Five-year survival (p. 48)
- Intention to treat (p. 49)
- Kaplan-Meier (product limit) method (p. 45)
- Life-table analysis (p. 45)
- Median survival (p. 46)
- Survival curve (p. 43)

CHAPTER 6

❖❖❖

Confidence Interval of Counted Data (Poisson Distribution)

> Not everything that can be counted counts; and not everything that counts can be counted.
>
> ALBERT EINSTEIN

When events occur randomly, independently of one another, and with an average rate that doesn't change over time, the number of events counted in a certain time interval follow a Poisson distribution. From the number of events actually observed, a confidence interval can be computed for the average number of events per unit of time. This also works for computing a CI for the average number of objects in a certain volume. Poisson distributions are common in many fields of science. In addition to teaching you about a useful distribution, this chapter also reinforces the concept of a confidence interval.

THE POISSON DISTRIBUTION

Some outcomes are expressed as the number of objects in some volume or the number of events in some time interval. Examples include the number of babies born in an obstetrics ward each day, the number of eosinophils seen in one microscope field, or the number of radioactive disintegrations detected by a scintillation counter in one minute.

As you'd expect, if you repeatedly sample time intervals (or objects in a volume), the number of events (or objects) will vary. This random distribution is called a *Poisson distribution*. Given the population's average number of occurrences in one unit of time (or space), the Poisson distribution predicts how often you'll observe any particular number of events or objects.

Figure 6.1 presents examples of the Poisson distribution. Background radiation depends on the types of radiation you measure, your location, and the sensitivity of the detector, but it ranges from a few to a few dozen counts per minute, on average. In any particular minute, the number of detected radiation counts may be higher or lower than the average, but the number of disintegrations counted will always be an integer.

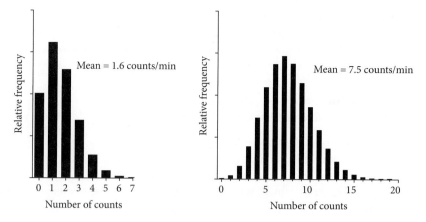

Figure 6.1. The Poisson distribution.
Assume that the average number of radiation counts is 1.6 counts/minute (left) or 7.5 counts/minute (right). In any particular minute, the count may be higher or lower than the average, as illustrated by these Poisson distributions. The horizontal axis lists the number of radioactive disintegrations actually counted (not converted to counts per minute), and the vertical axis plots the frequency.

Figure 6.1 illustrates the predictions of the Poisson distribution of background radiation. The left half of Figure 6.1 shows the predictions of a Poisson distribution with an average of 1.6 counts per minute. This Poisson distribution is notably asymmetrical. This makes sense, because the number of radioactive background counts cannot be less than zero but has no upper limit. The right half of Figure 6.1 shows the predictions of a Poisson distribution with an average of 7.5 counts per minute. This Poisson distribution is nearly symmetrical and almost looks like a Gaussian distribution (see Chapter 10).

ASSUMPTIONS: POISSON DISTRIBUTION

Calculations based on the Poisson distribution can be used for data expressed as the number of events per unit time or the number of objects in a certain volume.

Assumptions for number of events
Calculations about events assume the following:

- The event is clearly defined.
- Each event occurs randomly, independent of other events. This assumption might be violated if you tracked the number of babies born each day, because of the occurrence of twins and triplets. Instead, count the number of deliveries (counting twins and triplets as one delivery).
- The average rate doesn't change over time.
- Each event is counted only once. This assumption was violated in a study conducted to find out how often airplanes nearly collide. The investigators

surveyed pilots and copilots about the number of times they experienced a near miss. A single near miss, therefore, could be reported four times—by both pilots in both planes. For this reason, the initial analysis overestimated the number of near misses (*New Scientist*, 2007).

Assumptions for number of objects

Calculations about objects assume the following:

- Objects are randomly dispersed.
- Each object is only counted once. If you are counting cells, any clumping would make it hard to comply with this assumption.
- Objects are well defined, with no ambiguity about what to count (or what not to count). For example, if you are counting cells in a microscope field, there must be no ambiguity about differentiating cells from debris.

CONFIDENCE INTERVALS BASED ON POISSON DISTRIBUTIONS

The Poisson distribution can be used to compute a CI. As long as you know the actual number of objects counted in a volume, it is possible to compute a CI for the average number of objects in that volume. Similarly, if you know the number of events that occurred in one time interval, you can compute a CI for the average number of counts in that time interval. The calculations are explained later in this chapter. These examples are meant to explain the concept, not the calculations.

Raisins in bagels

Imagine that you carefully dissect a bagel and find 10 raisins. You assume that raisins are randomly distributed among bagels, that the raisins don't stick to each other to create clumps (perhaps a dubious assumption), and that the average number of raisins per bagel doesn't change over time (the recipe is constant). The 95% CI (determined using the Poisson distribution) ranges from 4.8 to 18.4 raisins per bagel (see Table 6.1). You are 95% certain that the average number of raisins per bagel lies in that range.

Radioactive counts

Imagine that you have counted 120 radioactive counts in one minute. Radioactive counts occur randomly and independently, and the average rate doesn't change (within a reasonable time frame much shorter than its half-life). Thus, the Poisson distribution is a reasonable model. The 95% CI for the average number of counts per minute is 99.5 to 143.5.

   Note that you must base this calculation on the actual number of radioactive disintegrations counted. If you counted the tubes for 10 minutes, the calculation must be based on the counts in 10 minutes and not on the calculated number of counts in 1 minute. This point is explained in more detail later in this chapter.

C	LOWER LIMIT	UPPER LIMIT
0	0.00	3.69
1	0.03	5.57
2	0.24	7.22
3	0.62	8.77
4	1.09	10.24
5	1.62	11.67
6	2.20	13.06
7	2.81	14.42
8	3.45	15.76
9	4.12	17.08
10	4.80	18.39
11	5.49	19.68
12	6.20	20.96
13	6.92	22.23
14	7.65	23.49
15	8.40	24.74
16	9.15	25.98
17	9.90	27.22
18	10.67	28.45
19	11.44	29.67
20	12.22	30.89
21	13.00	32.10
22	13.79	33.31
23	14.58	34.51
24	15.38	35.71
25	16.18	36.90
30	20.24	42.83
35	24.38	48.68
40	28.58	54.47
45	32.82	60.21
50	37.11	65.92
55	41.43	71.59
60	45.79	77.23
65	50.17	82.85
70	54.57	88.44
75	58.99	94.01
80	63.44	99.57
85	67.89	105.10
90	72.37	110.63
95	76.86	116.13
100	81.36	121.63

Table 6.1. The 95% CIs of a Poisson count (C).

Values calculated using the following Excel equations (where A1 is the cell containing C): lower limit = CHIINV(0.975, 2 * A1)/2; upper limit = CHIINV(0.025, 2*(A1 + 1))/2.

People-years

Exposure to an environmental toxin caused 1.6 deaths per 1,000 person-years exposure. What is the 95% CI? To calculate the CI, you must know the exact number of deaths that were observed in the study. This study observed 16 deaths in observations of 10,000 person-years (they might have studied 10,000 people for 1 year or 500 people for 20 years). If you set the count (C) to equal 16, the 95% CI for the number of deaths ranges from 9.15 to 25.98. That is the number of deaths per 10,000 person-years. Divide by 10 to return to the original units (per 1,000 person-years). The 95% confidence interval ranges from 0.92 to 2.6 deaths per 1,000 person-years exposure.

HOW TO CALCULATE THE POISSON CI

To compute the CI of a count, all you need is the count from one sample. If you have multiple samples, just add the counts from each sample to calculate a total count. It is essential that you base the calculations on the number of events (or objects) *actually observed*. Don't convert to a more convenient time scale or volume until after you have calculated the CI. If you attempt to calculate a CI based on the converted count, the results will be meaningless.

Table 6.1 shows you the confidence interval for many values of C and gives the formula to use within Excel. When C is large (greater than 25 or so), the following equation is a useful shortcut approximation for the 95% CI of C:

$$C - 1.96\sqrt{C} \text{ to } C + 1.96\sqrt{C}$$

If C is 25, the approximate 95% CI extends from 15.2 to 34.8, whereas the exact CI ranges from 16.2 to 36.9 (see Table 6.1).

If your count combined multiple samples, after calculating the confidence intervals using the total count divide both ends of the interval by the number of samples to calculate the CI of the number of counts per sample.

THE ADVANTAGE OF COUNTING FOR LONGER TIME INTERVALS (OR LARGER VOLUMES)

Figure 6.2 demonstrates the advantage of using a longer time interval. One tube containing a radioactive sample was counted repeatedly. The left side of Figure 6.2 illustrates radioactive decays counted in one-minute intervals. The right side of Figure 6.2 illustrates radioactive decays counted in 10-minute intervals. The graph plots counts per minute, so the number of radioactive counts counted in each 10-minute interval was divided by 10.0 after calculating the CI but before being graphed.

When computing the CIs for Poisson variables, it is essential to do the calculations with the number of objects or events actually counted. If you count 700 radioactive disintegrations in one minute, the 95% CI for the average number of disintegrations per minute will range from about 650 to 754 (using the free QuickCalc at www.graphpad.com/quickcalcs/confInterval1/, a tool that uses a more complicated and accurate equation than the one shown in the previous section). If you count 7,000 disintegrations in 10 minutes, the 95% CI for the average number of disintegrations per 10 minutes will range from 6,838 to 7,166. Divide those values by 10 to obtain the 95% CI for the average number of decays per minute, which ranges from 684 to 718. Counting for a longer period of time gives you a more precise assessment of the average number of counts per interval, and thus the CI is narrower.

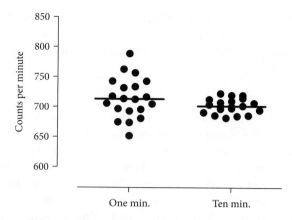

Figure 6.2. The advantages of counting radioactive samples for a longer time interval.

One radioactive sample was repeatedly measured for 1-minute intervals (left) and 10-minute intervals (right). The number of radioactive counts detected in the 10-minute samples was divided by 10, so both parts of the graph show counts per minute. By counting longer, there is less Poisson error. Thanks to Arthur Christopoulos for providing these data.

Let's revisit the example of raisins in bagels. Instead of counting raisins in one bagel, imagine that you dissected seven individual bagels and found counts of 9, 7, 13, 12, 10, 9, and 10 raisins. In total, there were 70 raisins in seven bagels, for an average of 10 raisins per bagel. The CI should be computed using the total counted. Given 70 objects counted in a certain volume, the 95% CI for the average number ranges from 54.57 to 88.44. This is the number of raisins per seven bagels. Divide by seven to express these results in more useful units. The 95% CI ranges from 7.8 to 12.6 raisins per bagel.

Q & A

What is the difference between the binomial and Poisson distributions?

Both the binomial and Poisson distributions are used for outcomes that are counted, but the two are very different. The binomial distribution describes the distribution of two possible outcomes. The Poisson distribution describes the possible number of objects you'll find in a certain volume or the number of events you'll observe in a particular time span.

Why use 1.96 in the equation in the previous section?

You'll learn about this in Chapter 10. In a Gaussian distribution, 95% of the values lie within 1.96 standard deviations of the mean. With large values of C, the Poisson distribution approximates a Gaussian distribution, with a standard deviation equal to the square root of C.

Is it possible to compute a CI when the observed number of counts is zero?

Yes. When you observe zero objects in a certain volume or zero events in a certain time, the 95% CI for the average number of objects in that volume (or events in that time interval) ranges from 0.0 to 3.69.

It is essential that you convert time (or volume) units only *after* computing the CI from the actual number of events or objects that were counted. If you first convert to counts per minute, you will get the wrong result when computing the CI.

CHAPTER SUMMARY

- When events occur randomly, independently of one another, and with an average rate that doesn't change over time, the number of events counted in a certain time interval follows a Poisson distribution.
- When objects are randomly distributed and not clumped, the number of objects counted in a certain volume follows a Poisson distribution.
- From the number of events actually observed (or the number of objects actually counted), a CI can be computed for the average number of events per unit of time (or number of objects per unit volume).
- When using a Poisson distribution to compute a CI, you must base the calculation on the actual number of objects (or events) counted. Any normalization to more convenient units must be done after you have computed the CI.

TERM INTRODUCED IN THIS CHAPTER

- Poisson distribution (p. 52)

PART C

Continuous Variables

CHAPTER 7

❖❖❖

Graphing Continuous Data

> When you can measure what you are speaking about and
> express it in numbers, you know something about it; but when
> you cannot measure it, when you cannot express it in numbers,
> your knowledge is of the meager and unsatisfactory kind.
>
> LORD KELVIN

When results are continuous (e.g., blood pressure, enzyme activity, IQ score, blood hemoglobin, oxygen saturation, temperature, etc.), the first and too-often neglected step is to visualize the data. This chapter shows how the actual distribution of values can be graphed without any statistical calculations.

CONTINUOUS DATA

When analyzing data, it is essential to choose methods appropriate for the kind of data with which you are working. To highlight this point, this book began with discussions of three kinds of data. Chapter 4 discussed data expressed as two possible outcomes, summarized as a proportion; Chapter 5 explained survival data and how to account for censored observations; and Chapter 6 discussed data expressed as the actual number of events counted in a certain time or as objects counted in a certain volume.

This chapter begins our discussion of continuous data, such as blood pressures, enzyme activity, weights, and temperature. In many scientific fields, continuous data are more common than other kinds of data.

THE MEAN AND MEDIAN

Mackowiak, Wasserman, and Levine (1992) measured body temperature from hundreds of healthy individuals to see what the normal temperature range really is. A subset of 12 of these values is shown in Table 7.1. These values, and the suggestion to use them to explain basic statistical principles, come (with permission) from Schoemaker (1996).

Calculating an arithmetic *mean* or average is easy: add up all the values and divide by the number of observations. The mean of the smaller (n = 12) subset is 36.77°C. If the data contain an outlier (a value far from the others), the mean

37.0
36.0
37.1
37.1
36.2
37.3
36.8
37.0
36.3
36.9
36.7
36.8

Table 7.1. The body temperature of 12 individuals in degrees centigrade.

These values are used as sample data.

won't be very representative (learn more about outliers in Chapter 25). For example, if the largest value (37.1) were mistakenly entered into the computer as 371 (that is, without the decimal point), the mean would increase to 64.6°C, which is larger than all the other values.

The mean is one way to quantify the middle, or central tendency, of the data, but it is not the only way. Here are some others:

- The *median* is the middle value. Rank the values from lowest to highest and identify the middle one. If there is an even number of values, average the two middle ones. For the n = 130 data set, the median is the average of the 65th and 66th ranked values, or 36.85°C. Compared to the mean, the median is not influenced by an outlier, and can be more useful with skewed distributions. For example, the mean cost of a US wedding in 2011 was $27,021, but the median price was $16,886 (Oremus 2013). A small proportion of really expensive weddings bring up the mean cost but don't affect the median cost.

- The *geometric mean* is calculated by transforming all values into their logarithms, computing the mean of these logarithms, and then taking the antilog of that mean (logarithms and antilogarithms are reviewed in Appendix E). Because a logarithm is only defined for values greater than zero, the geometric mean cannot be calculated if any values are zero or negative. None of the temperature values is zero or negative, so the geometric mean could be computed. However, the geometric mean would not be useful with the sample temperature data, because 0.0°C does not mean "no temperature." Temperature in degrees centigrade is not a ratio variable (defined in Chapter 8). Read more about the geometric mean in Chapter 11.

- The *harmonic mean* is calculated by first transforming each value to its reciprocal and then computing the (arithmetic) mean of those reciprocals. The harmonic mean is the reciprocal of that mean. The harmonic mean is not commonly used in the biological sciences. It can't be computed if any values are zero or negative.

- The *trimmed mean* is the mean of most of the values, ignoring the highest and lowest values. Olympic ice skating used to be scored this way, with the largest and smallest score eliminated before averaging the scores from all other judges. Sometimes several of the highest and lowest values are ignored.

- The *mode* is the value that occurs most commonly in the data set. It is not useful with continuous variables that are assessed with at least several digits of accuracy, because each value will be unique. The mode can be useful when working with variables that can only be expressed as integers. Note that the mode doesn't always assess the center of a distribution. Imagine, for example, a medical survey on which one of the questions is: "How many times have you had surgery?" In many populations, the most common answer will be zero, so that is the mode. In this case, some values will be higher than the mode, but none will be lower.

ERROR, BIAS, PRECISION, AND PERCENTILES

Error

In Table 7.1, the temperatures range from 36.0 to 37.3°C. Most of this variation is probably the result of biological variability. People (and animals, and even cells) are different from one another, and these differences are important! Moreover, people (and animals) vary over time because of circadian variations; aging; and alterations in activity, mood, and diet. In biological and clinical studies, much or most of the scatter is often caused by biological variation.

It is possible that some of the variation in the temperature values comes from imprecision, or experimental *error*. Reading many thermometers takes a bit of judgment, and the process can be prone to error. Many statistics books (especially those designed for engineers) implicitly assume that most variability is the result of imprecision. In medical studies, biological variation often contributes more variation than does experimental imprecision.

Mistakes and glitches can also contribute to variability. For example, it is possible that a value was written down incorrectly or that the thermometer wasn't positioned properly.

The word *error* is often used to refer to all three of these sources of variability. Note that this statistical use of the term is quite different from the everyday use of the word to mean a mistake. The terms *scatter* or *variability* are more understandable than the term *error*, but statistics books tend to use *error* when referring to any kind of variation not explained by treatment effects.

Bias

Biased measurements result from systematic errors. Bias can be caused by any factor that consistently alters the results: the proverbial thumb on the scale, defective thermometers, bugs in computer programs (maybe the temperature was measured in centigrade and the program that converted the values to Fahrenheit was buggy), the placebo effect, and so on. As used in statistics, the word *bias* refers to anything that leads to systematic errors, not only the preconceived notions of the experimenter. Biased data are not accurate.

Precision

Compare repeated measurements to see how precise the measurements are. *Precise* means the same thing as *repeatable* or *reproducible*. A method is precise when repeated measurements give very similar results. But note that a set of measurements could be quite precise without being accurate if the methodology is not working or is not calibrated properly. A measurement, or a method to obtain measurements, can be accurate and precise, accurate but not precise, precise but not accurate, or neither accurate nor precise.

Percentiles

You are probably familiar with the concept of a percentile. For example a quarter (25%) of the values in a set of data are smaller than the 25th percentile, and 75% of the values are smaller the 75th percentile (so 25% are larger).

The 50th percentile is the value in the middle. Half the values are larger and half are lower. If there is an even number of values, then the median is the average of the middle two values. The 50th percentile is identical to the median.

Calculating the percentiles is trickier than you'd guess. Eight different equations can be used to compute percentiles (Harter, 1984). All methods compute the same result for the median (50th percentile), but not necessarily for other percentiles. With large data sets, however, the results are all similar.

GRAPHING DATA TO SHOW SCATTER
OR DISTRIBUTION

Scatter plots show every value

Figure 7.1 plots the temperature data as a *column scatter plot*. The left side of the graph shows all 130 values. The right side of the graph shows the randomly selected subset of 12 values (Table 7.1), which we will analyze separately. Each value is plotted as a symbol. Within each half of the graph, symbols are moved to the right or left to prevent too much overlap; the horizontal position is arbitrary, but the vertical position, of course, denotes the measured value.

This kind of column scatter graph, also known as a *dot plot*, demonstrates exactly how the data are distributed. You can see the lowest and highest values and the distribution. There is no need to plot a mean and error bar, because each value is visible. A horizontal line is commonly drawn at the mean or median, and sometimes error bars are shown too (see Chapter 14).

With huge numbers of values, column scatter graphs get unwieldy with so many overlapping points. The left side of Figure 7.1, with 130 circles, is pushing

Figure 7.1. Column scatter graph of body temperatures.
(Left) The entire data set, with n = 130. (Right) A randomly selected subset (n = 12). In a column scatter graph, the vertical position of each symbol denotes its value. The horizontal position (within each lane) is adjusted (jittered) to prevent points from overlapping (too much).

the limits of this kind of graph. But the right side, with 12 circles, is quite clear and takes the same amount of space as a graph with mean and error bars, while showing more information.

Box-and-whiskers plots

A *box-and-whiskers plot* gives you a good sense of the distribution of data without showing every value (see Figure 7.2). Box-and-whiskers plots work great when you have too many data points to show clearly on a column scatter graph but don't want to take the space to show a full frequency distribution.

A horizontal line marks the median (50th percentile) of each group. The boxes extend from the 25th to the 75th percentiles and therefore contain half the values. A quarter (25%) of the values are higher than the top of the box, and 25% of the values are below the bottom of the box.

The whiskers can be graphed in various ways. The first box-and-whiskers plot in Figure 7.2 plots the whiskers down to the 5th and up to the 95th percentiles and plots individual dots for values lower than the 5th and higher than the 95th percentile. The other box-and-whiskers plot in Figure 7.2 plots the whiskers down to the smallest value and up to the largest, and so it doesn't plot any individual points. Whiskers can be defined in other ways as well.

Frequency distribution histograms

A frequency distribution lets you see the distribution of many values. Divide the range of values into a set of smaller ranges (bins) and then graph the number of values (or the fraction of values) in each bin. Figure 7.3 displays *frequency distribution histograms* for the temperature data. If you add the height of all

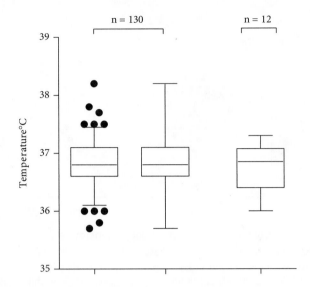

Figure 7.2. Box-and-whiskers plots.

(Left) Box-and-whiskers plots of the entire data set. The whiskers extend down to the 5th percentile and up to the 95th, with individual values showing beyond that. (Middle) The whiskers show the range of all the data. (Right) The box-and-whiskers plot the subset of 12 values.

the bars, you'll get the total number of values. If the graph plots fractions or percentages instead of number of values, then the sum of the height of all bars will equal 1.0 or 100%.

The trick in constructing frequency distributions is deciding how wide to make each bin. The three graphs in Figure 7.3 use different bin widths. The graph on the top left has too few bins (each bin covers too wide a range of values), so it doesn't really show you enough detail about how the data are distributed. The graph on the bottom has too many bins (each bin covers too narrow a range of values), so it shows you too much detail (in my opinion). The upper-right graph seems the most useful.

Watch out for the term *histogram*. It is usually defined as a frequency distribution plotted as a bar graph, as illustrated in Figure 7.3. Sometimes, however, the term histogram is used more generally to refer to any bar graph, even one that is not a frequency distribution.

Cumulative frequency distribution

One way to avoid choosing a bin width is to plot a *cumulative frequency distribution,* in which each Y value is the number of values less than X. The cumulative distribution begins at Y = 0 and ends at Y = n, the number of values in the data set. This kind of graph can be made without choosing a bin width. Figure 7.4 illustrates a cumulative frequency distribution of the temperature data.

Figure 7.3. Frequency distribution histograms of the temperature data with various bin widths.

With too few bins (top left), you don't get a sense of how the values vary. With many bins (bottom), the graph shows too much detail for most purposes. The top-right graph shows the number of bins that seems about right. Each bar plots the number of individuals whose body temperature is in a defined range (bin). The centers of each range, the bin centers, are labeled.

Figure 7.5 shows the same graph, but the Y-axis plots the percentage (rather than the number) of values less than or equal to each X value. The right side of Figure 7.5 illustrates the same distribution plotted with the Y-axis transformed in such a way that a cumulative distribution from a Gaussian distribution (see Chapter 10) becomes a straight line.

When graphing a cumulative frequency distribution, there is no need to decide on the width of each bin. Instead, each value can be individually plotted.

BEWARE OF DATA MASSAGE

Published graphs sometimes don't plot the data that were actually collected but instead plot the result of computations. Pay attention to the decisions and calculations done between data collection and graphing.

Beware of filtering out impossible values

Data sets are often screened to remove impossible values. Weights can't be negative. The year of death cannot be before the year of birth. The age of a child cannot

Figure 7.4. Cumulative frequency distribution.

Each circle illustrates the number of people whose temperature was less than (or equal to) the X value. With a cumulative distribution, there is no need to choose a bin width. This graph plots the distribution of 130 values. But since the data were recorded only to within 0.1°, there are only 21 unique values. Accordingly, the graph has 21 points.

Figure 7.5. Cumulative frequency distribution, shown as percentages.

(Left) The graph is the same as Figure 7.4, except that the Y-axis plots the percentage of values less than or equal to each X value (rather than the actual number). (Right) Plot of the data with the Y-axis transformed in such a way that a cumulative Gaussian distribution (to be explained in Chapter 10) becomes a straight line, as seen here.

be greater than the age of the mother. It makes no sense to run statistical analyses on obviously incorrect data. These values must be fixed (if the mistake can be traced) or removed and documented.

But beware! Eliminating "impossible" values can also prevent you from seeing important findings. In 1985, researchers first noticed the drop in ozone levels over Antarctica (which turned out to be caused by chlorofluorocarbons). But measurements from a satellite demonstrated no such drop. Why the discrepancy? It seems that the satellite had been reporting a drop in ozone levels for years, but these data were automatically removed because they were considered to be impossible values (Sparling, 2001). This story has been disputed, but it is illuminating even if false. One must be very careful when automatically filtering out "impossible" values.

Beware of adjusting data

The values analyzed by statistical tests are often not direct experimental measurements. It is common to adjust the data, so it is important to think about whether these adjustments are correct and whether they introduce errors. If the numbers going into a statistical test are questionable, then so are the statistical conclusions.

In some cases many adjustments are needed, and these adjustments can have a huge impact. As a case in point, NASA provides historical records of temperature. An examination of how temperature has changed over the past century has convinced most scientists that the world is getting warmer. The temperature record, however, requires many corrections. Corrections are needed to account for the fact that cities are warmer than surrounding areas (because of building heat), that different kinds of thermometers have been used at different historical times, that the time of day that temperatures have been taken is not consistent, and so on (Goddard, 2008). The net result of all these adjustments pushed earlier temperatures down by almost 1°C, which is about half of the observed temperature increase in the past century. These adjustments require judgment, and different scientists may do the adjustments differently. Anyone interpreting the data must understand the contribution of these adjustments to the overall observed effect, as well as the degree to which these adjustments could possibly be biased by the scientists' desire to make the data come out a certain way.

When interpreting published data, ask about how the data were adjusted before they were graphed or entered into a statistics program.

Beware of smoothing

When plotting data that change over time, it is tempting to remove much of the variability in order to make the overall trend more visible. This can be accomplished by plotting a *rolling average*, also called a *moving average* or *smoothed data*. For example, each point on the graph can be replaced by the average of it and the three nearest neighbors on each side. The number of points averaged to smooth the data can range from two to many. If more points are included in the rolling average (say 10 on each side, rather than 3), the curve will be smoother. Smoothing methods differ in how neighboring points are weighted.

Smoothed data should never be entered into statistical calculations. If you enter smoothed data into statistics programs, the results reported by most statistical tests will be invalid. Smoothing removes information, so most analyses of smoothed data are not useful.

Look ahead to Figure 33.4 on page 305 to see an example of how analysis of smoothed data can lead to a misleading conclusion.

Beware of variables that are the ratio of two measurements

Often the value that you care most about is the ratio of two values. For example, divide enzyme activity or the number of binding sites by the cell count or the protein concentration. Calculating the ratio is necessary to express the variable in a way that can be interpreted and compared—for example, as enzyme activity per milligram of protein. The numerator is usually what you are thinking about and what the experimenter spent a lot of time and effort to measure. The denominator can seem like a housekeeping detail. But, of course, the accuracy of the ratio depends on the accuracy of both the numerator and the denominator.

Beware of normalized data

Some scientists transform (normalize) data so that all values are between 0 and 100%. When you see these kinds of data, you should wonder about how the values defining 0% and 100% were chosen. Ideally, the definitions of 0% and 100% are based on theory, or on control experiments with plenty of precise replicates. If 0% and 100% are not clearly defined or seem to be defined sloppily, then the normalized values won't be very useful. Ask how many replicates there were of the control measurements that define 0 and 100%.

Q & A

Can the mean or median equal zero? Can it be negative?
> Yes. The mean (or median) can have any value, including negative values and zero.

Which is larger, the mean or median?
> It depends on the data. If the distribution of values is symmetrical, the mean and median will be similar. If the distribution is skewed to the right, with an excess of large values, then the mean will probably be larger than the median (von Hippel, 2005).

Are *mean* and *average* synonyms?
> Yes. As used by statisticians, they are interchangeable. Outside of statistics, some use ***average*** more generally, and it sometimes refers to the median.

In what units are the mean and median expressed?
> The mean and median are expressed in the same units as the data. So are the geometric mean and harmonic mean.

Can the mean and median be computed when n = 1? When n = 2?
> Computing a mean or median from two values is no problem. The whole idea of a mean or median makes no sense when there is only one value, but you could say that the mean and the median both equal that value.

Can the mean and median be calculated when some values are negative?
> Yes.

Can the geometric and harmonic means be calculated when some values are negative?
> No. They can only be calculated when all values are positive. The geometric and harmonic means cannot be computed if any values are negative or zero.

Is the 50th percentile the same as the median?
> Yes.

CHAPTER SUMMARY

- Many scientific variables are continuous.
- One way to summarize these values is to calculate the mean, median, mode, geometrical mean, or harmonic mean.
- When graphing this kind of data, consider creating a graph that shows the scatter of the data. Either show every value on a scatter plot or show the distribution of values with a box-and-whiskers plot or a frequency distribution histogram.
- It is often useful to filter, adjust, smooth, or normalize the data before further graphing and analysis. These methods can be abused. Think carefully about whether these methods are being used effectively and honestly.

TERMS INTRODUCED IN THIS CHAPTER

- Bias (p. 64)
- Box-and-whiskers plot (p. 65)
- Continuous data (p. 61)
- Cumulative frequency distribution (p. 66)
- Dot plot (or scatter plot) (p. 65)
- Error (p. 63)
- Frequency distribution histogram (p. 65)
- Geometric mean (p. 62)
- Harmonic mean (p. 62)
- Mean (p. 62)
- Median (p. 62)
- Mode (p. 63)
- Moving average (p. 69)
- Percentile (p. 64)
- Precision (p. 64)
- Rolling average (p. 69)
- Scatter plot (or dot plot) (p. 65)
- Smoothed data (p. 69)
- Trimmed mean (p. 63)

CHAPTER 8

❖❖❖

Types of Variables

Get your facts first, then you can distort them as you please.

MARK TWAIN

The past four chapters have discussed four kinds of data. This chapter reviews the distinctions among different kinds of variables. Much of this chapter is simply terminology, but these definitions commonly appear in exam questions.

INTERVAL VARIABLES

Chapter 7 used body temperature in degrees centigrade for its examples. This kind of continuous variable is termed an *interval variable* (but not a ratio variable). It is an interval variable because a difference (interval) of 1°C means the same thing all the way along the scale, no matter where you start.

Computing the difference between two values can make sense when using interval variables. The 10°C difference between the temperatures of 100 and 90°C has the same meaning as the difference between the temperatures of 90 and 80°C.

Calculating the ratio of two temperatures is not useful. The problem is that the definition of zero is arbitrary. A temperature of 0.0°C is defined as the temperature at which water freezes and certainly does not mean "no temperature." A temperature of 0.0°F is a completely different temperature (–17.8°C). Because the zero point is arbitrary (and doesn't mean no temperature), it would make no sense at all to compute ratios of temperatures. A temperature of 100°C is not twice as hot as 50°C.

Figure 8.1 illustrates average body temperatures of several species (Blumberg, 2004). The platypus has an average temperature of 30.5°C, whereas a canary has an average temperature of 40.5°C. It is incorrect to say that a canary has a temperature 33% higher than that of a platypus. If you did that same calculation using degrees Fahrenheit, you'd get a different answer.

Figure 8.1A is misleading. The bars start at zero, inviting you to compare their relative heights and think about the ratio of those heights. But that comparison is not useful. This graph is also not helpful because it is difficult to see the differences between values.

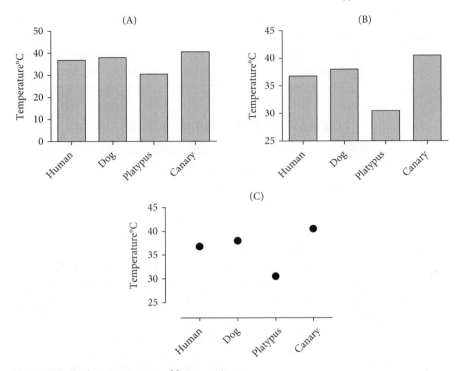

Figure 8.1. Body temperature of four species.
(A) is misleading. It invites you to compare the relative heights of the bars. But because a temperature of 0°C does not mean no temperature, the ratio of bar heights is not a meaningful value. (B) uses a different baseline to emphasize the differences. The bar for the canary is about three times as high as the bar for the platypus, but this ratio (indeed, any ratio) can be misleading. (C) illustrates the most informative way to graph these values.

Figure 8.1B uses a different baseline to demonstrate the differences. The bar for the canary is about three times as high as the bar for the platypus, but this ratio (indeed, any ratio) is not useful. Figure 8.1C illustrates the most informative way to graph these values. The use of points, rather than bars, doesn't suggest thinking in terms of a ratio. A simple table might be better than any kind of graph for these values.

RATIO VARIABLES

With a *ratio variable*, zero is not arbitrary. Zero height is no height. Zero weight is no weight. Zero enzyme activity is no enzyme activity. So height, weight, and enzyme activity are ratio variables.

As the name suggests, it can make sense to compute the ratio of two ratio variables. A weight of 4 grams is twice the weight of 2 grams, because weight is a ratio variable. But a temperature of 100°C is not twice as hot as 50°C, because temperature in degrees centigrade is not a ratio variable. Note, however, that temperature in Kelvin (°K) is a ratio variable, because 0.0°K really does mean (at least to a physicist) no temperature. Temperatures in Kelvin are far removed from temperatures that we ordinarily encounter, so they are rarely used in biology.

Like interval variables, you can compute the difference between ratio variables. Unlike interval variables, you can calculate the ratio of two ratio variables.

OTHER KINDS OF VARIABLES

The term *continuous variable* is applied to both interval and ratio variables. The next six chapters deal with continuous variables.

An *ordinal variable* expresses rank. The order matters, but not the exact value. For example, pain is expressed on a scale of 1 to 10. A score of 7 means more pain than a score of 5, which is more than a score of 3. But it doesn't make sense to compute the difference between two values, because the difference between 7 and 5 may not be comparable to the difference between 5 and 3. The values simply express an order. Another example would be movie or restaurant ratings from one to five stars.

To complete the list of terms describing kinds of variables, here are two more. Categorical outcomes with two possibilities (alive or dead, heads or tails, pass or fail) are called *binomial variables*. Categorical outcomes with more than two possible outcomes (not in any particular order) are *nominal variables*.

Table 8.1 summarizes which kinds of calculations are meaningful with which kinds of variables. It refers to the standard deviation and coefficient of variation, both of which are explained in Chapter 9, as well as the standard error of the mean (explained in Chapter 14).

NOT QUITE AS DISTINCT AS THEY SEEM

Note that the categories are nowhere near as distinct as they may sound (Velleman & Wilkinson, 1993). Here are some ambiguous situations:

- Color. In a psychological study of perception, different colors would be regarded as categories, so color would be a nominal variable. But monochromatic colors can be quantified by wavelength and thus can be considered a

OK TO COMPUTE	NOMINAL	ORDINAL	INTERVAL	RATIO
Frequency distribution	Yes	Yes	Yes	Yes
Median and percentiles	No	Yes	Yes	Yes
Add or subtract	No	No	Yes	Yes
Ratio	No	No	No	Yes
Mean, standard deviation, standard error of the mean	No	No	Yes	Yes
Coefficient of variation	No	No	No	Yes

Table 8.1. Calculations that are meaningful with various kinds of variables.
The standard deviation and coefficient of variation will be explained in Chapter 9, and the standard error of the mean will be explained in Chapter 14.

ratio variable (although a wavelength of zero is impossible.) Alternatively, you could rank the wavelengths and consider color to be an ordinal variable.

- The number of cells actually counted in a certain volume. This has all the properties of a ratio variable, except that the number must be an integer. This situation is similar to the examples outlined in Chapter 6.
- Effective concentrations $(EC)_{50}$: An EC_{50} is a measure of a drug's potency. It is the drug concentration that elicits 50% of the maximum response. An EC_{50} cannot equal zero, but it is very useful to calculate the ratios of two EC_{50} values. Therefore, it is sort of a ratio variable, but not quite.
- Percentages: Outcomes measured as ratio or interval variables are often transformed so that they are expressed as percentages. For example, a pulse rate (heartbeats per minute, a ratio variable) could be normalized to a percentage of the maximum possible pulse rate. A discrete outcome with a set of mutually exclusive categories can also be expressed as a percentage or proportion—for example, the percentage of transplanted kidneys that are rejected within a year of surgery. But these two situations are very different, and different kinds of statistical analyses are needed.

Q & A

Why is it important to remember these definitions?
> They appear commonly on statistics exams.

Does knowing these definitions help one choose an appropriate test?
> Maybe. It is important that every statistical test be designed to answer the scientific question at hand. Some people say that knowing the definitions of the various kinds of variables will help you choose the right test.

Why doesn't it make sense to compute a coefficient of variation (CV) of a continuous variable that is not a ratio variable?
> The CV (explained in Chapter 9) is the ratio SD/Mean. If the variable is a ratio variable, then the CV will be the same no matter what units are used. If the variable is not a ratio variable, then changing units (say from °C to °F) will change the value of the CV (which is therefore meaningless).

Why doesn't it make sense to compute the ratio of two continuous variables that are not ratio variables?
> For the same reason listed above. The ratio of two ratio variables expressed in the same units is a unitless ratio. If that ratio is 2.0, then one value is twice the other. . If the variable is not a ratio variable, then changing units (say from °C to °F) will change the value of the ratio (which is therefore meaningless).

CHAPTER SUMMARY

- Different kinds of variables require different kinds of analyses.
- Prior chapters have already discussed three kinds of data: proportions (binomial variables), survival data, and counts.
- This chapter explains how continuous data can be subdivided into interval and ratio (and ordinal) data.

- Interval variables are variables for which a certain difference (interval) between two values is interpreted identically no matter where you start. For example, the difference between 0 and 1 is interpreted the same as the difference between 999 and 1,000.
- Ratio variables are variables for which zero is not an arbitrary value. Weight is a ratio variable because weight = 0 means there is no weight. Temperature in Fahrenheit or centigrade is not a ratio variable, because 0° does not mean there is no temperature.
- It does not make any sense to compute a coefficient of variation, or to compute ratios, of continuous variables that are not ratio variables.
- The distinctions among the different kinds of variables are not quite as crisp as they sound.

TERMS INTRODUCED IN THIS CHAPTER

- Binomial variable (p. 74)
- Continuous variable (p. 74)
- Interval variable (p. 72)
- Nominal variable (p. 74)
- Ordinal variable (p. 74)
- Ratio variable (p. 73)

CHAPTER 9

❖❖❖

Quantifying Scatter

The average human has one breast and one testicle.

DES MCHALE

Chapter 7 demonstrated various ways to graph continuous data in order to make it easy to see the degree of variability. This chapter explains how variation can be quantified with the standard deviation, variance, coefficient of variation, interquartile range, and median absolute deviation.

INTERPRETING A STANDARD DEVIATION

The variation among values can be quantified as the *standard deviation* (abbreviated as SD in this book, and s in most others), which is expressed in the same units as the data. Let's first think about how to interpret the SD, and then look at how it is calculated.

You can interpret the SD using the following rule of thumb: about two-thirds of the observations in a population usually lie within the range defined by the mean minus 1 SD to the mean plus 1 SD. This definition is unsatisfying because the word *usually* is so vague. Chapter 10 will give a more rigorous interpretation of the SD for data sampled from a Gaussian distribution.

Let's turn to the larger (n = 130) sample in Figure 9.1. The mean temperature is 36.82°C and the SD is 0.41°C. Figure 9.1 plots the mean with error bars extending 1 SD in each direction (error bars can be defined in other ways, as you'll see in Chapter 14).

The range 36.4 to 37.2°C extends below and above the mean by 1 SD. The left side of Figure 9.1 is a column scatter graph of the body temperature data from Figure 7.1, and you can see that about two-thirds of the values are within that range.

Take a good look at Figure 9.1. Often you'll only see the mean and SD, as shown on the right, and you will need to imagine the actual scatter of the data. If you publish graphs, consider showing the actual data, as on the left side of Figure 9.1, rather than just the mean and the SD.

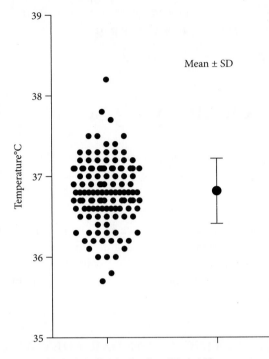

Figure 9.1. (Left) Each individual value. (Right) The mean and SD.

HOW IT WORKS: CALCULATING SD

It would seem that the simplest way to quantify variability would be to ask how far each value is from the mean (or median) of all values and to then report the average or median of those values. That strategy wouldn't work, however, because the positive deviations would balance out the negative deviations so that the average deviation would always equal zero. Another approach might be to take the average, or the median, of the absolute values of the deviations of each value from the mean. Indeed, *median absolute deviation* (MAD) is one way to quantify variability, and it is discussed later in this chapter. However, the most common way to quantify variation is by calculating the SD.

Of course, you'll use computer programs to compute the SD. Even so, the best way to understand the SD is to know how it is calculated.

1. Calculate the mean (average). For the n = 12 body temperature sample in Table 7.1, the mean is 36.77°C.
2. Calculate the difference between each value and the mean.
3. Square each of those differences.
4. Add up those squared differences. For the example data, the sum is 1.767.
5. Divide that sum by (n − 1), where n is the number of values. For the example, n = 12 and the result is 0.161. This value is called the *variance*.

6.   Take the square root of the value you calculated in Step 5. The result is the SD. For this example, the SD = 0.40°C.

These steps can be shown as an equation, where Y_i stands for one of the n values, and $\overline{Y}$ is the mean.

$$SD = \sqrt{\frac{\Sigma(Y_i - \overline{Y})^2}{n-1}}$$ **9.1**

WHY n − 1?

When calculating the SD, the sum of squares is divided by (n − 1). This is the definition of the *sample SD*, which is the best possible estimate of the SD of the entire population, as determined from one particular sample. Read on if you are curious to know why the denominator is (n − 1) rather than n.

When it may make sense to use n in the denominator
A denominator of n is sometimes used when the goal is to quantify the variation in a particular set of data without any extrapolation to make wider conclusions. The resulting SD is correct for those particular values, but it cannot be used to make general conclusions from the data. One example would be when a teacher wants to quantify variation among exam scores. The goal is not to make inferences about a larger population of hypothetical scores but simply to quantify the variation among those particular scores.

Why it usually makes sense to use n − 1 in the denominator
Most often, a scientist's goal is to compute SD from a sample of data and make inferences about the population from which the data were sampled. Depending on which values were randomly selected to be in the sample, the SD you calculate may be higher or lower than the true population SD, which you don't know. This is analogous to computing a mean from a sample of data. The result might be lower or higher than the true population mean, which you don't know. All you can do is use a method that will be correct on average in the long run.

When you go to calculate the SD using Equation 9.1, the numerator is the sum of the squares of the difference between each value and the mean of those values. Except for the rare case in which the sample mean happens to equal the population mean, the data will be closer to the sample mean than they will be to the true population mean (which you can't know). Therefore, the sum of squares will probably be a bit smaller (and can't be larger) than what it would be if you computed the difference between each value and the true population mean. Because the numerator is a bit too small, the denominator must be made smaller too.

But why n − 1? If you knew the sample mean and all but one of the values, you could calculate what that last value must be. Only n − 1 of the values are free to assume any value. Therefore, we calculate the average of the squared deviations by dividing by (n − 1) and say that there are n − 1 *degrees of freedom* (df). Many people find the concept of df to be quite confusing. Fortunately, being confused about df is not a big handicap! You can choose statistical tests and interpret the results with only a vague understanding of df.

The fine print

This SD computed with n − 1 is said to be the best possible estimate of the SD of the entire population, as determined from one particular sample. But best for what purpose? It is "best" because it will lead to accurate inferences when used to compute confidence intervals or P values. However, the SD computed with n − 1 as the denominator is *not* the most accurate estimate of the population SD. On the contrary, it is a biased estimate of the population SD. *Biased* means that on average, the computed SD will not equal (in this example, will be smaller than) the population SD, especially with small samples.

So why is the SD computed with (n − 1) used routinely? Because the sample variance (the square of the SD; discussed later in this chapter) computed using (n − 1) as the denominator is an unbiased estimate of the population variance, and the theory of confidence intervals and much of statistics depends on variances. This book, like most others, will ignore the fact that the sample SD is a biased estimate of the population SD, because that discrepancy doesn't affect any statistical conclusions.

SITUATIONS IN WHICH n CAN SEEM AMBIGUOUS

When calculating SD (and various other statistical tests), the definition of the sample size n can seem ambiguous in some situations.

Replicate measurements within repeat experiments

Table 9.1 presents a common situation. Data for five replicate measurements (rows) were collected from three animals (represented by the three columns). In total, 15 values were collected. But these are not 15 independent samples from one population. There are two sources of variability here: variation between animals

	ANIMAL A	ANIMAL B	ANIMAL C
	47.7	64.7	39.3
	43.1	65.4	40.0
	52.3	88.3	23.9
	55.2	64.0	36.6
	42.5	71.9	48.9
Mean	48.2	70.9	37.7

Table 9.1. What is n?

Data for five replicate measurements (rows) taken from three animals (columns). Because there are not 15 independent values here, it is not correct to compute an SD and CI using n = 15. Instead, average the values from each animal and then compute the SD and CI from those three means using n = 3.

(called *biological replicates*) and variation between replicate measurements in one animal (called *technical replicates*). Lumping these samples together would lead to invalid analyses. It is not correct to compute an SD (or CI; see Chapter 12) using n = 15. This mistake is called *pseudoreplication* (Lazic, 2010).

The simplest way to analyze such data is to average the values from each animal. For this example, there are three means (one for each animal). You would then compute the SD (and CI) from those three means using n = 3. The results can then be extrapolated to the population of animals you could have studied.

Representative experiments

Some investigators prefer to just show data from one "representative" experiment. The error bars in the table or graph are calculated only from the replicates in that one experiment. If the data were collected in triplicate, the graph might be labeled "n = 3," but in fact, all the data come from one experiment, not three. It can be useful to report the SD of replicate data within a single experiment as a way to demonstrate the precision of the method, and to spot experimental problems. But substantive results should be reported with data from multiple experiments (Vaux, Fidler, & Cumming, 2012).

Trials with one subject

Chapter 3 presented an example of an experiment conducted on one person with repeated samples (see Roberts, 2004); such experiments are sometimes called n = 1 studies. But despite the name, the n value used in statistical analysis is not 1 but rather is the number of values collected. The results can then be extrapolated to the population of results you could have collected from that one person.

SD AND SAMPLE SIZE

For the full body temperature data set (n = 130), as shown in Figure 7.1, SD = 0.41°C. For the smaller sample (n = 12, randomly sampled from the larger sample), SD = 0.40°C. Many find it surprising that the SD is so similar in samples of such different sizes. But, in fact, this is to be expected. The SD quantifies the variation within a population. As you collect larger samples, you'll be able to quantify the variability more precisely and more accurately, but collecting more data doesn't change the variability among the values.

OTHER WAYS TO QUANTIFY AND DISPLAY VARIABILITY

The SD is not the only way to quantify variability.

Coefficient of variation

For ratio variables, variability can be quantified as the *coefficient of variation* (CV), which equals the SD divided by the mean. If the CV equals 0.25, you know that the SD is 25% of the mean.

Because the SD and the mean are both expressed in the same units, the CV is a fraction with no units. Often the CV is expressed as a percentage.

For the preceding temperature example, the CV would be completely meaningless. Temperature is an interval variable, not a ratio variable, because zero is defined arbitrarily (see Chapter 8). A CV computed from temperatures measured in degrees centigrade would not be the same as a CV computed from temperatures measured in degrees Fahrenheit. Neither CV would be meaningful, because the idea of dividing a measure of scatter by the mean only makes sense with ratio variables, for which zero really means zero.

The CV is useful for comparing scatter of variables measured in different units. You could ask, for example, whether the variation in pulse rate is greater than or less than the variation in the concentration of serum sodium. The pulse rate and sodium are measured in completely different units, so comparing their standard deviations would make no sense. Comparing their coefficients of variation might be useful to someone studying homeostasis.

Variance

The variance equals the SD squared and so is expressed in the same units as the data, but squared. In the body temperature example, the variance is 0.16°C squared.

Statistical theory is based on variances, rather than SD, so mathematical statisticians routinely think about variances. Scientists analyzing data do not often have to encounter variances, not even when using analysis of variance (ANOVA), explained in Chapter 39.

Interquartile range

You probably are already familiar with the concept of percentiles. The 25th percentile is a value below which 25% of the values in your data set lie. The *interquartile range* is defined by subtracting the 25th percentile from the 75th percentile. Because both of these percentile values are expressed in the same units as the data, the interquartile range is also expressed in the same units.

For the body temperature data (n = 12 subset), the 25th percentile is 36.4°C and the 75th percentile is 37.1°C, so the interquartile range is 0.7°C. For the full (n = 130) data set, the 25th percentile is 36.6°C and the 75th percentile is 37.1°C, so the interquartile range is 0.5°C.

Five-number summary

The distribution of a set of numbers can be summarized with five values, known as the *five-number summary*: the minimum, the 25th percentile (first quartile), the median, the 75th percentile (the third quartile), and the maximum.

Median absolute deviation

The MAD is a simple way to quantify variation. It is the value such that half of the values are closer to the median than it and half are farther away.

The best way to understand the MAD is to understand the "recipe" for how it is calculated. First, compute the median of all values. The median is the 50th percentile. Then calculate how far each value is from the median of all values. Regardless of whether the value is greater than or less than the median, express the distance between it and the median as a positive value (that is, take the absolute value of the difference between the value and the median). Now find the median of that set of differences. The result is the MAD.

For the body temperature data (n = 12 subset), the median is 36.85°C and the MAD is 0.2°C. For the full (n = 130) data set, the median is 36.8°C and the MAD is 0.3°C.

Note two points of confusion:

- There are two distinct computations of the median. First, you compute the median (the 50th percentile) of the actual data. Then you compute the median of the absolute values of the distances of the data from the 50th percentile.
- Some investigators take the mean (rather than the median) of the set of differences and call the result the *mean absolute deviation*, also abbreviated MAD.)

Half of the values are within 1 MAD of the median. Therefore, a symmetrical range that extends one MAD in each direction from the median will contain about half of the values. The interquartile range also contains half of the values. The distinction is that the interquartile range can be asymmetrical around the median. Like the interquartile range, but unlike the SD, computation of the MAD is resilient to the presence of outliers (see Chapter 25).

Q & A

Can the SD ever equal zero? Can it be negative?
The SD will equal zero if all the values are identical. The SD can never be a negative number.

In what units do you express the SD?
The SD is expressed in the same units as the data.

Can the SD be computed when n = 1? When n = 2?
The SD quantifies variability, so it cannot be computed from a single value. It can be computed from two values (n = 2).

Is the SD the same as the standard error of the mean?
No. They are very different. See Chapter 14.

Can the SD be computed if the data clearly do not come from a Gaussian distribution?
Yes. The SD can be computed from any set of values. The Gaussian distribution is explained in Chapter 10. If the data do not come from a Gaussian population, you cannot necessarily use the rule of thumb that roughly two-thirds of the values are expected to be within 1 SD of the mean.

Is the SD larger or smaller than the CV?

The SD is expressed in the same units as the data. The CV is a unitless ratio, often expressed as a percentage. Because the two are not expressed in the same units, it makes no sense to ask which is larger.

Is the SD larger or smaller than the variance?

The variance is the SD squared, so it is expressed in different units. It makes no sense to ask which is larger.

Will all programs compute SD the same way?

The only ambiguity is whether to use n or (n − 1) in the denominator.

Will all programs compute the same value for the interquartile range?

No. Eight different equations can be used to compute percentiles (Harter, 1984), and the difference between these methods can be especially noticeable with small samples.

CHAPTER SUMMARY

- The most common way to quantify scatter is with a standard deviation.
- A useful rule of thumb is that about two-thirds of the observations in a population usually lie within the range defined by the mean minus 1 SD to the mean plus 1 SD.
- Other variables used to quantify scatter are the variance (SD squared), the coefficient of variation (which equals the SD divided by the mean), the interquartile range, and the median absolute deviation.
- While it is useful to quantify variation, it is often easiest to understand the variability in a data set by seeing a graph of every data point or a graph of the frequency distribution.

TERMS INTRODUCED IN THIS CHAPTER

- Biological replicates (p. 84)
- Coefficient of variation (CV) (p. 81)
- Degrees of freedom (df) (p. 80)
- Five-number summary (p. 82)
- Interquartile range (p. 82)
- Median absolute deviation (MAD) (p. 82)
- Pseudoreplication (p. 81)
- Sample standard deviation (p. 79)
- Standard deviation (SD) (p. 77)
- Technical replicates (p. 81)
- Variance (p. 82)

❖❖❖

CHAPTER 10

The Gaussian Distribution

Everybody believes in the normal approximation, the experimenters because they think it is a mathematical theorem, the mathematicians because they think it is an experimental fact.

G. LIPPMAN

Many statistical methods assume that data follow a Gaussian distribution. This chapter briefly explains the origin and use of the Gaussian distribution. Chapter 24 will explain how to test for deviations from a Gaussian distribution.

THE NATURE OF THE GAUSSIAN DISTRIBUTION

The Gaussian bell-shaped distribution is the basis for much of statistics. It arises when many random factors create variability. This happens because random factors tend to offset each other. Some will push the value higher and others will pull it lower. The effects usually partly cancel one another, so many values end up near the center (the mean). Sometimes many random factors will tend to work in the same direction, pushing a value away from the mean. Only rarely do the majority of random factors all work in the same direction, pushing that value far from the mean. Thus, many values are near the mean, some values are farther from the mean, and very few are quite far from the mean. When you plot the data on a frequency distribution, the result is a symmetrical, bell-shaped distribution, idealized as the *Gaussian distribution.*

Variation among values will approximate a Gaussian distribution when there are many sources of variation, so long as the various contributors to that variation are added up to get the final result and the sample size is large. As you have more and more sources of scatter, the predicted result approaches a Gaussian distribution.

Interpreting many commonly used statistical tests requires an assumption that the data were sampled from a population that follows a Gaussian distribution. This is often a reasonable assumption. For example, in a laboratory experiment, variation between experiments might be caused by several factors: imprecise weighing of reagents, imprecise pipetting, the random nature of radioactive decay, nonhomogenous suspensions of cells or membranes, and so on. Variation

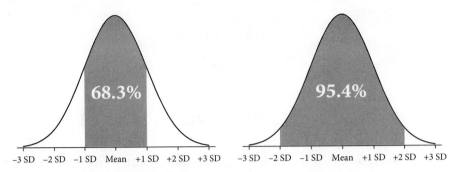

Figure 10.1. Ideal Gaussian distributions.
The horizontal axis plots various values, and the vertical axis plots their relative abundance.
The area under the curve represents all values in the population. The fraction of that area
within a range of values tells you how common those values are. (Left) About two-thirds
of the values are within 1 SD of the mean. (Right) Slightly more than 95% of the values are
within 2 SD of the mean.

in a clinical value might be caused by many genetic and environmental factors.
When scatter is the result of many independent causes, the distribution will tend
to follow a bell-shaped Gaussian distribution.

SD AND THE GAUSSIAN DISTRIBUTION

Figure 10.1 illustrates an ideal Gaussian distribution. The horizontal axis shows
various values that can be observed, and the vertical axis quantifies their relative
frequency. The mean, of course, is the center of the Gaussian distribution. The
Gaussian distribution is high near the mean, because that is where most of
the values are. As you move away from the mean, the distribution follows its
characteristic bell shape. The distribution is symmetrical, so the median and the
mean are identical.

The SD is a measure of the spread or width of the distribution. The area under
the entire curve represents the entire population. The left side of Figure 10.1
shades the area under the curve within 1 SD of the mean. You can see that the
shaded portion is about two-thirds (68.3%) of the entire area, demonstrating that
about two-thirds of the values in a Gaussian population are within 1 SD of the
mean. The right side of Figure 10.1 demonstrates that about 95% of the values in
a Gaussian population are within 2 SD of the mean (the actual multiplier is 1.96).

Scientific papers and presentations often show the mean and SD, but not
the actual data. If you assume that the distribution is approximately Gaussian,
you can re-create the distribution in your head. Going back to the sample body
temperature data in Figure 9.1, what could you infer if you knew only that the
mean is 36.82°C and its SD is 0.41°C (n = 130)? If you assume a Gaussian distribution, you could infer that about two-thirds of the values lie between 36.4 and
37.2°C and that 95% of the values lie between 36.0 and 37.6°C. If you look back
at Figure 9.1, you can see that these estimates are fairly accurate.

THE STANDARD NORMAL DISTRIBUTION

When the mean equals 0 and the SD equals 1.0, the Gaussian distribution is called a *standard normal curve*. Figure 10.1 would be a standard normal distribution if the labels went from –3 to +3 without using the labels SD and mean.

All Gaussian distributions can be converted to a standard normal distribution. To do this, subtract the mean from each value and divide the difference by the SD.

$$z = \frac{\text{Value} - \text{Mean}}{\text{SD}}$$

The variable z is the number of SD away from the mean. When $z = 1$, a value is 1 SD above the mean. When $z = -2$, a value is 2 SD below the mean. Table 10.1 tabulates the fraction of a normal distribution between $-z$ and $+z$ for various values of z.

If you work in pharmacology, don't confuse this use of the variable z with the specialized Z-factor used to assess the quality of an assay used to screen drugs (Zhang, Chung, & Oldenburg, 1999). The two are not related.

THE NORMAL DISTRIBUTION DOES
NOT DEFINE NORMAL LIMITS

A Gaussian distribution is also called a *normal distribution*. This is another case in which statistics has endowed an ordinary word with a special meaning. Don't mistake this special use of the word *normal* with its more ordinary meaning to describe commonly observed value or a value of a lab test or clinical measurement that does not indicate disease.

Here is a simple, but flawed, approach that is often used to determine the normal range of a measured value (in this case body temperature): Assume that the population's body temperatures (from the examples of the previous chapters) follow a Gaussian distribution and define 5% of the population to be abnormal.

z	PERCENTAGE OF STANDARD NORMAL DISTRIBUTION BETWEEN –z AND z
0.67	50.00
0.97	66.66
1.00	68.27
1.65	90.00
1.96	95.00
2.00	95.45
2.58	99.00
3.00	99.73

Table 10.1. The standard normal distribution.

Table 10.1 is best understood by example. The range between z = –1 and z = +1 contains 68.27% of a standard normal distribution. Calculate additional values using this Excel formula: =1 – 2*(1-NORMSDIST(z))

Using the sample mean and SD from our study, the normal range can be (incorrectly) defined as the mean plus or minus 1.96 SD, that is, $36.816 \pm (1.96 \times 0.405)$, which ranges from 36.0 to 37.6°C.

There are a number of problems with this approach:

- It really doesn't make sense to define normal and abnormal just in terms of the distribution of values in the general population. We know that high body temperature can be an indication of infection or inflammatory disease. So, the question we really want to answer is, When is a temperature high enough that it needs investigation? The answer to this question requires scientific or clinical context. It cannot be answered by statistical calculations.
- The definitions of *normal* and *abnormal* really should depend on other factors such as age and sex. What is abnormal for a 25-year-old may be normal for an 80-year-old.
- In many cases, we don't really want a crisp boundary between normal and abnormal. It often makes sense to label some values as "borderline."
- Even if the population is approximately Gaussian, it is unlikely to follow a Gaussian distribution exactly. The deviations from a Gaussian distribution are likely to be most apparent in the tails (extreme values) of the distribution, where the abnormal values lie.
- This approach defines just as many people as having abnormally high values as those having abnormally low values. But in fact having a temperature in the lowest 2.5% is not an indication that something is wrong requiring medical attention. There is no reason to define the normal limits symmetrically around the mean or median.

Defining the normal limits of a clinical measurement is not straightforward and requires clinical thinking. Simple statistics rules based on the mean, SD, and Gaussian distribution are rarely useful except as a starting point before defining smarter limits.

WHY THE GAUSSIAN DISTRIBUTION IS SO CENTRAL TO STATISTICAL THEORY

The Gaussian distribution plays a central role in statistics because of a mathematical relationship known as the *central limit theorem*. You'll need to read a more theoretical book to really understand this theorem, but the explanation below gives you the basics.

To understand this theorem, follow this imaginary experiment:

1. Create a population with a known distribution that is not Gaussian.
2. Randomly pick many samples of equal size from that population.
3. Tabulate the means of these samples.
4. Graph the frequency distribution of those means.

The central limit theorem says that if your samples are large enough, the distribution of means will approximate a Gaussian distribution, even though the population is not Gaussian. Because most statistical tests (such as the t test and analysis of variance) are concerned only with differences between means, the central limit theorem explains why these tests work well even when the populations are not Gaussian.

Q & A

Who was Gauss?

Karl Gauss was a mathematician (one of the greatest of all time) who used this distribution in 1809 to analyze astronomical data. Although his name is now attached to the distribution, others (Laplace and de Moivre) actually used it earlier.

Is the Gaussian distribution the same as a normal distribution?

Yes, the two terms are used interchangeably.

Are all bell-shaped distributions Gaussian?

As you can see in Figure 10.1, the Gaussian distribution is bell shaped. But not all bell-shaped curves are Gaussian.

Will numerous sources of scatter always create a Gaussian distribution?

No. A Gaussian distribution is formed only when each source of variability is independent and additive with the others and no one source dominates. Chapter 11 discusses what happens when the sources of variation multiply.

CHAPTER SUMMARY

- The Gaussian bell-shaped distribution is the basis for much of statistics. It arises when many random factors create variability.
- With a Gaussian distribution, about two-thirds of the values are within 1 SD of the mean, and about 95% of the values are within 2 SD of the mean.
- The Gaussian distribution is also called a normal distribution. But this use of *normal* is very different than the usual use of that word to mean ordinary or abundant.
- The central limit theorem explains why Gaussian distributions are central to much of statistics. Basically this theorem says that the distribution of many sample means will tend to be Gaussian, even if the data are not sampled from a Gaussian distribution.

TERMS INTRODUCED IN THIS CHAPTER

- Central limit theorem (p. 88)
- Gaussian distribution (p. 85)
- Normal distribution (p. 87)
- Standard normal distribution (p. 87)
- z (p. 87)

CHAPTER 11

❖❖❖

The Lognormal Distribution
and Geometric Mean

42.7 percent of all statistics are made up on the spot.

STEVEN WRIGHT

Lognormal distributions are not an obscure mathematical quirk. On
the contrary, they are very common in many fields of science. This
chapter explains how lognormal distributions arise and how to analyze
lognormal data. Read this chapter to avoid making the common mistake
of choosing analyses that assume sampling from a Gaussian distribution when working with data that are actually sampled from a lognormal
distribution.

EXAMPLE: RELAXING BLADDERS

And you thought statistics was never relaxing!

Frazier, Schneider, and Michel (2006) measured the ability of isoprenaline
(a drug that acts much like the neurotransmitter norepinephrine) to relax the
bladder muscle. The results are expressed as the EC50, which is the concentration
required to relax the bladder halfway between its minimum and maximum
possible relaxation. The graph on the left side of Figure 11.1 illustrates the data
plotted on a linear scale.

The distribution is far from symmetrical and is quite skewed. One value is far
from the rest and almost looks like a mistake.

THE ORIGIN OF A LOGNORMAL DISTRIBUTION

Chapter 10 explained that a Gaussian distribution arises when variation is caused
by many factors that are additive. Some factors push a value higher and some pull
it lower, and the cumulative result is a symmetrical bell-shaped distribution that
approximates a Gaussian distribution.

But some factors act in a multiplicative, rather than additive, manner. If a
factor works multiplicatively, it is equally likely to double a value as to cut it
in half. If that value starts at 100 and is multiplied by 2, it ends up at 200. If it

90

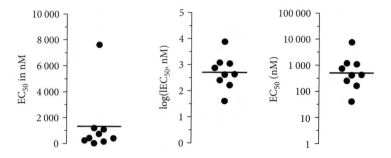

Figure 11.1. Lognormal data.

These data demonstrate the EC50 of isoprenaline for relaxing bladder muscles (Frazier, Schneider, & Michel, 2006). Each dot indicates data from the bladder of a different animal. The EC50 is the concentration required to relax the bladder halfway between its minimum and maximum possible relaxation, measured in nanomoles per liter (nM). The graph on the left plots the original concentration scale. The data are far from symmetrical, and the highest value appears to be an outlier. The middle graph plots the logarithm (base 10) of the EC50 and is symmetrical. The graph on the right plots the raw data on a logarithmic axis. This kind of graph is a bit easier to read.

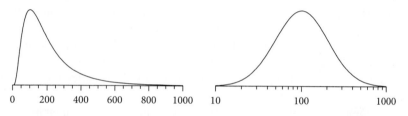

Figure 11.2. Lognormal distribution.

(Left) Lognormal distribution. The distribution appears Gaussian when plotted on a logarithmic axis (right) or when all values are transformed to their logarithm.

is divided by 2, it ends up at 50. Consequently, that factor is equally likely to increase a value by 100 or decrease it by 50. The effect is not symmetrical.

If many factors act in a multiplicative manner, the resulting distribution is asymmetrical, as shown in Figure 11.2 (left). This distribution is called a *lognormal distribution*. Logarithms? How did logarithms get involved? Briefly, it is because the logarithm of the product of two values equals the sum of the logarithm of the first value plus the logarithm of the second. So logarithms convert multiplicative scatter (lognormal distribution) to additive scatter (Gaussian). Thus, the logarithms of the values follow a Gaussian distribution when the raw data are sampled from a lognormal distribution.

HOW TO ANALYZE LOGNORMAL DATA

If you transform each value sampled from a lognormal distribution to its logarithm, the distribution becomes Gaussian. Logarithms (and antilogarithms) are reviewed in Appendix E.

The middle graph in Figure 11.1 plots the logarithm of the EC50 values. Note that the distribution is symmetrical.

The graph on the right in Figure 11.1 illustrates an alternative way to plot the data. The axis has a logarithmic scale. Note that every major tick on the axis represents a value 10 times higher than the previous value. The distribution of data points is identical to that in the middle graph, but the graph on the right is easier to comprehend because the Y values are in natural units of the data, rather than logarithms.

GEOMETRIC MEAN

The mean of the data from Figure 11.1 is 1,333 nM. This is illustrated as a horizontal line in the left panel of Figure 11.1. The mean is larger than all but one of the values, so it is not a good measure of the central tendency of the data.

The middle panel plots the logarithms of the values on a linear scale. The horizontal line is at the mean of the logarithms, 2.71. About half of the values are higher and half are smaller.

The right panel in Figure 11.1 uses a logarithmic axis. The values are the same as those of the graph on the left, but the spacing of the values on the axis is logarithmic. The horizontal line is at the antilog of the mean of the logarithms. This graph uses logarithms base 10, so the antilog is computed by calculating $10^{2.71}$, which equals 513. The value 513 nM is called the *geometric mean*.

To compute a geometric mean, first transform all the values to their logarithms and then calculate the mean of those logarithms. Finally, transform that mean of the logarithms back to the original units of the data.

COMMON MISTAKES: LOGNORMAL DISTRIBUTIONS

Mistake: Being inconsistent with the use of common and natural logarithms

To compute the geometric mean, you need to first convert all the values to their logarithms, then compute the mean of those logarithms, and finally compute the antilogarithm of that mean. It is critical that the same base be used for computing the logarithm and antilogarithm. If you use log10 common logarithms, compute the antilogarithm by taking 10 to the power of the mean. If you use natural logarithms, compute the antilogarithm using the exp() function.

Mistake: Converting data to logarithms when some values are zero or negative

The logarithm is simply not defined for zero or negative values. If you transform a set of values to their logarithms, any values are zero or negative will be lost to the analysis.

Q & A

Where can I review logarithms and antilogarithms?
> In Appendix E.

Lognormal or log-normal?
> Both forms are commonly used. In this text, we use "lognormal."

Are values in a lognormal distribution always positive?
> Yes. The logarithm of zero and negative numbers is simply not defined. Distributions that contain zero or negative values cannot be treated as lognormal distributions.

Can the geometric mean be computed if any values are zero or negative?
> No.

When computing the geometric mean, should I use natural logarithms or logarithms base 10?
> It doesn't matter, as long as you are consistent. More often, scientists use logarithms base 10, so the reverse transform is the power of 10. The alternative is to use natural logarithms, so the reverse transform is taking e to that power. Whichever log base is used, the geometric mean will have exactly the same value. See Appendix E.

Are lognormal distributions common?
> Yes, they occur commonly (Limpert, Stahel, & Abbt, 2001). For example, the potency of a drug (assessed as EC50, IC50, Km, Ki, etc.) is almost always lognormal. For this reason, it makes sense to compare treatment groups using ratios rather than differences, and to summarize data with a geometric mean. Other examples of lognormal distributions are the blood serum concentrations of many natural or toxic compounds.

Should I use a logarithmic axis when plotting data from a lognormal distribution?
> Yes, this can make it easier to understand. See Figure 11.1 (right).

What happens if you analyze data from a lognormal distribution as if they were sampled from a Gaussian distribution?
> It depends on the details of the data and the sample size. It is likely that the results of any statistical analysis that assumes a Gaussian distribution will be misleading.

Since the lognormal distribution is asymmetrical, is it hard to correctly spot outliers?
> Yes. An easy mistake would be to view the left graph of Figure 11.1 and conclude that these values are sampled from a Gaussian distribution but are contaminated by the presence of a single outlier. Running an outlier test would confirm that conclusion. But that would be misleading. Outlier tests assume that all the values (except for any outliers) are sampled from a Gaussian distribution. When presented with data from a lognormal distribution, outlier tests are likely to incorrectly flag very high values as outliers, when, in fact, those high values are expected in a lognormal distribution. Chapter 25 shows some examples.

What units are used to express the geometric mean?
> The same units that are used with the values being analyzed. Thus, the mean and geometric mean are expressed in the same units.

Is the geometric mean larger or smaller than the regular mean?
> The geometric mean is always smaller. (One trivial exception: if all the values are identical, then the mean and geometric mean will be equal.)

CHAPTER SUMMARY

- Lognormal distributions are very common in many fields of science.
- Lognormal distributions arise when multiple random factors are multiplied together to determine the value. In contrast, Gaussian distributions arise when multiple random factors are added together.

- Lognormal distributions have a long right tail (are said to be skewed to the right).
- You may get misleading results if you make the common mistake of choosing analyses that assume sampling from a Gaussian distribution when in fact your data are actually sampled from a lognormal distribution.
- In most cases, the best way to analyze lognormal data is to take the logarithm of each value and then analyze those logarithms.

TERMS INTRODUCED IN THIS CHAPTER

- Geometric mean (p. 92)
- Lognormal distribution (p. 90)

CHAPTER 12

❖❖❖

Confidence Interval of a Mean

It is easy to lie with statistics. It is hard to tell the truth without it.

ANDREJS DUNKELS

Chapters 4 through 6 explained confidence intervals of proportions, survival fractions, and counts. This chapter extends those concepts to the confidence interval of a mean, a calculation that depends on the size of the sample and the variability of the values (expressed as the standard deviation).

INTERPRETING A CI OF A MEAN

For our ongoing n = 130 body temperature example (see Figure 7.1), any statistics program will calculate that the 95% CI of the mean ranges from 36.75 to 36.89°C. For the smaller n = 12 subset, the 95% CI of the mean ranges from 36.51 to 37.02°C.

Note that there is no uncertainty about the sample mean. We are 100% sure that we have calculated the sample mean correctly. Any errors in recording the data or computing the mean will not be accounted for in computing the CI of the mean. By definition, the CI is always centered on the sample mean. The population mean is not known and can't be known. However, given some assumptions discussed below, we can be 95% sure that the calculated interval contains it.

What exactly does it mean to be "95% sure"? When you have measured only one sample, you don't know the value of the population mean. The population mean either lies within the 95% CI or it doesn't. You don't know, and there is no way to find out. If you calculate a 95% CI from many independent samples, the population mean will be included in the CI in 95% of the samples but will be outside of the CI in the other 5% of the samples. Using data from one sample, therefore, you can say that you are 95% confident that the 95% CI includes the population mean.

The correct syntax is to express the CI as "36.75 to 36.89" or as "[36.75, 36.89]." It is considered bad form to express the CI as "36.75–36.89," because the hyphen would be confusing when the values are negative. Although it seems sensible to express the CI as "36.82 ± 0.07," that format is rarely used.

Figure 12.1. The 95% CI does not contain 95% of the values, especially when the sample size is large.

WHAT VALUES DETERMINE THE CI OF A MEAN?

The CI of a mean is computed from four values:

- The sample mean. Our best estimate of the population mean is the sample mean. Accordingly, the CI is centered on the sample mean.
- The SD. If the data are widely scattered (large SD), then the sample mean is likely to be farther from the population mean than if the data are tight (small SD). The width of the CI, therefore, is proportional to the sample SD.
- Sample size. Our sample has 130 values, so the sample mean is likely to be quite close to the population mean, and the CI will be very narrow. With tiny samples, the sample mean is likely to be farther from the population mean, so the CI will be wider. The width of the CI is inversely proportional to the square root of the sample size. If the sample were four times larger, the CI would be half as wide (assuming the same SD). Note that the CI from the n = 12 sample is wider than the CI for the n = 130 sample (see Figure 12.1).
- Degree of confidence. Although CIs are typically calculated for 95% confidence, any value can be used. If you wish to have more confidence (e.g., 99% confidence), you must generate a wider interval. If you are willing to accept less confidence (e.g., 90% confidence), you can generate a narrower interval.

ASSUMPTIONS: CI OF A MEAN

To interpret a CI of a mean, you must accept the following assumptions:

Assumption: Random (or representative) sample

The 95% CI is based on the assumption that your sample was randomly selected from the population. In many cases, this assumption is not strictly true. You can still interpret the CI as long as you assume that your sample is representative of the population.

In clinical studies, it is not feasible to randomly select patients from the entire population of similar patients. Instead, patients are selected for the study because they happened to be at the right clinic at the right time. This is called a *convenience sample* rather than a *random sample*. For statistical calculations to be meaningful, we must assume that the convenience sample adequately represents the population, and that the results are similar to what would have been observed had we used a true random sample.

This assumption would be violated in the body temperature example if the people who participated chose to join a study of body temperature because they knew (or suspected) that their own temperature was consistently higher or lower than that of most other people.

Assumption: Independent observations

The 95% CI is only valid when all subjects are sampled from the same population and each has been selected independently of the others. Selecting one member of the population should not change the chance of selecting any other person. This assumption would be violated if some individuals' temperatures were measured twice and both values were included in the sample. The assumption would also be violated if several of the subjects were siblings, because it is likely that genetic factors affect body temperature.

Assumption: Accurate data

The 95% CI is only valid when each value is measured correctly. This assumption would be violated if subjects didn't place the thermometer in their mouths correctly, or if the thermometer was misread.

Assumption: Assessing an event you really care about

The 95% CI allows you to extrapolate from the sample to the population for the event that you tabulated. But sometimes you might really care about a different event. In this example, what you really want to know is the core body temperature. Instead, temperature under the tongue was measured. The difference in this example is trivial, but it is always worth thinking about the distinction between what is measured and what one really wants to know. Chapter 45 gives two examples in which drugs changed a surrogate marker as desired—fewer arrhythmias in one example, higher "good cholesterol" in the other—but increased the number of deaths.

Assumption: The population is distributed in a Gaussian manner, at least approximately

The most common method of computing the CI of a mean is based on the assumption that the data are sampled from a population that follows a Gaussian distribution. This assumption is important when the sample is small but doesn't matter much when samples are large. How big must a sample be for us to disregard this assumption? It depends on the shape of the non-Gaussian distribution. See the discussion of the central limit theorem at the end of Chapter 10.

An alternative resampling method of computing the 95% CI that does not assume sampling from a Gaussian distribution is explained in Chapter 13.

What if the assumptions are violated?

In many situations, these assumptions are not strictly true. The patients in your study may be more homogeneous than the entire population of patients. Measurements made in one lab will have a smaller SD than measurements made in other labs at other times. More generally, the population you really care about may be more diverse than the population from which the data were sampled. Furthermore, the population may not be Gaussian. If any assumption is violated, the CI will probably be too optimistic (too narrow). The true CI (taking into account any violation of the assumptions) is likely to be wider than the calculated CI.

HOW TO CALCULATE THE CI OF A MEAN

Although computers will do the calculations, it is easier to understand a CI of a mean once you know how it is computed. The CI is centered on the sample mean (m). To calculate the width requires taking account of the SD (s), the number of values in the sample (n), and the degree of confidence you desire (usually 95%).

Use Table 12.1 (reprinted in Appendix D) to determine the value of a constant based on the sample size and the degree of confidence you desire. This value is called a *constant from the t distribution* (to be explained in Chapter 13), denoted in this book by the nonstandard abbreviation t*. For the body temperature example of the right side of Figure 12.1, n = 12, so df = 11 and t* (for 95% confidence) = 2.201.

Calculate the margin of error of the CI, W, which is half the width of the CI:

$$W = \frac{t^* \cdot s}{\sqrt{n}}$$

For the example data, s = 0.40°C, so W = 0.254 (for 95% confidence). The CI covers this range:

m − W to m + W

For the example data, m = 36.77°C, so the 95% CI extends from 36.52 to 37.02°C. If you want 90% confidence, t* = 1.7959 (n = 12, so df = 11), W = 0.207, and the 90% CI extends from 36.56 to 36.98°C.

DESIRED CONFIDENCE LEVEL					DESIRED CONFIDENCE LEVEL			
df	80%	90%	95%	99%	df	80%	90%	95%	99%
1	3.0777	6.3138	12.7062	63.6567	27	1.3137	1.7033	2.0518	2.7707
2	1.8856	2.9200	4.3027	9.9248	28	1.3125	1.7011	2.0484	2.7633
3	1.6377	2.3534	3.1824	5.8409	29	1.3114	1.6991	2.0452	2.7564
4	1.5332	2.1318	2.7764	4.6041	30	1.3104	1.6973	2.0423	2.7500
5	1.4759	2.0150	2.5706	4.0321	35	1.3062	1.6896	2.0301	2.7238
6	1.4398	1.9432	2.4469	3.7074	40	1.3031	1.6839	2.0211	2.7045
7	1.4149	1.8946	2.3646	3.4995	45	1.3006	1.6794	2.0141	2.6896
8	1.3968	1.8595	2.3060	3.3554	50	1.2987	1.6759	2.0086	2.6778
9	1.3830	1.8331	2.2622	3.2498	55	1.2971	1.6730	2.0040	2.6682
10	1.3722	1.8125	2.2281	3.1693	60	1.2958	1.6706	2.0003	2.6603
11	1.3634	1.7959	2.2010	3.1058	65	1.2947	1.6686	1.9971	2.6536
12	1.3562	1.7823	2.1788	3.0545	70	1.2938	1.6669	1.9944	2.6479
13	1.3502	1.7709	2.1604	3.0123	75	1.2929	1.6654	1.9921	2.6430
14	1.3450	1.7613	2.1448	2.9768	80	1.2922	1.6641	1.9901	2.6387
15	1.3406	1.7531	2.1314	2.9467	85	1.2916	1.6630	1.9883	2.6349
16	1.3368	1.7459	2.1199	2.9208	90	1.2910	1.6620	1.9867	2.6316
17	1.3334	1.7396	2.1098	2.8982	95	1.2905	1.6611	1.9853	2.6286
18	1.3304	1.7341	2.1009	2.8784	100	1.2901	1.6602	1.9840	2.6259
19	1.3277	1.7291	2.0930	2.8609	150	1.2872	1.6551	1.9759	2.6090
20	1.3253	1.7247	2.0860	2.8453	200	1.2858	1.6525	1.9719	2.6006
21	1.3232	1.7207	2.0796	2.8314	250	1.2849	1.6510	1.9695	2.5956
22	1.3212	1.7171	2.0739	2.8188	300	1.2844	1.6499	1.9679	2.5923
23	1.3195	1.7139	2.0687	2.8073	350	1.2840	1.6492	1.9668	2.5899
24	1.3178	1.7109	2.0639	2.7969	400	1.2837	1.6487	1.9659	2.5882
25	1.3163	1.7081	2.0595	2.7874	450	1.2834	1.6482	1.9652	2.5868
26	1.3150	1.7056	2.0555	2.7787	500	1.2832	1.6479	1.9647	2.5857

Table 12.1. Critical values of the t distribution for computing CIs used in Equation 12.1.
When computing a CI of a mean, df equals (n − 1). More generally, df equals n minus the number of parameters being estimated.

MORE ABOUT CONFIDENCE INTERVALS

One-Sided CIs

CIs are almost always calculated and presented as described above: as an interval defined by two limits. But it is also possible to create *one-sided confidence intervals*.

For the example, the 90% CI extends from 36.56 to 36.98°C. Because this is a 90% two-sided CI, there is room for 5% error on each end. There is a 5% chance that the upper limit (36.98) is less than the population mean. There is also a 5% chance that the lower limit (36.56) is greater than the population mean. That leaves a 90% (100% minus 5% minus 5%) chance that the interval 36.56 to 36.98°C includes the true population mean.

But what if we are clinically interested only in fevers and want to know only the upper confidence limit? Because there is a 5% chance that the population mean is greater than 36.98°C (see previous paragraph), that leaves a 95% chance that the true population mean is less than 36.98°C. This is a one-sided 95% CI.

More precisely, you could say that the range from minus infinity to 36.98°C is 95% likely to contain the population mean.

CI of the SD

A CI can be determined for nearly any value you calculate from a sample of data. With the n = 12 body temperature data, the SD was 0.40°C. It isn't done often, but it is possible to compute a 95% CI of the SD itself. In this example, the 95% CI of the SD extends from SD = 0.28°C to SD = 0.68°C (calculated using a Web-based calculator at www.graphpad.com/quickcalcs/CISD1/).

The interpretation is straightforward. Given the same assumptions as those used when interpreting the CI of the mean, we are 95% sure that the calculated interval contains the true population SD. With the n = 130 body temperature data, the SD is 0.41°C. Because the sample size is so much larger, the sample SD is a more precise estimate of the population SD, and the CI is narrower, ranging from SD = 0.37°C to SD = 0.47°C.

Table 12.2 presents the CI for a standard deviation for various sample sizes.

CI of a Geometric Mean

Chapter 11 explained how to compute the geometric mean. This process is easily extended to computing the 95% CI of the geometric mean. The first step in computing a geometric mean is to transform all the values to their logarithms. Then the mean of the logarithms is computed and reverse transformed to its antilog. The result is the geometric mean.

To create a CI of the geometric mean, compute the CI of the mean of the logarithms and then reverse transform each confidence limit.

For the EC50 example in Chapter 11, the mean of the logarithms is 2.71, and the 95% CI of this mean extends from 2.22 to 3.20. Reverse transform each of these values, and the 95% CI of the geometric mean (which equals 512 nM) ranges from 166 to 1585 nM. Note that the CI of the geometric mean is not symmetrical around the geometric mean.

The CI and the SEM

Note that since the standard error of the mean (SEM; explained in Chapter 14) is computed from the standard deviation and sample size, the equation for W shown above can be simplified.

$$W = \frac{t^* \cdot s}{\sqrt{n}}$$

$$SEM = \frac{s}{\sqrt{n}}$$

$$W = t^* \cdot SEM$$

The first line repeats the definition of W from above. The second line defines the SEM (see Chapter 14 for details). The third line simply combines the two.

n	95% CI OF SD
2	0.45 to 31.9 · SD
3	0.52 to 6.29 · SD
5	0.60 to 2.87 · SD
10	0.69 to 1.83 · SD
25	0.78 to 1.39 · SD
50	0.84 to 1.25 · SD
100	0.88 to 1.16 · SD
500	0.94 to 1.07 · SD
1,000	0.96 to 1.05 · SD

Table 12.2. The CI of an SD.

Assuming random sampling from a Gaussian population, this table gives the 95% CI for the population SD, given the sample SD and sample size (n). These values were computed using equations from Sheskin (2007, page 197).

Q & A

Why 95% confidence?

CIs can be computed for any degree of confidence. By convention, 95% CIs are presented most commonly, although 90 and 99% CIs are sometimes published.

Is a 99% CI wider or narrower than a 95% CI?

To be more confident that your interval contains the population value, the interval must be made wider. Thus, 99% CIs are wider than 95% CIs, and 90% CIs are narrower than 95% CIs (see Figure 12.2).

Figure 12.2. If you want to be more confident that a CI contains the population mean, you must make the interval wider.

The 99% interval is wider than the 95% interval, which is wider than the 90% CI.

Does a CI quantify variability?

No. The CI of the mean tells you how precisely you have determined the population mean. It does not tell you about scatter among values.

When can I use the rule of thumb that a 95% CI equals the mean plus or minus 2 SD?

Never! In a Gaussian distribution, you expect to find about 95% of the individual values within 2 SD of the mean. But the idea of a CI is to define how precisely you know the population mean. For that, you need to take into account sample size.

When can I use the rule of thumb that a 95% CI equals the mean plus or minus 2 SEM?

When n is large. Chapter 14 explains the SEM.

Do 95% of values lie within the CI?

No! The CI quantifies how precisely you know the population mean. With large samples, you can estimate the population mean quite accurately, so the CI is quite narrow and includes only a small fraction of the values (see Figure 12.1).

If I collect more data, should I expect 95% of the new values to lie within this CI?

No! The CI quantifies how precisely you know the population mean. With large samples, you can estimate the population mean quite accurately, so the CI is quite narrow and will include only a small fraction of the values you sample in the future.

Is the 95% CI of the mean always symmetrical around the mean?

It depends on how it is computed. But with the standard method, explained in the previous section, the CI always extends equally above and below the mean.

How is a confidence interval distinct from confidence limits?

The confidence interval is a range of values. The two values that define that range are called *confidence limits*.

CHAPTER SUMMARY

- A confidence interval of the mean shows you how precisely you have determined the population mean.
- If you compute 95% confidence intervals from many samples, you expect that 95% will include the true population mean and 5% will not. You'll never know whether or not a particular confidence interval includes the population mean or not.
- The CI of a mean is computed from the sample mean, the sample SD, and the sample size.
- The CI does not display the scatter of the data. In most cases, the majority of the data values will lie outside the CI.
- If you desire more confidence (99% rather than 95%), the CI will be wider.
- Larger samples have narrower confidence intervals than smaller samples with the same SD.
- Interpreting the CI of a mean requires accepting a list of assumptions.

TERMS INTRODUCED IN THIS CHAPTER

- Confidence limit (p. 102)
- Constant from the t distribution (p. 98)
- Convenience sample (p. 97)
- One-sided confidence interval (p. 100)
- Random sample (p. 97)

CHAPTER 13

❖❖❖

The Theory of
Confidence Intervals

> Confidence is what you have before you understand the problem.
>
> WOODY ALLEN

The statistical theory underlying confidence intervals is hard to understand. Probability theory starts with a known population and then computes the probabilities of obtaining various possible samples. Statistical analysis flips this theory: it starts with data and then computes the likelihood that the data were sampled from various populations. This chapter tries to give you a general sense of how the theory is flipped. It includes more equations than the rest of the book. Later chapters do not assume you have mastered this one.

CI OF A MEAN VIA THE t DISTRIBUTION

What is the t distribution?

Chapter 12 demonstrated how to calculate a CI of a mean using a method that required us to find a value from the *t distribution*. A full understanding of the t distribution requires a mathematical approach beyond the scope of this book, but the following provides an overview.

Let's assume we know that a population follows a Gaussian distribution. The mean of this population is designated by the Greek letter μ (mu). Let's assume we also know the standard deviation of this population, designated by the Greek letter σ (sigma).

Using a computer program, one can randomly choose values from this defined population. Pick n values and compute the mean of this sample, which we'll abbreviate m. Also compute the SD of this sample, which we'll abbreviate s. This sample was chosen randomly, so m and s won't equal the population values μ and σ.

Repeat these computations thousands of times. For each random sample, compute the ratio t:

$$t = \frac{m - \mu}{s / \sqrt{n}} \qquad\qquad \textbf{13.1}$$

104

Because these are simulated data from a hypothetical population, the value of μ is known and constant, as is n. For each sample, compute m and s, and then use the equation above to compute t.

For each random sample, the sample mean is equally likely to be larger or smaller than the population mean, so the t ratio is equally likely to be positive or negative. It will usually be fairly close to zero, but it can also be far away. How far? It depends on sample size (n), the standard deviation (s), and chance. Figure 13.1 illustrates the distribution of t computed for a sample size of 12. Of course, there is no need to actually perform these simulations. Mathematical statisticians have derived the distribution of t using calculus.

Sample size, n, is included in the equation that defines t, so you might expect the distribution of t to be the same regardless of sample size. In fact, the t distribution depends on sample size. With small samples, the curve in Figure 13.1 will be wider; with large samples, the curve will be narrower. With huge samples, the curve in Figure 13.1 will become indistinguishable from the Gaussian distribution shown in Figure 10.1.

The critical value of t

The area under the curve in Figure 13.1 represents all possible samples. Chop off the 2.5% tails on each side (shaded) to determine a range of t ratios that includes 95% of the samples. For this example with n = 12 and df = 11, the t ratio has a value between –2.201 and 2.201 in 95% of the samples (see Table 12.1). Let's define the variable t* to equal 2.201 (this abbreviation is not standard).

The value of t* depends on sample size and the degree of confidence you want (here we chose the standard, 95%). Note that its value does not depend on the actual data we are analyzing. You can look up this value in Appendix D.

The flip!

Here comes the most important step, in which the math is flipped around to compute a CI.

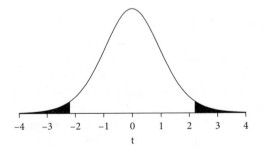

Figure 13.1. The t distribution for 11 df.

The shaded areas in the two tails each represent 2.5% of the area under the curve, which leaves 95% of the area unshaded. Because 95% of the time a value sampled from this t distribution is between –2.201 and 2.201, a 95% CI is computed by setting t* equal to 2.201.

In any one sample, the sample mean (m) and standard deviation (s) are known, as is the sample size (n). What is not known is the population mean. That is why we want to generate a CI.

Rearrange Equation 13.1 to solve for μ (that is, put μ on the left side of the equation):

$$\mu = m \pm t^* \cdot \frac{s}{\sqrt{n}}$$

For the example, n = 12, m = 36.77°C, and s = 0.40°C. The t ratio has a 95% probability of being between –2.201 and +2.201, so t* = 2.201. Plug in those values and compute the population mean twice. The first time use the minus sign, and the second time do the calculations again using the plus sign. The results are the limits of the 95% CI, which ranges from 36.52 to 37.02°C.

How it works

The t distribution is defined by assuming the population is Gaussian and investigating the variation among the means of many samples. The math is then flipped to make inferences about the population mean from the mean and SD of one sample

This method uses probability calculations that are logically simple and flips them to answer questions about data analysis. Depending on how you look at it, this flip is either really obvious or deeply profound. Thinking about this kind of inverse probability is tricky and has kept statisticians and philosophers busy for centuries.

CI OF A MEAN VIA RESAMPLING

The resampling approach

One of the assumptions of the usual (t distribution) method for computing a CI of a mean is that the values were sampled from a Gaussian distribution. What if you can't support that Gaussian assumption?

Resampling is an alternative approach to statistics that does not assume a Gaussian (or any other) distribution.

Create many pseudosamples via resampling

The first step is to generate many more pseudosamples via resampling.

The data set we are analyzing has 12 values, so each of our "resampled samples" will also have 12 values. To find the first value, pick a random integer between 1 and 12. If it is 3, pick the third value from the original sample. To find the second value, again pick a random integer between 1 and 12 and then pick the corresponding value. It might be the third value again, but it is more likely to be one of the other 11 values. Repeat this process again until you have a new *pseudosample* with 12 values.

Repeat this many times to create, say, 500 pseudosamples. These new samples are created by sampling from the same set of values (12 in this example). Those same 12 values appear over and over, and no other values ever appear. But the samples aren't all identical. In each sample, some values are repeated and others are left out.

For each of the 500 new samples, compute the mean (because we want to know the CI of the population mean). Next, determine the 2.5th and 97.5th percentiles of that list of means. For this example, the 2.5th percentile is 36.55°C and the 97.5th percentile is 36.97°C. Because the difference between 97.5 and 2.5 is 95, we can say that 95% of the means of the pseudosamples are between 36.55 and 36.97°C.

The flip!

Statistical conclusions require flipping the logic from the distribution of multiple samples to the CI of the population mean. With the resampling approach, the flip is simple. The range of values that contains 95% of the resampled means (36.55 to 36.97°C) is also the 95% CI of the population mean.

For this example, the resampled CI is nearly identical to the CI computed using the conventional method (which assumes a Gaussian distribution). But the resampling method does not require the assumption of a Gaussian distribution and is more accurate when that assumption is not justified. The only assumption used in the resampling approach is that the values in the sample vary independently and are representative of the population.

That's it?

This resampling method seems too simple to be useful! It is surprising that randomly resampling from the same set of values gives useful information about the population from which that sample was drawn. But it does create useful results. Plenty of theoretical and practical (simulation) work has validated this approach, which some statisticians think should be widely used.

Learn More

To learn more about the resampling approach—also called the *bootstrapping* or *computer-intensive method*—start by reading books by Wilcox (2010) and Manly (2006). The method explained here is called a *percentile-based resampling confidence interval.* Fancier methods create slightly more accurate CIs.

A huge advantage of the resampling approach is that it is so versatile. It can be used to obtain the CI of the median, the interquartile range, or almost any other parameter. Resampling approaches are extensively used in the analysis of genomics data.

CI OF A PROPORTION VIA RESAMPLING

Let's revisit the voting example from Chapter 4. When polled, 33 of 100 people said they would vote a certain way. Our goal is to find the 95% CI for the true population proportion.

The resampling approach is easier to use for proportions than for means. The sample proportion is 33% with n = 100. To resample, simply assume the population proportion is 0.33, draw numerous random samples, and find the range of proportions that includes 95% of those samples.

Figure 13.2 illustrates the answer. If 33% of the entire population of voters is in favor of your candidate, then if you collect many samples of 100 voters,

Observed proportion (33/100)

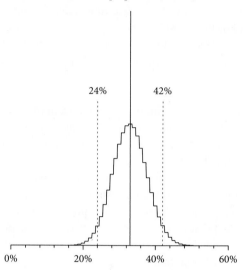

24%   42%

0%   20%   40%   60%

**Figure 13.2. A resampling approach to computing the CI of a
proportion.**

If the true proportion of "success" (in this example, voting for your
candidate) were 33%, the graph illustrates the probability distribution
of the success rate in many samples of 100 people. After removing
the 2.5% tails on each side, you can see that 95% of samples will have
success rates between 24 and 42%. Surprisingly, you can flip the logic
and use this graph as a way to derive CIs.

in 95% of those samples, between 24 and 42% of people will be in favor of your
candidate.

That range tells you about multiple samples from one population. We want
to know what can be inferred about the population from one sample. It turns out
that that same range is the CI. Given our one sample, we are 95% sure the true
population value is somewhere between 24 and 42%.

With continuous data, the resampling approach is more versatile than the t
distribution approach, because it doesn't assume a Gaussian (or any other) distribution. With binomial data, there is no real advantage to the resampling approach,
except that it is a bit easier to understand than the approach based on the binomial
distribution (see the next section).

CI OF A PROPORTION VIA BINOMIAL DISTRIBUTION

When polled, 33 of 100 people said they would vote a certain way. The goal is
to find two confidence limits, L (lower) and U (upper), so that the 95% CI of the
proportion ranges from L to U. This section demonstrates one approach, summarized in Figure 13.3. As mentioned in Chapter 4, there are several approaches to
computing CIs of proportions, and they don't give exactly the same result.

Observed proportion (33/100)

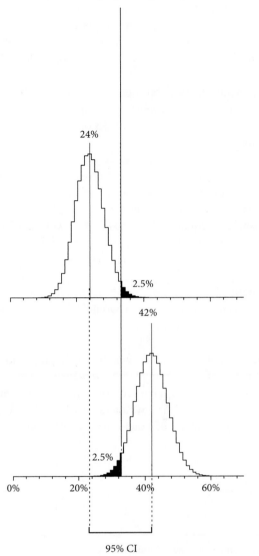

95% CI

Figure 13.3. How the CI of a proportion is derived.

The solid vertical line is at 33%, which is the sample proportion (33/100). The top probability distribution demonstrates the probability of various outcomes if the true population proportion were 24%. This distribution is centered on 24%, but the right tail of the distribution indicates that there is a 2.5% chance that the observed proportion in one particular sample will be 33% or higher. The bottom distribution indicates the probability of various outcomes if the population proportion were 42%. The left tail demonstrates that there is a 2.5% chance that the observed proportion in one particular sample will be 33% or lower. Subtracting 2.5% from 100% twice, we are left with a 95% interval ranging from 24 to 42%.

The lower 95% CI (L) can be determined by an indirect approach. We can use probability equations to answer the question, If the population proportion equals L, what is the probability that a sample proportion (n = 100) will equal 0.33 (as we observed in our sample) or more? To compute a 95% CI, each tail of the distribution must be 2.5%, so we want to find the value of L that makes the answer to that question 2.5%.

Determining the upper 95% confidence limit of a proportion (U) uses a similar approach. If the population proportion equals U, what is the probability that a sample proportion (n = 100) will equal 0.33 or less? We want the answer to be 2.5% and solve for U.

Using the binomial distribution to solve for L and U is not a straightforward task. A brute force approach is to try lots of possible values of L or U in order to find values for which the cumulative binomial distribution gives a result of 2.5%. Excel's Solver can automate this process (Winston, 2004).

For this example, L equals 0.24, or 24%. If fewer than 24% of voters were truly in favor of your candidate, there would be less than a 2.5% chance of randomly choosing 100 subjects of which 33% (or more) are in favor of your candidate.

For the example, U equals 0.42, or 42%. If more than 42% of voters really were in favor of your candidate, there would be less than a 2.5% chance of randomly choosing 100 subjects of which 33% (or less) were in favor of your candidate.

So far, we have been asking about which sample proportions are likely given known population proportions (L and U). Now we want to make inferences about the population from one sample. That requires a flip in reasoning. Here it comes.

Calculate 100% − 2.5% − 2.5%, which leaves 95%. Therefore, there is a 95% chance that the true percentage of voters in favor of your candidate is between 24 and 42%.

Q & A

How can resampling from one sample tell you anything useful about the population from which that sample was drawn?

It is amazing that resampling works. It is also called bootstrapping, which comes from the phrase "pulling oneself up by the bootstrap," which is impossible. It is not obvious that bootstrap/resample methods produce valid inferences about the population the sample was drawn from, but this has been proven by both mathematical proofs and simulations.

Is it correct to say that there is a 95% chance that the population mean (or proportion) lies within the computed confidence interval?

No. The population mean (or proportion) has a fixed value (that we never know). So it is not appropriate to ask about the chance that the population mean (or proportion) has any particular value. It is what it is with no chance involved. In contrast, the computed confidence intervals depend on which sample of data you happened to choose, so vary from sample to sample based on random sampling. It is correct to say that there is a 95% chance that a 95% confidence interval includes the population mean (or proportion).

Why the focus on parameters like the mean? I don't know want to know the precision of the mean. I want to know about the distribution of values in the population.

> The 95% prediction interval answers your question. It is the range of values that is 95% sure to contain 95% of the values in the entire population. I'm not sure why, but prediction intervals are not often used in most fields of biology. Prediction intervals are much wider than confidence intervals.

CHAPTER SUMMARY

- You can understand confidence intervals without understanding how they are computed. This chapter gives you a peek at how the math works. You don't need to understand this chapter to understand how to interpret confidence intervals.
- The math works by flipping around (solving) equations that predict samples from a known population in order to let you make inferences about the population from a single sample.
- Depending on your frame of mind, this flip can seem really obvious or deeply profound.
- An alternative approach is to use resampling (also called bootstrapping) methods. The idea is that analyzing a set of bootstrapped samples created from the original data will generate a confidence interval for the population mean or proportion.

TERMS INTRODUCED IN THIS CHAPTER

- Bootstrapping (p. 107)
- Computer-intensive method (p. 107)
- Percentile-based resampling confidence interval (p. 107)
- Prediction intervals (p. 111)
- Pseudosample (p. 106)
- Resampling (p. 106)
- t distribution (p. 104)

CHAPTER 14

Error Bars

The only role of the standard error... is to distort and conceal the data. The reader wants to know the actual span of the data; but the investigator displays an estimated zone for the mean.

A. R. FEINSTEIN

Scientific papers often tabulate or graph results as mean ± SD or mean ± SEM. This chapter explains the difference, as well as how to compute the SD from the SEM and sample size.

SD VERSUS SEM

What is the SD?
The standard deviation, abbreviated SD or s, was explained in Chapters 9 and 10. It is expressed in the same units as the data and quantifies variation among the values. If the data are sampled from a Gaussian distribution, you expect about two-thirds of the values to lie within 1 SD of the mean.

What is the SEM?
The ratio of the SD divided by the square root of the sample size (n) is called the *standard error of the mean*, abbreviated SEM:

$$SEM = \frac{SD}{\sqrt{n}}$$

Often the SEM is referred to as the *standard error*, with the word *mean* missing but implied. Like many statistical terms, this jargon can be confusing. The SEM has nothing to do with standards or errors, and standard errors can be computed for values other than the mean (e.g., the standard error of the best-fit value of a slope in linear regression; see Chapter 33).

The SEM does not quantify the variability among values
Note that the SEM does not directly quantify scatter or variability among values in the population. Many scientists misunderstand this point. The SEM will be small when the SD is large, provided that the sample size is also large. With huge samples, the SEM is always tiny. For the n = 130 sample of the body temperature data, the SEM equals 0.0359°C.

The SEM quantifies how precisely you know the population mean

Imagine taking many samples of sample size n from your population. These sample means will not be identical, so you could quantify their variation by computing the SD of the set of means. The SEM computed from one sample is your best estimate of what the SD among sample means would be if you collected an infinite number of samples.

Chapter 12 defined the margin of error (W) of a CI of a mean as follows:

$$W = \frac{t^* \cdot SD}{\sqrt{n}}$$

Substituting the definition of the SEM,

$$W = t^* \cdot SEM$$

So the width of a CI of the mean is proportional to the SEM.

How to compute the SD from the SEM

Many scientists present the SEM in papers and presentations. Remember this equation so you can easily calculate the SD:

$$SD = SEM \cdot \sqrt{n} \qquad\qquad \textbf{14.1}$$

For the n = 12 body temperature sample, the SEM equals 0.1157°C. If that was all you knew, you could compute the SD. Multiply 0.1157 times the square root of 12, and the SD equals 0.40°C. This is an exact calculation, not an approximation.

WHICH KIND OF ERROR BAR SHOULD I PLOT?

An *error bar* displays variation or uncertainty. Your choice of error bar depends on your goals.

Goal: To show the variation among the values

If each value represents a different individual, you probably want to show the variation among values. Even if each value represents a different lab experiment, it often makes sense to show the variation.

With fewer than 100 or so values, you can create a scatter plot that shows every value. What better way to show the variation among values than to show every value? If your data set has more than 100 or so values, a scatter plot becomes messy. Alternatives include a box-and-whiskers plot, a frequency distribution (histogram), or a cumulative frequency distribution (see Chapter 7).

What about plotting the mean and SD? The SD does quantify scatter, so this is indeed one way to graph variability. But since the SD is only one value, plotting it is a pretty limited way to show variation. A graph showing mean and SD *error bars* is less informative than any of the other alternatives. The advantages of plotting a mean with SD error bars (as opposed to a column scatter graph, a box-and-whiskers plot, or a frequency distribution) are that the graph is simpler and the approach is conventional in many fields.

Of course, if you do decide to show SD error bars, be sure to say so in the figure legend, so they won't be confused with SEM error bars.

If you are creating a table, rather than a graph, there are several ways to show the data, depending on how much detail you want to show. One choice is to show the mean as well as the smallest and largest values. A more compact choice is to tabulate the mean ± SD.

Goal: To show how precisely you have determined the population mean

If your goal is to compare means with a t test (Chapter 30) or ANOVA (Chapter 39), or to show how closely your data come to the predictions of a model, you are likely trying to show how precisely the data define the mean, rather than to show variability among values. In this case, the best approach is to plot the 95% CI of the mean (or perhaps a 90 or 99% CI).

What about the SEM? Graphing the mean with SEM error bars is a method commonly used to show how well you know the mean. The only advantages of SEM error bars versus CI error bars is that SEM error bars are shorter and are more conventional in some fields. But it is harder to interpret SEM error bars than it is to interpret a CI.

Whatever error bars you choose to show, be sure to state your choice.

Goal: To create persuasive propaganda

Simon (2005) "advocated" the following approach in his excellent blog. Of course he meant it as a joke!

- When you are trying to emphasize small and unimportant differences in your data, show your error bars as SEM and hope that your readers think they are SD. (The readers will believe that the standard deviation is much smaller than it really is so will incorrectly conclude that the difference among means is large compared to variation among replicates.)
- When you are trying to cover up large differences, show the error bars as SD and hope that your readers think they are SEM. (The readers will believe that the standard deviation is much larger than it really is so will incorrectly conclude that the difference among means is small compared to the variation among replicates.)

THE APPEARANCE OF ERROR BARS

Frazier, Schneider, and Michel (2006) measured how well the neurotransmitter norepinephrine relaxes bladder muscles. Chapter 11 examined the concentrations required to get half-maximal bladder relaxation (EC_{50}) as an example of lognormal data. Now let's look at the maximal relaxation (abbreviated % E_{max}) that can be achieved with large doses of norepinephrine in old rats. Figure 14.1 (left) illustrates the raw data plotted as a column scatter plot.

Figure 14.1. (Left) The actual data. (Right) Five kinds of error bars, each representing a different way to portray variation or precision.

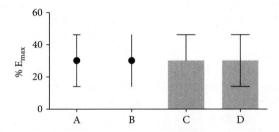

Figure 14.2. Four different styles for plotting the mean and SD.

These methods all plot the same values, and there is no real reason to prefer one style over another.

Figure 14.1 (right) shows the same data with several kinds of error bars representing the SD, SEM, 95% CI, range, and interquartile range. When you create graphs with error bars, be sure to state clearly how they were computed.

Figure 14.2 shows that error bars (in this case, representing the SD) can be plotted with or without horizontal caps, and (with bar graphs) in one direction or both directions. The choice is a matter of preference and style. When error bars with caps are placed on bars and only extend above the bar (bar C in Figure 14.2), the resulting plot is sometimes called a *dynamite plunger plot*, sometimes shortened to *dynamite plot*.

HOW ARE SD AND SEM RELATED TO SAMPLE SIZE?

If you increase the sample size, is the SEM expected to get larger, get smaller, or stay about the same?

It is expected to get smaller. The SEM quantifies how precisely the population mean has been determined. Larger samples have smaller SEMs,, because the mean of a large sample is likely to be closer to the true population mean than is the mean of a small sample.

Of course, in any particular experiment, you can't be sure that increasing n will decrease SEM. It is a general rule, but it is also a matter of chance. It is possible, but unlikely, that the SEM might increase if you collect a larger sample size.

If you increase the sample size, is the SD expected to get larger, get smaller, or stay about the same?

The SD quantifies the scatter of the data. Whether increasing the size of the sample increases or decreases the scatter in that sample depends on the random selection of values. The SD, therefore, is equally likely to get larger or to get smaller as the sample size increases.

That statement, however, has some fine print: for all sample sizes, the sample variance (i.e., the square of the sample SD) is the best possible estimate of the population variance. If you were to compute the sample variance from many samples from a population, on average, it would be correct. Therefore, the sample variance is said to be *unbiased*, regardless of n. In contrast, when n is small, the sample SD tends to slightly underestimate the population SD. Increasing n, therefore, is expected to increase the SD a bit. Because statistical theory is based on the variance rather than the SD, it is not worth worrying about (or correcting for) the bias in the sample SD. Few statistics books even mention it. The expected increase in the sample SD as you increase sample size is tiny compared to the expected decrease in the SEM.

COMMON MISTAKES: ERROR BARS

Mistake: Plotting mean and error bar instead of plotting a frequency distribution

Figure 14.3 shows the number of citations (references in other papers within five years of publication) received by 500 papers randomly selected from the journal *Nature* (Colquhoun, 2003). The first (leftmost) bar shows that there are 71 (Y value) papers that had between 0 and 20 citations (since the X value is centered at 10 and each bar represents a bin width of 20 citations). The last (rightmost) bar shows that a single paper (of the 500 studied) had 2,364 citations. This frequency distribution is easy to understand even though it is very asymmetrical, even more asymmetrical than a lognormal distribution (see Chapter 11). With a glance at the graph, you can see that many papers are only cited a few times (or not at all), while only a few are heavily cited.

The mean number of citations is approximately 115, with a standard deviation of approximately 157 ("approximately" because I don't have access to the raw data, but only to the frequency distribution). Viewing a graph showing that the mean is 115 and the SD is 157 would not help you understand the data. Seeing

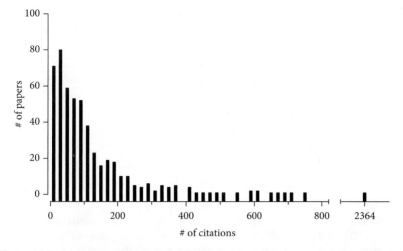

Figure 14.3. A very asymmetrical distribution

This figure, remade from Colquhoun (2003), shows the number of citations (X-axis) received (within five years of publication) by 500 papers randomly chosen from the journal *Nature*. The bin width is 20 citations. The first bar shows that 74 papers received between 0 and 20 citations, the second bar shows that 80 papers received between 21 and 40 citations, and so on.

that the mean is 115 and the SEM is 7 would be less helpful and might be quite misleading.

The median number of citations is approximately 65, and the interquartile range runs from about 20 to 130. Reporting the median with that range would be more informative than reporting the mean and SD or SEM. However, it seems to me that the best way to summarize these data is to show a frequency distribution. No numerical summary really does a good job.

Mistake: Assuming that all distributions are Gaussian

What would you think about the distribution of a data set if you were told only that the mean is 101 and the SD is 43 (or if you saw the bar graph of Data Set A in Figure 14.4)? You'd probably imagine that the data look something like Data Set B, and that would often be correct. But Data Sets C and D of Figure 14.4 also share that same mean and SD (and SEM). The distributions of values in Data Sets C and D are extremely different from the distribution of values in Data Set B. It can be useful to summarize data as mean and SD (or SEM), but it can also be misleading. Look at the raw data when possible.

When graphing your own data, keep in mind that it is easy to show the actual data on dot plots (column scatter graphs) with up to 100 or even a few hundred values. With more points, it is easy to plot a box-and-whiskers graph or a frequency distribution histogram. Don't plot mean and an error bar without thinking first about alternative ways to plot the data.

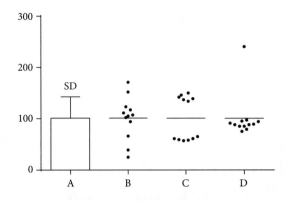

Figure 14.4. The mean and SD can be misleading.
All four data sets in the graph have approximately the same mean (101) and SD (43). If you were only told those two values or only saw the bar graph of Data Set A, you'd probably imagine that the data look like Data Set B. Data Sets C and D have very different distributions, but the same mean, SD and SEM as Data Set B.

Mistake: Plotting a mean and error bar without defining how the error bars were computed

If you make a graph showing means and error bars (or a table with means plus or minus error values), it is essential that you state clearly what the error bars are—SD, SEM, CI, range, or something else? Without a clear legend or a statement in the methods section, the graph or table is ambiguous.

Q & A

What does the SD quantify?
The SD quantifies scatter—how much the values vary from one another.

What does the SEM quantify?
The SEM quantifies how precisely you know the true mean of the population. The value of the SEM depends upon both the SD and the sample size.

Are the SD and SEM expressed in the same units?
Yes. Both are expressed in the same units as the data.

Which is smaller?
The SEM is always smaller than the SD.

When can I use the rule of thumb that the 95% CI of a mean equals the mean plus or minus 2 SEMs?
With large samples. Calculating a CI requires using a multiplier from the t distribution, abbreviated (in this book) as t*. With large samples, t* has a value close to 2.0. Hence the rule of thumb that a 95% CI of a mean extends about 2 SEM in each direction from the mean. With small sample sizes, this approximate CI will be too narrow because the correct multiplier is greater than 2.0. For example, if n = 3, the distance a 95% CI extends in each direction from the mean equals 4.30 times the SEM.

Graphs often show an error bar extending 1 SEM in each direction from the mean.
What does that denote?

> The range defined by the mean plus or minus 1 SEM has no simple interpretation. With large samples, that range is a 68% CI of the mean. But the degree of confidence depends on sample size. When n = 3, that range is only a 58% CI. You can think of the mean plus or minus 1 SEM as covering about a 60% CI.

Why are SEM error bars used so often?

> SEM error bars are popular because they are always smaller than the SD, and because they are a compact way to show how precisely you know the mean.

What can I learn by observing whether two error bars overlap or not?

> Not much. See Chapter 30.

CHAPTER SUMMARY

- The SD quantifies variation.
- The SEM quantifies how precisely you know the population mean. It does not quantify variability. The SEM is computed from the SD and the sample size.
- The SEM is always smaller than the SD. Both are measured in the same units as the data.
- Graphs are often plotted as means and error bars, which are usually either the SD or the SEM.
- Consider plotting each value, or a frequency distribution, before deciding to graph a mean and error bar.
- You can calculate the SD from the SEM and n using Equation 14.1.
- Show the SD when your goal is to show variability and show the SEM (or, better, the confidence interval) when your goal is to show how precisely you have determined the population mean.
- The choice of SD or SEM is often based on traditions in a particular scientific field. Often, SEMs are chosen simply because they are smaller.
- Graphs with error bars (or tables with error values) should always contain labels, whether the error bars are SD, SEM, or something else.

TERMS INTRODUCED IN THIS CHAPTER

- Dynamite plunger plot (p. 115)
- Error bar (p. 115)
- Standard error of the mean (SEM) (p. 112)

P Values and Statistical Significance

CHAPTER 15

Introducing P Values

> Imperfectly understood CIs are more useful and less dangerous
> than incorrectly understood P values.
>
> HOENIG AND HEISEY (2001)

Many statistical analyses are reported with P values, which are explained in this essential chapter. P values are often misunderstood because they answer a question you probably never thought to ask. This chapter explains what P values are and the many incorrect ways in which they are interpreted.

INTRODUCING P VALUES

You've probably already encountered P values. They appear in many statistical results, either as a number ("P = 0.0432") or as an inequality ("P < 0.01"). They are also used to make conclusions about statistical significance, as explained in Chapter 16.

P values are difficult to understand and are often misinterpreted. This chapter starts with several examples to explain what P values are and then identifies the common incorrect beliefs many have about P values.

EXAMPLE 1: COIN FLIPPING

You flip a coin 20 times and observe 16 heads and 4 tails. Because the probability of heads is 50%, you'd expect about 10 heads in 20 flips. How unlikely is it to find 16 heads in 20 flips? Should you suspect that the coin is unfair?

Calculating the P value

To evaluate our suspicion that the coin might be unfair, we need to define what a fair coin is. A coin flip is fair when each coin flip has a 50% chance of landing on heads and a 50% chance of landing on tails, and each result is accurately recorded.

What probability do we want to compute? We could compute the chance of observing 16 heads in 20 flips, but that isn't sufficient. We would have been even more suspicious that the coin flipping wasn't fair had the coins landed on heads

HEADS	TAILS	PROBABILITY (%)
0	20	0.000
1	19	0.002
2	18	0.018
3	17	0.109
4	16	0.462
5	15	1.479
6	14	3.696
7	13	7.393
8	12	12.013
9	11	16.018
10	10	17.620
11	9	16.018
12	8	12.013
13	7	7.393
14	6	3.696
15	5	1.479
16	4	0.462
17	3	0.109
18	2	0.018
19	1	0.002
20	0	0.000

Table 15.1. Probabilities when flipping a fair coin 20 times.

When you flip a fair coin, there is a 50% chance of heads and a 50% chance of tails. This table shows all the probabilities of flipping a fair coin 20 times. This table was created using the free binomial distribution calculator at www.graphpad.com/quickcalcs/probability1/.

The values in the third column can also be computed using the Excel function Binomdist. For example, the value for 10 heads is computed as =BINOMDIST(10, 20, 0.5, FALSE). The first value is the number of "successes" (heads in this case); the second value is the number of tries (20 in this case), the third number is the fraction success for each try (0.5 for a coin flip), and the last value is False because you don't want a cumulative probability. Newer version of Excel also offer the function Binom.Dist which works the same way but is said to be more accurate.

17 times, or 18 times, and so on. Similarly, we would have been suspicious if the coin had landed on tails 16 or more times (and so on heads 4 or fewer times). Putting all this together, we want to answer the following question: If the coin tosses are random and the answers are recorded correctly, what is the chance that when you flip the coin 20 times, you'll observe either 16 or more heads or 4 or fewer heads (meaning 16 or more tails)?

Table 15.1 shows all the possible probabilities when you flip a fair coin 20 times. This book doesn't explain how to compute these answers, but the table legend points you to a Web page and an Excel function that does these calculations.

To answer the current question, we need to sum the probability of seeing 0, 1, 2, 3, 4, 16, 17, 18, 19, or 20 heads in 20 coin flips. The answer is 0.0118, or 1.18%. This result is called a *P value*. It is always reported as a fraction rather than a percentage.

Interpreting the P value

The P value answers this question: Assuming the coin tosses are random and the answers are recorded correctly, what is the chance that when you flip a coin 20 times, you'll observe results as extreme (or more extreme) than those you observed in the previous example—that is, that you would observe 16 or more, or 4 or fewer, heads? The answer, the P value, is 0.0118. You'd only see results this far from an even split of heads to tails in 1.18% of 20-coin runs.

Should you conclude that the coin is unfair? There are two possible explanations for your observations. One possibility is that it is a coincidence. The other possibility is that the coin flipping wasn't fair. Which is it? All statistics can do is tell you how rare it is that such a coincidence would happen (1.18% of the time). What you conclude depends on the context. Consider these three scenarios:

- You've examined the coin and are 100% certain it is an ordinary coin, and you did the coin flips and recorded the results yourself, so you are sure there is no trickery. In this case, you can be virtually 100% certain that this streak of many heads is just a chance event. The calculation shown above won't change that conclusion.
- If the coin flips were part of a magic show, you can be pretty sure that trickery invalidates the assumptions that the coin is fair, the tosses are random, and the results are accurately recorded. You'll conclude that the coin flipping was not fair.
- If everyone in a class of 200 students had flipped the coin 20 times and the run of 16 heads was the most heads obtained by any student, you wouldn't conclude that the coin was unfair. It is not surprising to see a coincidence that happens 1.18% of the time when you run the test (coin flipping) 200 times.

EXAMPLE 2: ANTIBIOTICS ON SURGICAL WOUNDS

Heal and colleagues (2009) tested whether applying an antibiotic (chloramphenicol) to surgical wounds would reduce the incidence of wound infection. They randomly assigned 972 surgical patients to receive either an antibiotic ointment or an ointment with no active medication.

Infections occurred in 6.6% of the patients who received the antibiotic, compared with 11.0% of the patients who received inactive ointment. In other words, infections occurred in about 40% fewer patients who received the antibiotic. Does this mean that the antibiotic worked in this situation? Or could the results be due to a coincidence of random sampling?

To assess this question, we first must tentatively assume that the risk of infections is the same in patients who receive an antibiotic ointment and those who receive an inactive ointment, and that the discrepancy in the incidence of infections observed in this study is the result of chance. Then we can ask the following question: If the risk of infection overall is identical in the two groups, what is the chance that random sampling would lead to a difference in incidence rates equal to or greater than the difference actually observed in this study?

The authors did the calculation and published the answer, which is 0.010. Since this P value is low, and since the hypothesis that an antibiotic would prevent infection is quite sensible, the investigators concluded that the antibiotic worked to prevent wound infection.

EXAMPLE 3: ANGIOPLASTY AND MYOCARDIAL INFARCTION

Cantor and colleagues (2009) studied the optimal way to treat patients who had experienced myocardial infarctions (heart attacks) but were admitted to a hospital that could not do percutaneous coronary intervention (PCI, also known as angioplasty). They randomized 1,059 such patients into two groups. One group was immediately transferred to a different hospital and received PCI. The other half was given standard therapy at the admitting hospital. The researchers assessed the fraction of patients who died or experienced worsening heart disease (another myocardial infarction or worsened congestive heart failure) within 30 days of being admitted to the hospital.

By the end of 30 days, one (or more) of these outcomes had occurred in 11.0% of the patients who were transferred to a different hospital and 17.2% of the patients who received standard therapy. The risk ratio was $11.0/17.2 = 0.64$. That means that the risk of death or worsening heart disease in patients who were transferred was 64% of the risk for patients who stayed at the original hospital.

Is this a real difference? Or is it just a coincidence of random sampling, such that sicker or unluckier patients in this study tended to be assigned to stay in the same hospital? To answer this question, we first must tentatively *assume* that the risk of death or worsening heart disease was identical in the two populations, and that the difference observed in this study was a coincidence of random sampling. Given that assumption, we can ask the question, What is the chance that the risk ratio would have been 0.64 or less in a study this size? The investigators state that the P value is 0.004. You don't need to know how they calculated the P value in order to interpret it.

There are two possibilities. One is that the sicker or unluckier patients in this study just happened to be assigned to stay in the same hospital. That coincidence would happen in 0.4% , which is 1 in 250, of studies of this size. The other possibility is that patients who are transferred to a hospital that can do angioplasty on average have better outcomes than those who stay at a hospital that cannot do angioplasty. Since the P value is so low, and since that hypothesis makes sense, the investigators concluded that transferring patients is the safer protocol.

LINGO: NULL HYPOTHESIS

In all three examples, the P value definition begins with a tentative assumption.

- In the coin-flipping example, the assumption is that the coin is fair so that the observed deviation from an outcome of half heads and half tails occurred randomly.
- In the antibiotic example, the assumption is that the antibiotic used does not actually prevent infection in the experimental situation, so that the observed difference in infection rates was due to random sampling of subjects.
- In the angioplasty example, the assumption is that patients transferred to another hospital have the same risk of death or worsening heart disease as do those who remain in the existing hospital, so that the observed difference in risk of death or heart disease was a coincidence.

In each case, the assumption is called a *null hypothesis*. The definition of every P value begins by assuming a null hypothesis is true. If you can't state what the null hypothesis is, you can't understand the P value.

P VALUES ARE NOT VERY REPRODUCIBLE

P values are less reproducible than most people would guess. To demonstrate this, I simulated multiple data sets randomly sampled from the same populations. The top portion of Figure 15.1 shows four simulated experiments. The four P values vary considerably. The bottom part of Figure 15.1 shows the distribution of P values from 2,500 such simulated experiments. Leaving out the 2.5% highest and lowest P values, the middle 95% of the P values range from 0.0001517 to 0.6869—a range covering more than three orders of magnitude!

This example uses an experiment in which the median P value is about 0.05. Simulation with a smaller SD resulted in smaller P values, but the range still covered more than three orders of magnitude. Boos and Stefanski (2011) have shown that P values from repeated simulated experiments vary over more than three orders of magnitude in a number of situations.

Cumming (2008; 2011) systematically studied the reproducibility of P values. If the P value from one particular experiment equals 0.05, he asked what the P value from a repeat experiment is likely to be (taking into account only random sampling and assuming there are no glitches with either experiment). Surprisingly, the results don't depend on sample size. If the first P value equals 0.05, there is

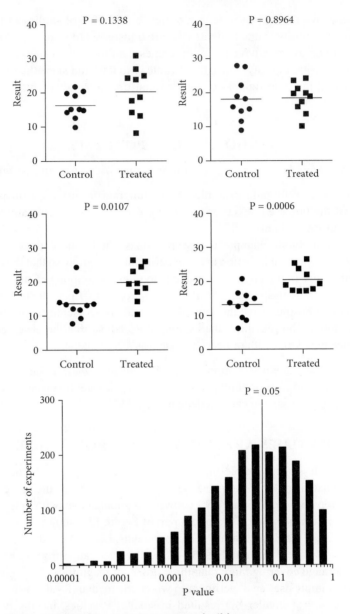

Figure 15.1. P values are not very reproducible.

The top panels show four pairs of data sets simulated (using GraphPad Prism) from Gaussian distributions with an SD equal to 5.0. The population means of the two populations differ by 5.0. Ten values were simulated in each group, and an unpaired t test was used to compare the sample means. Note how far apart the four P values are, even though the data in each of the four experiments were sampled from the same populations. The histogram at the bottom shows the distribution of P values from 2,500 simulated data sets sampled from those same populations. Note that the X-axis is logarithmic. The P values span a very wide range.

a 20% chance that the P value from a repeat experiment will be greater than 0.38 and a 5% chance it will be greater than 0.82.

WHY P VALUES ARE CONFUSING

P values are confusing for several reasons:

- The null hypothesis is usually the opposite of the hypothesis the experimenter expects or hopes to be true.
- In many situations, you know before collecting any data that the null hypothesis is almost certainly false. The null hypothesis is that in the populations being studied there is zero difference between the means, zero correlation, or a relative risk or odds ratio of 1.0000. People rarely conduct experiments or studies in which it is even conceivable that the null hypothesis is true. The difference or correlation or association in the overall population might be trivial, but it almost certainly isn't zero. So the P value asks a question that it rarely makes sense to ask.
- Clinicians and scientists find it strange to calculate the probability of obtaining results that weren't actually obtained. The math of theoretical probability distributions is beyond the easy comprehension of most scientists.
- The logic of P values goes in a direction that seems backward. In conducting a study, you observe a sample of data and ask about the populations from which the data are sampled. The definition of the P value starts with an assumption about the population (the null hypothesis) and asks about possible data samples. It's backwards! Thinking about P values seems quite counterintuitive, except maybe to lawyers or Talmudic scholars used to this sort of argument by contradiction.
- The P value answers a question that few scientists ever think to ask!

ONE- OR TWO-TAILED P VALUE?

Example 1: Coin flipping
Let's return to the coin-flipping example from the beginning of the chapter. The P value we've already defined is known as a *two-tailed P value*. It answers the question, If the null hypothesis (fair coin) were true, what is the chance of observing 16 or more, or 4 or fewer, heads in 20 flips? It is two-tailed because it asks about both extremes: lots of heads or very few heads. This two-tailed P value is 0.0118.

To define a *one-tailed P value*, you must predict which way the data will go before collecting them. Perhaps someone told you they suspected that this coin comes up heads too often and so the question you want to ask is whether that claim is true. In that case, the P value answers the question:, If the null hypothesis were true, what is the chance of observing 16 or more heads in 20 flips? Find out by adding up the bottom five rows of Table 15.1. The answer, the one-tailed P value, equals 0.0059. Note that this is half the two-tailed P value.

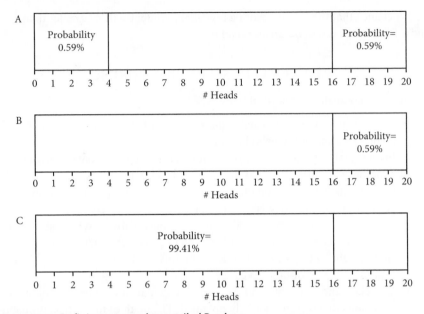

Figure 15.2. Defining one- and two-tailed P values

You flip a coin 20 times, see 16 heads, and want to compute a P value testing the null hypothesis that the coin flips are fair. Part A defines the two-tailed P value, the probability (under the null hypothesis) of observing 16 or more heads, plus the probability (the other tail) of observing 4 or fewer heads. Part B shows the one-tailed P value if you had predicted that the coin would land on heads more than half the time. Since the actual result (16 heads) corresponds to the hypothesis, the P value is less than 0.50. Part C shows the one-tailed P value if you had predicted that the coin would land on heads less than half the time. Since the actual data (16 heads) was opposite to the prediction, the P value is greater than 0.50. The P values in the figure are shown as percentages to make them more compact. However, P values are almost always presented as proportions, not percentages.

What if someone had suggested that this coin was biased the other way, with tails coming up too often? In that case, the one-tailed P value would answer the question, If the null hypothesis were true, what is the chance of observing 16 or fewer heads in 20 flips? Find out the answer by adding up the top 17 rows in Table 15.1 (which starts with 0 heads). The result, the one-tailed P value, is 0.9940. You expected to have fewer heads than tails, but in fact, the results went the other way, so the one-tailed P value is high.

Figure 15.2 shows the definitions of the two-tailed P value and the two distinct one-tailed P values. Note that the two one-tailed P values add up to 1.00.

Example 2: Antibiotics on Surgical Wounds

The null hypothesis for the antibiotics example is that the antibiotic ointment does not change the risk of infection. In the patients studied, there were 40% fewer infections among those treated with the antibiotic compared to those treated with the inactive ointment.

The two-tailed P value answers the question, If the null hypothesis were true, what is the chance that random sampling would lead to either a 40% reduction (or more) in infection rate in those treated with antibiotics or a 40% reduction (or more) in those treated with inactive ointment? The answer, as calculated by the investigators and included in the published paper, is 0.010.

To compute a one-tailed P value, you must make a prediction about the direction of the findings. Assume you had predicted that there would be fewer infections in the antibiotic-treated patients. In that case, the one-tailed P value answers the question, If the null hypothesis were true, what is the chance that random sampling would lead to a 40% reduction (or more) in infection rate in those treated with antibiotics? The answer is 0.005, which is half the two-tailed P value.

What if you had predicted that there would be more infections in the patients treated with the antibiotic than in those treated with the inactive ointment? Sure, that is a pretty unlikely prediction for this example, but go with it. Now the one-tailed P value answers the question, If the null hypothesis were true, what is the chance that random sampling would lead to a 40% reduction (or less) in infection rate in those treated with the inactive ointment? Since the prediction runs opposite to the data, this one-tailed P value, 0.995, is high.

What is the advantage of a one-tailed P value?

When planning a study, choosing a one-tailed P value gives you a more focused hypothesis and so reduces the necessary sample size. A study designed to use a one-tailed P value requires about 20% fewer subjects than the same study designed to use a two-tailed P value (Moyé & Tita, 2002). Even though that is not a huge difference, it does decrease cost and risk and so is a reason to prefer one-tailed P values (Knottnerus & Bouter, 2001).

Once the data are collected, the appeal of a one-tailed P value is clear. Assuming you predicted the direction of the effect correctly, a one-tailed P value equals one-half of the two-tailed P value (this is not always exactly true, but it is true for most statistical tests).

When is a one-tailed P value appropriate?

Use of a one-tailed P value requires planning. If you haven't given any thought as to whether you will use one- or two-tailed P values until after you have begun to analyze the data, it is too late. Use two-tailed P values. You should only report a one-tailed P value when you have predicted which treatment group will have the larger mean (or proportion, or survival time, etc.) before you collect any data and have recorded this prediction (so you can't be tempted to change your "prediction" after viewing the data).

Here is an example in which one could argue that it is appropriate to choose a one-tailed P value: imagine you are testing whether a new antibiotic impairs kidney function, as measured by serum creatinine. Many antibiotics poison kidney cells, resulting in reduced glomerular filtration and increased serum creatinine. As far as I know, no antibiotic is known to decrease serum creatinine.

Before collecting any data, it would not be unreasonable to state that there are two possibilities: either the drug will not change the mean serum creatinine of the population, or else it will increase the mean serum creatinine in the population. Accordingly, it makes sense to calculate a one-tailed P value that tests the null hypothesis that the drug does not increase the creatinine level. If the creatinine level goes down, no matter how much, you'd attribute that decrease to chance.

The problem with one-tailed P values

Before committing to a decision to use a one-tailed P value, think about what you would do if the data surprised you and the treatment effect went in the direction opposite to what you predicted. In this case, the one-tailed P value would be larger than 0.50.

Would you be tempted to switch to a two-tailed P value? Would you be tempted to change your "prediction" about the direction of the effect? If you answer yes to either of these questions, then you should plan to use a two-tailed P value.

One famous example of this problem is the Cardiac Arrhythmia Suppression Trial (CAST) study of the use of anti-arrhythmic drugs following myocardial infarction (heart attack). These drugs suppress premature ventricular contractions (extra heart beats), which can occur after a heart attack and were assumed to also prevent fatal arrhythmias (including sudden death) and thus prolong life after a heart attack. For many years, these drugs were the standard therapy for anyone who survived a heart attack. The CAST study tested the hypothesis that anti-arrhythmic drugs prolonged life after a heart attack (CAST Investigators, 1989). The people planning the study expected that the drugs would either prolong the life of the patients or have no effect, and so they chose to calculate and report a one-tailed P value.

The results were a surprise to the cardiologists who designed the study. Patients treated with anti-arrhythmic drugs turned out to be four times more likely to die than patients given a placebo. The results went in the direction opposite to what was predicted.

When in doubt, use two-tailed P values

When in doubt, I suggest that you use a two-tailed P value for the following reasons:

- The relationship between P values and CIs is more straightforward with two-tailed P values.
- Some tests compare three or more groups, in which case the concept of one or two tails makes no sense.
- The practical effect of choosing a one-tailed P value is to make it seem as though the evidence is stronger (because the P value is lower). But this is not due to any data collected in the experiment, but rather to a choice (or perhaps a belief) of the experimenter (Goodman, 2008).

- Some reviewers or editors may criticize any use of a one-tailed P value, no matter how well you justify it.

THERE IS MUCH MORE TO STATISTICS THAN P VALUES

Many statistical analyses generate both P values and CIs. Many scientists report the P value and ignore the CI. I think that this is a mistake.

Interpreting P values is tricky, requiring one to define a null hypothesis and think about the distribution of the results of hypothetical experiments. Interpreting CIs, in contrast, is quite simple. You collect some data and do some calculations to quantify a treatment effect. Then you report that value (perhaps a difference or a relative risk) along with a CI that quantifies the precision of that value. This approach is very logical and intuitive. You report what you calculated along with an interval that shows how sure you are.

The underlying theory is identical for CIs and P values. If both are interpreted correctly, the conclusions will be identical. But what a big "if"! P values are often interpreted incorrectly (by people who haven't read this book, of course).

When reading papers, I suggest that you don't let yourself get distracted by P values. Instead, look for a summary of how large the effect was and its confidence interval.

When writing a paper, first make sure that the results are shown clearly. Then compute confidence intervals to quantify the precision of the effect sizes. Consider stopping there. Only add P values to your manuscript when you have a good reason to think that they will make the paper easier to understand.

COMMON MISTAKES: P VALUES

Many of the points listed below come from Kline (2004) and Goodman (2008).

The most common misconception about P values

Many people misunderstand what a P value means. Let's assume that you read that investigators compared a mean in groups given two alternative treatments and reported that the P value equals 0.03. Here are two correct definitions of this P value:

- If the two population means are identical (the null hypothesis is true), there is a 3% chance of observing a difference as large as you observed (or larger).
- Random sampling from identical populations would lead to a difference smaller than you observed in 97% of experiments and larger than you observed in 3% of experiments.

Many people want to believe the definition below, which is appears crossed out to emphasize that it is wrong:

~~There is a 97% chance that there is a real difference between populations and a 3% chance that the difference is a random coincidence.~~

The P value is not the probability that the result was due to sampling error

The P value is computed assuming the null hypothesis is true. In other words, the P value is computed based on the assumption that the difference was due to randomness in selecting subjects—that is, to sampling error. Therefore, the P value cannot tell you the probability that the result is due to sampling error.

The P value is not the probability that the null hypothesis is true

The P value is computed *assuming* that the null hypothesis is true, so it cannot be the probability that it is true.

The probability that the alternative hypothesis is true is not (1.0 minus the P value)

If the P value is 0.03, it is very tempting to think that if there is only a 3% probability that my difference would have been caused by random chance, then there must be a 97% probability that it was caused by a real difference. But this is wrong!

What you *can* say is that if the null hypothesis were true, then 97% of experiments would lead to a difference smaller than the one you observed and 3% of experiments would lead to a difference as large or larger than the one you observed.

Calculation of a P value is predicated on the assumption that the null hypothesis is true. P values cannot tell you whether this assumption is correct. Instead, the P value tells you how rarely you would observe a difference as large or larger than the one you observed if the null hypothesis were true.

The probability that the results will hold up when the experiment is repeated is not (1.0 minus the P value)

If the P value is 0.03, it is tempting to think that this means there is a 97% chance of getting similar results in a repeated experiment. Not so. The P value does not itself quantify reproducibility.

A high P value does not prove that the null hypothesis is true

A high P value means that if the null hypothesis were true, it would not be surprising to observe the treatment effect seen in a particular experiment. But that does not prove that the null hypothesis is true. It just says the data are consistent with the null hypothesis.

"P = 0.05" is not the same as "P < 0.05"

"P = 0.05" means what it says: that the P value equals 0.05. In contrast, "P < 0.05" means that P is less than 0.05.

A P value does not always have to be written as an inequality

A P value is a number, and it should be expressed as such. When very small, it can make sense to simply say, for example, that $P < 0.0001$. But it is more informative to state the P value as a number (e.g., 0.0345) than to just state some value it is less than (e.g., $P < 0.05$).

Q & A

Can P values be negative?

No. P values are fractions, so they are always between 0.0 and 1.0.

Can a P value equal 0.0?

A P value can be very small but can never equal 0.0. If you see a report that a P value equals 0.0000, that really means it is less than 0.0001.

Can a P value equal 1.0?

A P value would equal 1.0 only in the rare case in which the treatment effect size in your sample precisely equals the one defined by the null hypothesis. When a computer program reports that the P value is 1.0000, it may really mean that the P value is greater than 0.9999.

Should P values be reported as fractions or percentages?

By tradition, P values are always presented as fractions and never as percentages.

Is a one-tailed P value always equal to half the two-tailed P value?

Not always. Some distributions are asymmetrical. For example, a one-tailed P value from a Fisher's exact test (see Chapter 27) is usually not exactly equal to half the two-tailed P value. With some data, in fact, the one- and two-tailed P values can be identical. This event is rare.

Even if the distribution is symmetrical (as most are), the one-tailed P value is only equal to half the two-tailed value if you correctly predicted the direction of the difference (correlation, association, etc.) in advance. If the effect actually went in the opposite direction to your prediction, the one-tailed P will not be equal to half the two-tailed P value. If you were to calculate this value, it would be greater than 0.5 and greater than the two-tailed P value.

Can a one-tailed P value have a value greater than 0.50?

Yes, this happens when the direction of the treatment effect is opposite to the prediction.

Is a P value always associated with a null hypothesis?

Yes. If you can't state the null hypothesis, then you can't interpret the P value.

Shouldn't P values always be presented with a conclusion about whether the results are statistically significant?

No. A P value can be interpreted on its own. In some situations, it can make sense to go one step further and report whether or not the results are statistically significant (as will be explained in Chapter 16). But this step is optional.

Should it be P or p? Italicized or not? Hyphenated or not?

Different books and journals use different styles. This book uses "P value," but "*p*-value" is probably more common.

I chose to use a one-tailed P value, but the results came out in the direction opposite to my prediction. Can I report the one-tailed P value calculated by a statistical program?

Probably not. Most statistical programs don't ask you to specify the direction of your hypothesis and report the one-tailed P value assuming you correctly predicted the direction of the effect. If your prediction was wrong, then the correct one-tailed P value equals 1.0 minus the P value reported by the program.

Why do some investigators present the negative logarithm of P values?

When presenting many tiny P values, some investigators (especially those doing GWAS; see Chapter 22) present the negative logarithm of the P value to avoid the confusion of dealing with tiny fractions. For example, if the P value is 0.01, its logarithm (base 10) is -2 so the negative logarithm is 2. If the P value is 0.00000001, its logarithm is -8, so the negative logarithm is 8.

Is the null hypothesis ever true?

Rarely.

CHAPTER SUMMARY

- P values are frequently reported in scientific papers, so it is essential that every scientist understand exactly what a P value is and is not.
- All P values are based on a null hypothesis. If you cannot state the null hypothesis, you can't understand the P value.
- A P value answers the following general question: If the null hypothesis is true, what is the chance that random sampling would lead to a difference (or association, etc.) as large or larger than that observed in this study?
- P values are calculated values between 0.0 and 1.0. They can be reported and interpreted without ever using the word "significant."
- When interpreting published P values, note whether they are calculated for one or two tails. If the author doesn't say, the result is somewhat ambiguous.
- If you repeat an experiment, expect the P value to be very different. P values are much less reproducible than you would guess.
- One-tailed P values have their advantages, but you should always use a two-tailed P value unless you have a really good reason to use a one-tailed P value.
- There is much more to statistics than P values. When reading scientific papers, don't get mesmerized by P values. Instead, focus on what the investigators found and how large the effect size is.

TERMS INTRODUCED IN THIS CHAPTER

- Null hypothesis (p. 127)
- One-tailed P value (p. 129)
- P value (p. 125)
- Two-tailed P value (p. 129)

CHAPTER 16

❖❖❖

Statistical Significance and Hypothesis Testing

> For the past eighty years, it appears that some of the sciences have
> made a mistake by basing decisions on statistical significance.
>
> ZILIAK AND McCLOSKEY (2008)

The phrase "statistically significant" is commonly misunderstood, because this use of the word "significant" has almost no relationship to the conventional meaning of the word as used to describe something that is important or consequential. This chapter explains that the concept is quite straightforward once you get past the confusing terminology.

STATISTICAL HYPOTHESIS TESTING

Making a decision is the primary goal when analyzing some kinds of data. In a pilot experiment of a new drug, the goal may be to decide whether the results are promising enough to merit a second experiment. In a Phase III drug study, the goal may be to decide whether the drug should be recommended. In quality control, the goal is to decide whether a batch of product can be released.

Statistical hypothesis testing automates decision-making. First, define a threshold P value for declaring whether a result is statistically significant. This threshold is called the *significance level* of the test; it is denoted by α (*alpha*) and is commonly set to 0.05. If the P value is less than α, conclude that the difference is *statistically significant* and decide to reject the null hypothesis. Otherwise, conclude that the difference is not statistically significant and decide to not reject the null hypothesis.

Note that statistics gives the term *hypothesis testing* a unique meaning that is very different than the meaning most scientists think of when they go about testing a scientific hypothesis.

ANALOGY: INNOCENT UNTIL PROVEN GUILTY

Scientist as juror

The steps that a jury must follow to determine criminal guilt are very similar to the steps that a scientist follows when using hypothesis testing to determine statistical significance.

137

A juror starts with the presumption that the defendant is innocent. A scientist starts with the presumption that the null hypothesis of "no difference" is true.

A juror bases his or her decision only on factual evidence presented in the trial and should not consider any other information, such as newspaper stories. A scientist bases his or her decision about statistical significance only on data from one experiment, without considering what other experiments have concluded.

A juror reaches the verdict of guilty when the evidence is inconsistent with the assumption of innocence. Otherwise, the juror reaches a verdict of not guilty. When performing a statistical test, a scientist reaches a conclusion that the results are statistically significant when the P value is small enough to make the null hypothesis unlikely. Otherwise, a scientist concludes that the results are not statistically significant.

A juror does not have to be convinced that the defendant is innocent to reach a verdict of not guilty. A juror reaches a verdict of not guilty when the evidence is consistent with the presumption of innocence, when guilt has not been proven. A scientist reaches the conclusion that results are not statistically significant whenever the data are consistent with the null hypothesis. The scientist does not have to be convinced that the null hypothesis is true.

A juror can never reach a verdict that a defendant is innocent. The only choices are guilty or not guilty. A statistical test never concludes the null hypothesis is true, only that there is insufficient evidence to reject it (but see Chapter 21).

A juror must try to reach a conclusion of guilty or not guilty and can't conclude, "I am not sure." Similarly, each statistical test leads to a crisp conclusion of statistically significant or not statistically significant. A scientist who strictly follows the logic of statistical hypothesis testing cannot conclude, "Let's wait for more data before deciding."

Scientist as journalist

Jurors aren't the only people who evaluate evidence presented at a criminal trial. Journalists also evaluate evidence presented at a trial, but they follow very different rules than jurors. A journalist's job is not to reach a verdict of guilty or not guilty, but rather to summarize the proceedings.

Scientists in many fields are often more similar to journalists than jurors. If you don't need to make a clear decision based on one P value, you don't need to use the term *statistically significant* or use the rubric of statistical hypothesis testing.

EXTREMELY SIGNIFICANT? BORDERLINE SIGNIFICANT?

Tiny P values

Is a result with P = 0.004 more statistically significant than a result with P = 0.04? By the strict rules of statistical hypothesis testing, the answer is no. Once you

SYMBOLS	PHRASE	P VALUE
ns	Not significant	P > 0.05
*	Significant	P < 0.05
**	Highly significant	P < 0.01
***	Extremely significant	P < 0.001

Table 16.1. Using asterisks to denote statistical significance.
The scheme shown here is commonly used but is not universal. If you
see graphs decorated with asterisks to denote statistical significance,
be sure to check the legend to see how the authors defined the
symbols.

have established a significance level, every result is either statistically significant
or not statistically significant. Because the goal of statistical hypothesis testing is
to make a crisp decision, only two conclusions are needed.

Most scientists are less rigid and refer to *very significant* or *extremely significant* results when the P value is tiny. When showing P values on graphs, investigators commonly use asterisks to create a scale such as that used in Michelin Guides
or movie reviews (see Table 16.1). When you read this kind of graph, make sure
that you look at the key that defines the symbols, because different investigators
use different threshold values.

Borderline statistical significance

If you follow the strict paradigm of statistical hypothesis testing and set α to its
conventional value of 0.05, then a P value of 0.049 denotes a statistically significant difference and a P value of 0.051 does not. This arbitrary distinction is
unavoidable, because the whole point of statistical hypothesis testing is to reach a
crisp conclusion from every experiment, without exception.

When a P value is just slightly greater than α, some scientists avoid the phrase
"not significantly different" and instead refer to the result as "marginally significant" *or* "almost statistically significant." Some get more creative and use
phrases like "a clear trend," "barely missed statistical significance," "not exactly
significant," "not currently significant," "on the cusp of significance," "provi-
sionally significant," "trending towards significance," "verging on significance,"
"weakly significant," and so on. Hankin (2013) lists 468 such phrases he found in
published papers. Rather than deal with linguistic tricks, it is better in many cases
to just report the actual P value and not worry about whether it is above or below
some arbitrary threshold value.

When a two-tailed P value is between 0.05 and 0.10, you may be tempted to
switch to a one-tailed P value. Because the one-tail P value (with rare exceptions)
equals half the two-tailed P value (as explained in Chapter 15), it will be smaller
than 0.05. Your results will have become statistically significant, as if by magic.
Obviously, this is not an appropriate reason to choose a one-tailed P value! The
choice should be made before the data are collected.

	DECISION: REJECT NULL HYPOTHESIS	DECISION: DO NOT REJECT NULL HYPOTHESIS
Null hypothesis is true	Type I error	(No error)
Null hypothesis is false	(No error)	Type II error

Table 16.2. Definition of Type I and Type II errors.

LINGO: TYPE I AND TYPE II ERRORS

Statistical hypothesis testing makes a decision based on the results of one comparison. When you make this decision, there are two kinds of mistakes you can make (see Table 16.2).

Type I error

When there really is no difference (or association, or correlation) between the populations, random sampling can lead to a difference (or association, or correlation) large enough to be a statistically significant. This is a *Type I error*. It occurs when you decide to reject the null hypothesis when in fact the null hypothesis is true. It is a false positive.

Type II error

When there really is a difference (or association, or correlation) between the populations, random sampling (and small sample size) can lead to a difference (or association, or correlation) small enough to be not statistically significant. This is a *Type II error*. It occurs when you decide not to reject the null hypothesis when in fact the null hypothesis is false. It is a false negative.

A worse kind of error

What if you conclude that a difference (or association, or correlation) is statistically significant but are wrong about the direction of the effect? You are correct that the null hypothesis truly is wrong, so you haven't made a Type I error. But you've got the direction of the effect backwards. In your sample, the new drug works better than the standard drug but in fact (if you had infinite sample size), it works (on average) worse. Or the data in your sample show that people taking a certain drug lose weight, when in fact (if you had an infinite sample) that drug causes people (on average) to gain weight. Gelman and Tuerlinckx (2000) call this a *Type S error*, because the sign of the difference is backwards. Hsu (1996) calls this a *Type III error*. Neither term is widely used.

You won't know

Type I and II (and III) errors are theoretical concepts. When you analyze any particular set of data, you don't know whether the populations are identical. You only know the data in your particular samples. You will never know whether you made one of these errors as part of any particular analysis.

CHOOSING A SIGNIFICANCE LEVEL

The distinction between the P value and α

The P value and α are not the same:

- The significance level α is chosen by the experimenter as part of the experimental design before collecting any data. If the null hypothesis is true, α is the probability of rejecting the null hypothesis.
- A P value is computed from the data. You reject the null hypothesis (and conclude that the results are statistically significant) when the P value from a particular experiment is less than the significance level α set in advance.

The trade-off

A result is deemed statistically significant when the P value is less than a significance level (α) set in advance. So you need to choose that level. By tradition, α is usually set to equal 0.05, but you can choose whatever value you want. When choosing a significance level, you confront a trade-off.

If you set α to a very low value, you will make few Type I errors. That means that if the null hypothesis is true, there will be only a small chance that you will mistakenly call a result statistically significant. However, there is also a larger chance that you will not find a significant difference, even if the null hypothesis is false. In other words, reducing the value of α will decrease your chance of making a Type I error but increase the chance of a Type II error.

If you set α to a very large value, you will make more Type I errors. If the null hypothesis is true, there is a large chance that you will mistakenly conclude that the effect is statistically significant. But there is a small chance of missing a real difference. In other words, increasing the value of α will increase your chance of making a Type I error but decrease the chance of a Type II error. The only way to reduce the chances of both a Type I error and a Type II error is to collect bigger samples (see Chapter 26).

Example: Detecting spam

Table 16.3 recasts Table 16.2 in the context of detecting spam (junk) email. Spam filters use various criteria to evaluate the probability that a particular email is spam and then use this probability to decide whether to deliver that message to the inbox or to the spam folder.

The null hypothesis is that an email is good (not spam). A Type I error occurs when a good email is mistakenly sent to the spam mailbox, where it probably will never be seen. A Type II error occurs when spam is delivered to the inbox.

	DECISION: DELETE AS JUNK	DECISION: PLACE IN INBOX
Good email	Type I error	(No error)
Spam	(No error)	Type II error

Table 16.3. Type I and Type II errors in the context of spam filters.

	VERDICT: GUILTY	VERDICT: NOT GUILTY
Did not commit the crime	Type I error	(No error)
Did commit the crime	(No error)	Type II error

Table 16.4. Type I and Type II errors in the context of trial by jury in a criminal case.

When deciding how aggressively to define spam, people designing the spam filters have to decide on the relative consequences of the two types of errors. Some spam-filtering software lets the user adjust how aggressively it defines spam, so that each user can choose whether he or she would rather have more spam in the inbox or more good mail in the spam box.

Example: Trial by jury

Table 16.4 continues the analogy between statistical significance and the legal system. The relative consequences of Type I and Type II errors depend on the type of trial.

In the United States (and many other countries), a defendant in a criminal trial is considered innocent until proven guilty "beyond a reasonable doubt." This system is based on the principle that it is better to let many guilty people go free than to falsely convict one innocent person. The system is designed to avoid Type I errors in criminal trials, even at the expense of many Type II errors. You could say that α is set to a very low value.

In civil trials, the court or jury rules in favor of the plaintiff if the evidence shows that the plaintiff is "more likely than not" to be right. The thinking is that it is no worse to falsely rule in favor of the plaintiff than to falsely rule in favor of the defendant. The system attempts to equalize the chances of Type I and Type II errors in civil trials.

Example: Detecting athletes that dope

When an athletic association tests an athlete for using a banned drug (doping), there are two kinds of mistakes it can make (Fung, 2011). A false positive (Type I error) is concluding that the athlete had used a banned drug when in fact he or she hadn't done so. This can ruin an athlete's reputation and career. A false negative (Type II error) is concluding that an athlete had not been using a banned drug when she or he actually had done so. This is unfair to other athletes who follow the rules. The officials for each sport need to decide where to draw the line.

Example: Clinical trial for a me-too drug

A clinical trial tests the use of a drug to treat a disease (like hypertension) for which there are already effective existing therapies. If the results are statistically significant, you will market the drug. If the results are not statistically significant, work on the drug will cease. In this case, the null hypothesis is that the drug doesn't work at all, and a Type I error results in treating some future patients with a drug that is actually ineffective. A Type II error results in aborting development of a good new drug for a condition that can be treated adequately with existing drugs. Thinking scientifically (not commercially), it makes sense to minimize the risk of a Type I error, even at the expense of a higher chance of a Type II error.

Example: Clinical trial for a novel drug

This example also concerns a clinical trial, but for a drug that purportedly treats a disease for which there is no good existing therapy. If the results are statistically significant, you will market the drug. If the results are not statistically significant, work on the drug will cease. In this case, a Type I error results in treating future patients with a useless drug, instead of nothing. A Type II error results in canceling development of a good drug for a condition that is currently not treatable. It makes sense to set α to a relatively high value because a Type I error isn't so bad, but a Type II error would be awful. It might make sense to set α to 0.10 or even higher.

Of course, this example and the last one are simplistic. Difficult decisions about drug development rarely depend on a single conclusion about statistical significance. The scenarios above are designed to help you understand the trade-offs between Type I and Type II errors, and not to provide guidance in how to make those difficult decisions.

Example: Particle physics and the five-sigma cut-off

A large international group of physicists announced in 2012 the discovery of the Higgs boson. The results were not announced in terms of a P value or statistical significance. Instead, the group announced that two separate experiments met the five-sigma threshold (Lamb, 2012).

Meeting the *five-sigma threshold* means that the results would occur by chance alone as rarely as a value sampled from a Gaussian distribution would be five standard deviations (sigma) away from the mean. How rare is that? The probability is 0.0000003, or 0.00003%, or about one in 3.5 million, as computed with the Excel function "=1- normsdist(5)." In other words, the one-tailed P value is less than 0.0000003. If the Higgs boson doesn't exist, that probability is the chance that random sampling of data would give results as striking as the physicists observed. Two separate groups obtained data with this level of evidence.

The standard of statistical significance in most fields is that the two-tailed P is less than 0.05. About 5% of a Gaussian distribution is more than two standard deviations away from the mean (in either direction). So the conventional definition of statistical significance can be called a *two-sigma threshold*.

The tradition in particle physics is that the threshold to report "evidence of a particle" is $P < 0.003$ (three sigma), and the standard to report a "discovery of a particle" is $P < 0.0000003$ (five sigma). These are one-tailed P values.

The point of this example is to demonstrate that different scientific fields have different traditions for defining statistical significance.

STATISTICAL SIGNFIICANCE VS. SCIENTIFIC SIGNIFICANCE

A result is said to be statistically significant when the calculated P value is less than an arbitrary preset value of α. A conclusion that a result is statistically significant does not mean the difference is large enough to be interesting or intriguing enough to be worthy of further investigations, so does not mean the finding is scientifically or clinically significant.

To avoid the ambiguity in the word *significant*, never use that word by itself. Always use the phrase "statistically significant" when referring to statistical hypothesis testing and phrases such as "scientifically significant" or "clinically significant" when referring to the size of a difference (or association).

Even better, avoid the word "significant" altogether. It is not needed. When writing about statistical hypothesis testing, state the P value and whether or not the null hypothesis can be rejected. When discussing the importance or impact of data, banish the word "significant," and instead use words such as *consequential, eventful, meaningful, momentous, large, big, important, substantial, remarkable, valuable, worthwhile, impressive, major*, or *prominent*.

COMMON MISTAKES: STATISTICAL HYPOTHESIS TESTING

Mistake: Believing that statistical hypothesis testing is an essential part of all statistical analyses

The whole point of statistical hypothesis testing is to make a crisp decision from one result. Make one decision when the results are statistically significant and another decision when the results are not statistically significant. This situation is common in quality control but rare in exploratory basic research.

In many scientific situations, it is not necessary—and can even be counterproductive—to reach a crisp conclusion that a result is either statistically significant or not statistically significant. P values and CIs can help you assess and present scientific evidence without ever using the phrase "statistically significant."

When reading the scientific literature, don't let a conclusion about statistical significance stop you from trying to understand what the data really show. Don't let the rubric of statistical hypothesis testing get in the way of clear scientific thinking. Because statistical significance is often shown on graphs and tables as asterisks or stars, the process of overemphasizing statistical significance is sometimes referred to as *stargazing*. To avoid stargazing, ignore statements of statistical significance when you first read a paper and focus instead on the size of the effect (difference, association, etc.) and its confidence interval.

Mistake: Believing that if a result is statistically significant, the effect must be large

The conclusion that something is statistically significant applies to the strength of the evidence, not the size of the effect. Statistical hypothesis testing answers the question, Is there significant evidence of a difference? It does not answer the question, Is there evidence of a significant difference? (Marsupial, 2013). If the sample size is large, even tiny, inconsequential differences will be statistically significant.

Once some people hear the word *significant*, they often stop thinking about what the data actually show. Belief in the usefulness of statistical significance has even been called a cult (Ziliak & McCloskey, 2008). Be sure you know what a conclusion that a result is statistically significant does not mean. This conclusion:

- Does *not* mean the difference is large enough to be interesting
- Does *not* mean the results are intriguing enough to be worthy of further investigations
- Does *not* mean that the finding is scientifically or clinically significant

Mistake: P-hacking to obtain statistical significance

One problem with statistical hypothesis testing is that investigators may be tempted to work hard to push their P value to be low enough to be declared significant. Gotzsche (2006) used a clever approach to quantify this. He figured that if results were presented honestly, the number of P values between 0.04 and 0.05 would be similar to the number of P values between 0.05 and 0.06. However, in 130 abstracts of papers published in 2003, he found five times more P values between 0.04 and 0.05 than between 0.05 and 0.06.

Masicampo and Lalande (2012) found similar results. They reviewed papers in three highly regarded, peer-reviewed psychology journals. The distribution of 3,627 P values shows a clear discontinuity of 0.05. There are "too many" P values just smaller than 0.05 and too many greater than 0.05. They conclude, "The number of P values in the psychology literature that barely meet the criterion for statistical significance (i.e., that fall just below .05) is unusually large, given the number of P values occurring in other ranges."

Some of the investigators probably did the analyses properly and simply didn't write up results when the P value was just a bit higher than 0.05. And some may have had the borderline results rejected for publication. You can read more about publication bias in Chapter 43. It also is likely, I think, that many investigators cheat a bit by:

- Tweaking. The investigators may have played with the analyses. If one analysis didn't give a P value less than 0.05, then they tried a different one. Perhaps they switched between parametric and nonparametric analysis or tried including different variables in multiple regression models. Or perhaps they reported only the analyses that gave P values less than 0.05 and ignored the rest.
- Dynamic sample size. The investigators may have analyzed their data several times. Each time, they may have stopped if the P value was less than 0.05 but collected more data when the P value was above 0.05. This approach would yield misleading results, as it is biased toward stopping when P values are small.

- Slice and dice. The investigators may have analyzed various subsets of the data and only reported the subsets that gave low P values.
- Selectively report results of multiple outcomes. If several outcomes were measured, the investigators may have chosen to only report the ones for which the P value was less than 0.05.
- Play with outliers. The investigators may have tried using various definitions of outliers, reanalyzed the data several times, and only reported the analyses with low P values.

Simmons, Nelson, & Simonsohn (2012) coined the term *P-hacking* to refer to attempts by investigators to lower the P value by trying various analyses or by analyzing subsets of data.

Q & A

Is it possible to report scientific data without using the word *significant*?
Yes. Report the data, along with CIs and perhaps P values. Decisions about statistical significance are often not helpful.

Is the concept of statistical hypothesis testing about making decisions or about making conclusions?
Decision-making. The system of statistical hypothesis testing makes perfect sense when it is necessary to make a crisp decision based on one statistical analysis. If you have no need to make a decision from one analysis, then it may not be helpful to use the term *statistically significant*.

But isn't the whole point of statistics to decide when an effect is statistically significant?
No. The goal of statistics is to quantify scientific evidence and uncertainty.

Why is statistical hypothesis testing so popular?
There is a natural aversion to ambiguity. The crisp conclusion "The results are statistically significant" is more satisfying to many than the wordy conclusion "Random sampling would create a difference this big or bigger in 3% of experiments if the null hypothesis were true."

Who invented the threshold of $P < 0.05$ as meaning statistically significant?
That threshold, like much of statistics, came from the work of Ronald Fisher.

Are the P value and α the same?
No. A P value is computed from the data. The significance level α is chosen by the experimenter as part of the experimental design before collecting any data. A difference is termed statistically significant if the P value computed from the data is less than the value of α set in advance.

Is α the probability of rejecting the null hypothesis?
Only if the null hypothesis is true. In some experimental protocols, the null hypothesis is often true (or close to it). In other experimental protocols, the null hypothesis is almost certainly false. If the null hypothesis is actually true, α is the probability that you will reject the null hypothesis, thus making a Type I error.

If I perform many statistical tests, is it true that the conclusion "statistically significant" will be incorrect 5% of the time?
No! That would only be true if the null hypothesis is, in fact, true in every single experiment. It depends on the scientific context.

My two-tailed P value is not low enough to be statistically significant, but the one-tailed P value is. What do I conclude?

> Stop playing games with your analysis. It is only OK to compute a one-tailed P value when you decided to do so as part of the experimental protocol (see Chapter 15).

Isn't it possible to look at statistical hypothesis testing as a way to choose between alternative models?

> Yes. See Chapter 35.

What is the difference between Type I and Type II errors?

> Type I errors reject a true null hypothesis. Type II errors accept a false null hypothesis.

CHAPTER SUMMARY

- Statistical hypothesis testing reduces all findings to two conclusions, "statistically significant" or "not statistically significant."
- The distinction between the two conclusions is made purely based on whether the computed value of a P value is less than or not less than an arbitrary threshold, called the significance level.
- A conclusion of statistical significance does not mean the difference is large enough to be interesting, does not mean the results are intriguing enough to be worthy of further investigations, and does not mean that the finding is scientifically or clinically significant.
- The concept of statistical significance is often overemphasized. The whole idea of statistical hypothesis testing only is useful when a crisp decision needs to be made based on one analysis.
- Avoid the use of the term *significant* when you can. Instead, talk about whether the P value is low enough to reject the null hypothesis, and whether the effect size (or difference, or association) is large enough to be important.
- Some scientists use the phrase "very significant" or "extremely significant" when a P value is tiny and "borderline significant" when a P value is just a little bit greater than 0.05.
- While it is conventional to use 0.05 as the threshold P value (called α) that separates statistically significant from not statistically significant, this value is totally arbitrary.
- Ideally, one should choose α based on the consequences of Type I (false positive) and Type II (false negative) errors.
- It is important to distinguish the P value from the significance level α. A P value is computed from data, while the value of α is (or should be) decided as part of the experimental design.
- There is more—a lot more—to statistics than deciding if an effect is statistically significant or not statistically significant.

TERMS DEFINED IN THIS CHAPTER

- Alpha (α) (p. 137)
- Five-sigma threshold (p. 143)

- P-hacking (p. 145)
- Significance level (p. 145)
- Stargazing (p. 144)
- Statistical hypothesis testing (p. 144)
- Statistically significant (p. 137)
- Type I error (p. 140)
- Type II error (p. 140)
- Type III error (p. 140)
- Type S error (p. 140)

CHAPTER 17

❖❖❖

The Relationship Between Confidence Intervals and Statistical Significance

> Reality must take precedence over public relations, for Nature cannot be fooled.
>
> RICHARD FEYNMAN

This book has presented the concepts of CIs and statistical hypothesis testing in separate chapters. The two approaches are based on the same assumptions and the same mathematical logic. This chapter explains how they are related.

CIs AND STATISTICAL HYPOTHESIS TESTING ARE CLOSELY RELATED

CIs and statistical hypothesis testing are based on the same statistical theory and assumptions, so they are closely linked.

The CI computes a range that 95% of the time will contain the population value (given some assumptions).

The hypothesis testing approach computes a range that you can be 95% sure would contain experimental results if the null hypothesis were true. Any result within this range is considered not statistically significant, and any result outside this range is considered statistically significant.

WHEN A CI INCLUDES THE NULL HYPOTHESIS

Figure 17.1 shows the results of comparing observed body temperatures (n = 12) with the hypothetical mean value of 37°C (see Chapter 7).

The bottom bar of Figure 17.1 shows the 95% CI of the mean. It is centered on the sample mean. It extends in each direction a distance computed by multiplying the SEM by the critical value of the t distribution from Appendix D (for 95% confidence).

The top bar in Figure 17.1 shows the range of results that would not be statistically significant (P > 0.05). It is centered on the null hypothesis, in this case, that the mean body temperature is 37°C. It extends in each direction exactly

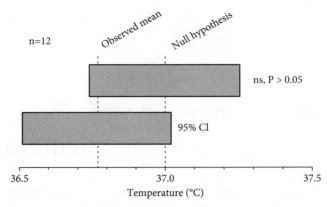

Figure 17.1. Comparing observed body temperatures (n = 12) with the hypothetical mean value of 37°C.

The top bar shows the range of results that are not statistically significant (P > 0.05). The bottom bar shows the 95% CI of the mean and is centered on the sample mean. The lengths of the two bars are identical. Because the 95% CI contains the null hypothesis, the zone of statistically not significant results must include the sample mean.

the same distance as the other bar, equal to the product of the SEM times the critical value of the t distribution.

The two bars have identical widths but different centers. In this example, the 95% CI contains the null hypothesis (37°C). Therefore, the zone of statistically not significant results must include the sample result (in this case, the mean).

WHEN A CI DOES NOT INCLUDE
THE NULL HYPOTHESIS

Figure 17.2 shows the n = 130 data but is otherwise similar to Figure 17.1.

The bottom bar of Figure 17.2 shows the 95% CI of the mean. It is centered on the sample mean. It extends in each direction a distance computed by multiplying the SEM by the critical value of the t distribution from Appendix D (for 95% confidence). This bar is much shorter than the corresponding bar in Figure 17.1. The main reason for this is that the SEM is much smaller because of the much larger sample size. Another reason is that the critical value from the t distribution is a bit smaller (1.98 instead of 2.20) because the sample size (and number of degrees of freedom) is larger.

The top bar in Figure 17.2 shows the range of results that would not be statistically significant (P > 0.05). It is centered on the null hypothesis, in this case, that the mean body temperature is 37°C. It extends in each direction exactly the same distance as the other bar, equal to the product of the SEM times the critical value of the t distribution.

In this example, the 95% CI does not contain the null hypothesis. Therefore, the zone of statistically not significant results cannot include the sample mean.

Figure 17.2. Comparing observed body temperatures (n = 130) with the hypothetical mean value of 37°C.

Because the sample size is larger, the CI is narrower than it is in Figure 17.1, as is the zone of possible results that would not be statistically significant. Because the 95% CI does not contain the null hypothesis, the zone of not statistically significant results cannot contain the sample mean. Whether the two boxes overlap is not relevant.

A RULE THAT LINKS CONFIDENCE INTERVALS AND STATISTICAL SIGNIFICANCE

Here is a general rule that you should remember:

- If a 95% CI does not contain the value of the null hypothesis, then the result must be statistically significant, with $P < 0.05$.
- If a 95% CI does contain the value of the null hypothesis, then the result must not be statistically significant, with $P > 0.05$.

There is nothing special about 95% CIs and a significance level of 5%. The rule also works like this: if the 99% CI does not contain the null hypothesis value, then the P value must be less than 0.01.

In the examples, the outcome was the difference between the sample mean and a hypothetical population mean. This rule works for many other kinds of outcomes. For example:

- If the CI for the difference between two means does not include zero (the null hypothesis), then the result must be statistically significant ($P < 0.05$).
- If the CI for the ratio of two proportions does not include 1.0 (the null hypothesis), then the result must be statistically significant ($P < 0.05$).
- If you are comparing a set of percentages against a hypothetical value of 100 and the 95% CI of the mean of the percentages does not include 100, then the discrepancy must be statistically significant ($P < 0.05$).

Q & A

If the 95% confidence interval just barely reaches the value that defines the null hypothesis, what can you conclude about the P value?

> If the 95% CI includes the value that defines the null hypothesis, you can conclude that the P value is greater than 0.05. If the 95% CI excludes the null hypothesis value, you can conclude that the P value is less than 0.05. So if the 95% CI ends right at the value that defines the null hypothesis, then the P value must equal 0.05.

If the 95% CI is centered on the value that defines the null hypothesis, what can you conclude about the P value?

> The observed outcome equals the value that defines the null hypothesis. In this case, the two-tailed P value must equal 1.000.

The 99% CI includes the value that defines the null hypothesis, but the P value is reported to be < 0.05. How is this possible?

> If the 99% CI includes the value that defines the null hypothesis, then the P value must be greater than 0.01. But the P value was reported to be less than 0.05. You can conclude that the P value must be between 0.01 and 0.05.

The 99% CI includes the value that defines the null hypothesis, but the P value is reported to be < 0.01. How is this possible?

> It is inconsistent. Perhaps the CI or P value was calculated incorrectly. Or perhaps the definition of the null hypothesis included in the 99% CI is not the same definition used to compute the P value.

CHAPTER SUMMARY

- Confidence intervals and statistical hypothesis testing are closely related.
- If the 95% confidence interval includes the null hypothesis, then the P value must be less than 0.05.
- If the 95% confidence interval does not include the null hypothesis, then the P value must be less than 0.05.
- The same rule works for 99% confidence intervals and $P < 0.01$, or 90% confidence intervals and $P < 0.10$.

CHAPTER 18

⋅⋅⋅⋅⋅
⋅⋅⋅⋅

Interpreting a Result That Is Statistically Significant

Facts do not "speak for themselves," they are read in the light of theory.

STEPHEN JAY GOULD

When you see a result that is "statistically significant," don't stop thinking. That phrase is often misunderstood, as this chapter explains. All it means is that the calculated P value is less than a preset threshold you set. This means that the results would be surprising (but not impossible) if the null hypothesis were true. A conclusion of statistical significance does not mean the difference is large enough to be interesting or worthy of follow-up, nor does it mean that the finding is scientifically or clinically significant.

SAMPLE SIZE MATTERS WHEN INTERPRETING P VALUES

A P value quantifies the probability of observing a difference or association as large (or larger) than actually observed if the null hypothesis were true. Table 18.1 demonstrates that the relationship between P value and the size of the observed effect depends on sample size. Each row in the table represents a simulated experiment that compares an observed proportion (the outcome tabulated by the experimenter) in control and treated groups. In each case, the observed proportion in the control group was 10%. The proportions in the expected group were chosen so that the P values in all the rows is about the same. In the top row, n=10. The proportion in the treated group was 80%, so the ratio of the two proportions (tabulated as the Relative Risk, see Chapter 27) is 8.0. In bottom row, n=10000 per group. The observed proportion in the treated group was only 11.2%, so the Relative Risk was 1.12. The P value was the same in both cases, 0.006.

Focus on the last row of Table 18.1. The effect is tiny, 11.2% vs. 10.0% of the subjects had the outcome being studied. Even though the P value is small, it is unlikely that such a small effect would be worth pursuing. It depends on the scientific or clinical situation, but there are few fields where such a tiny difference is important. With huge sample sizes, it takes only a tiny effect to produce such a small

SAMPLE SIZE (PER GROUP)	P VALUE	RELATIVE RISK	95% CI
10	0.006	8.00	1.21 to 52.7
100	0.005	2.60	1.32 to 5.11
1000	0.006	1.41	1.11 to 1.79
10000	0.006	1.12	1.03 to 1.21

Table 18.1. Sample size matters when interpreting P values

Four simulated experiments are tabulated, each with almost the same P value determined by comparing two proportions. In each experiment the proportion in the control group was 10%, which was compared to the proportion observed in the treated group (see Chapter 27). In addition to the sample size and P value, this table shows the relative risk (the ratio of the proportion in the treated group divided by the ratio in the control group) and its 95% CI. The P values for the top three rows were calculated by Fishers's exact test. The P values for the bottom rows was calculated using a chi-square test. (Why is the P value 0.005 in the second row, and not 0.0006 like the others? Because with n=100 and setting the control proportion to 10%, it is impossible to get a P value of 0.006.)

P value. In contrast, with tiny samples, it takes a huge effect to yield a small P value. It is essential when reviewing scientific results to look at more than just the P value.

DISTINGISHING THE SIGNIFICANCE LEVEL (α) FROM THE FALSE DISCOVERY RATE (FDR)

Table 18.2 shows the results of many hypothetical statistical analyses, each analyzed to reach a decision to reject or not reject the null hypothesis. The top row tabulates results for experiments in which the null hypothesis is really true. The second row tabulates experiments in which the null hypothesis is not true. When you analyze data, you don't know whether the null hypothesis is true, so you could never create this table from an actual series of experiments.

Distinguish two quantities:

- The *significance level* (α) equals A/(A+B).
- The *false discovery rate* (FDR) equals A/(A+C).

The significance level (α) is the answer to these two equivalent questions:

- If the null hypothesis is true, what is the probability of incorrectly rejecting that null hypothesis?
- Of all experiments you could conduct when the null hypothesis is actually true, in what fraction will you reach a conclusion that the results are statistically significant?

The *false discovery rate* (FDR) is the answer to these two equivalent questions:

- If a result is statistically significant, what is the probability that the null hypothesis is really true?
- Of all experiments that reach a statistically significant conclusion, in what fraction is the null hypothesis true?

	DECISION: REJECT NULL HYPOTHESIS	DECISION: DO NOT REJECT NULL HYPOTHESIS	TOTAL
Null hypothesis is true	A(Type I error)	B	A +B
Null hypothesis is false	C	D (Type II error)	C + D
Total	A + C	B + D	A + B + C + D

Table 18.2. The results of many hypothetical statistical analyses, each analyzed to reach a decision to reject or not reject the null hypothesis.

The top row tabulates results for experiments in which the null hypothesis is really true. The second row tabulates experiments in which the null hypothesis is not true. When you analyze data, you don't know whether the null hypothesis is true, so you could never create this table from an actual series of experiments. A, B, C, and D are integers (not proportions) that count the number of analyses.

THE PRIOR PROBABILITY INFLUENCES THE FALSE DISCOVERY RATE

The FDR is influenced by the context of the experiment (the prior probability), as demonstrated by the following two extreme examples.

An extreme example in which you know that every null hypothesis is true

In a randomized clinical trial, it is common to compare two groups at baseline by measuring, for example, the hematocrit (blood count), fasting glucose, blood pressure, and weight. Then each subject is randomly assigned to receive one of two (or more) treatments. After all the subjects are entered into the study, it is common to compare all those baseline variables between the groups.

Table 18.3 shows what happens if you compare 20 different variables in this way. Since the measurements are made before any treatments are given (and possibly before the subject is even randomly assigned to a treatment), you absolutely, positively know that both samples are drawn from the same population. In other words, you know that all 20 null hypotheses are true. By chance, you'd expect to find a statistically significant finding 5% of the time when the null hypothesis is true, so you would expect (on average, if you were to do this many times) that one of the 20 comparisons would yield a statistically significant result. Since every null hypothesis is true, every time the result is statistically significant, that conclusion will be a Type I (false positive) error. Of all the results that are statistically significant, not a single one is due to a real difference between populations. In other words, the false discovery rate must be 100%.

Does it even make sense to compare relevant variables at the beginning of a clinical trial? Sure. If a clinically important variable differs substantially between the two randomly chosen groups before any treatment or intervention, it will be impossible to interpret the study results. You won't be able to tell if any difference in the outcome variable is due to treatment, or due to unlucky randomization

	DECISION: REJECT NULL HYPOTHESIS	DECISION: DO NOT REJECT NULL HYPOTHESIS	TOTAL
Populations are the same (null hypothesis is true)	1	19	20
Populations differ (null hypothesis is false)	0	0	0
Total	1	19	20

Table 18.3. Distribution of results when comparing 20 variables collected in two randomly assigned groups before treatment.

giving you very different kinds of people in the two groups. When comparing groups in a randomized trial before treatment, ignore the P values and ask whether the difference in relevant variables prior to treatment are large enough to matter, defining "large enough" physiologically or clinically (but not based on a P value).

An extreme example in which you know that every null hypothesis is false

Now let's look at the other extreme case, in which you know for sure that every hull hypothesis is false. Imagine you are running a screening assay to try to find a drug that works on a particular pathway. You screen tens of thousands of compounds, and none (so far) work. To prove to yourself (and others) that your assay would detect an active drug, you run a positive control every day that tests a drug known to work on that pathway. You know what to expect and therefore run this positive control to demonstrate that your experimental system is working each day. This is important, because you want to make sure that all the negative results you obtain are really because you are screening inactive drugs, and not because the experimental system is simply not working on a given day (maybe you left out one ingredient).

Table 18.4 shows what happens when you run this positive control 100 times. The drug always provokes a big response, and the P value is always less than 0.05. That means that when you run the positive control 100 times, you'll get a statistically significant result 100 times, without a single false positive. A false positive (Type I error) is impossible, because you are repeatedly testing a drug that is known to work. Therefore, the FDR is 0%.

ACCOUNTING FOR PRIOR PROBABILITY INFORMALLY

The extreme examples in the previous section demonstrate that your interpretation of a statistically significant result depends on what you know before you collect the data—that is, the *prior probability* (see Chapter 2). Understanding the results requires combining the P value obtained from the experiment with the prior probability based on the context of the experiment.

Experienced scientists often account for the prior probability informally, without performing any additional calculations, and without mentioning Bayes

	DECISION: REJECT NULL HYPOTHESIS	DECISION: DO NOT REJECT NULL HYPOTHESIS	TOTAL
Drug is ineffective	0	0	0
Drug works	100	0	100
Total	100	0	100

Table 18.4. Distribution of results running positive control experiments 100 times.

(see below). When interpreting a result that is statistically significant, here are three scenarios that take into account prior probability:

- This study tested a hypothesis that is biologically sound and supported by previous data. The P value is 0.04. I have a choice of believing that the results occurred by a coincidence that will happen 1 time in 25 under the null hypothesis, or of believing that the experimental hypothesis is true. Because the experimental hypothesis makes so much sense, I'll believe it. The null hypothesis is probably false.

- This study tested a hypothesis that makes no biological sense and has not been supported by any previous data. The P value is 0.04, which is lower than the usual threshold of 0.05, but not by very much. I have a choice of believing that the results occurred by a coincidence that will happen 1 time in 25 under the null hypothesis, or of believing that the experimental hypothesis is true. Because the experimental hypothesis is so unlikely to be correct, I think that the results are the result of coincidence. The null hypothesis is probably true.

- This study tested a hypothesis that makes no biological sense and has not been supported by any previous data. I'd be amazed if it turned out to be true. The P value is incredibly low (0.000001). I've looked through the details of the study and cannot identify any biases or flaws. These are reputable scientists, and I believe that they've reported their data honestly. I have a choice of believing that the results occurred by a coincidence that will happen 1 time in a million under the null hypothesis, or of believing that the experimental hypothesis is true. Although the hypothesis seems crazy to me, the data force me to believe it. The null hypothesis is probably false.

You *should* interpret experimental data in the context of scientific theory and previous data. That's why different people can legitimately reach different conclusions from the same data.

ACCOUNTING FOR PRIOR PROBABILITY FORMALLY

Example: Screening for new drugs

Imagine that you work at a drug company and are screening drugs as possible treatments for hypertension. You test the drugs in a group of animals. You have decided that you are interested in a mean decrease in blood pressure of 10 mmHg, and you are using large enough samples so that there is an 80% chance of finding a statistically significant difference ($P < 0.05$) if the true difference between population means is 10 mmHg. (You will learn how to calculate the sample size in Chapter 26.)

You test a new drug and find a statistically significant drop in mean blood pressure. You know that there are two possibilities. Either the drug really works to lower blood pressure or the drug doesn't alter blood pressure at all and you just happened to get lower pressure readings on the treated animals. How likely are the two possibilities?

Because you set α to 0.05, you know that 5% of studies done with inactive drugs will demonstrate a statistically significant drop in blood pressure. But that isn't the question you are asking. You want to know the answer to a different question: In what fraction of experiments in which you observe a statistically significant drop in pressure is the drug really effective? In other words, what is the false discovery rate? The answer is not necessarily 5%.

The answer depends on what you knew about the drug before you started the experiment, expressed as the prior probability that the drug works. This point is illustrated in the following three examples.

Drug A. Prior probability = 10%
This drug is known to weakly block angiotensin receptors, but the affinity is low and the drug is unstable. From your experience with such drugs, you estimate that there is about a 10% chance that it will depress blood pressure. In other words, the prior probability that the drug works is 10%. What will happen if you test many such drugs? The answer is shown in Table 18.5.

These are the steps you need to follow to create the table:

1. We will predict the results of 1,000 experiments with 1,000 different drugs, so set the grand total to 1,000. This number is arbitrary, because the important conclusions will be expressed as ratios.
2. Of those 1,000 drugs we screen, we expect that 10% will really work. In other words, the prior probability equals 10%. So we place 10% of 1,000, or 100, as the total of the second row, leaving 900 for the sum of the first row.
3. Of the 100 drugs that really work, we will obtain a statistically significant result in 80% (because our experimental design has 80% power; see

	DECISION: REJECT NULL HYPOTHESIS	DECISION: DO NOT REJECT NULL HYPOTHESIS	TOTAL
Drug is ineffective	45	855	900
Drug works	80	20	100
Total	125	875	1,000

Table 18.5. Distribution of results with 80% power, 5% significance level, and prior probability = 10%.

Chapter 20). So we place 80% of 100, or 80, into the bottom left cell of the table. This leaves 20 experiments with a drug that really works, but P > 0.05, so we conclude that the drug is not effective.

4. Of the 900 drugs that are really ineffective, we will by chance obtain a statistically significant reduction in blood pressure in 5% (because we set α equal to 0.05). Thus, the top-left cell is 5% × 900, or 45. That leaves 855 experiments in which the drug is ineffective, and we correctly observe no statistically significant difference.

5. Determine the column totals by addition.

Of 1,000 tests of different drugs, we expect to obtain a statistically significant difference (P < 0.05) in 125 of them. Of those, 80 drugs are really effective and 45 are not. When you see a statistically significant result for any particular drug (for which the prior probability is 10%), you can conclude that there is a 64% chance (80/125) that the drug is really effective and a 36% chance (45/125) that it is really ineffective. In other words, you expect the FDR to equal 36%.

Drug B. Prior probability = 80%
Here the pharmacology is much better characterized. Drug B blocks the right kinds of receptors with reasonable affinity and the drug is chemically stable. From your experience with such drugs, you estimate that the prior probability that the drug is effective equals 80%. What will happen if you test 1,000 such drugs? The calculations are similar to those for Drug A, and the results are shown in Table 18.6.

If you test 1,000 drugs like this one, you expect to see 650 statistically significant results. Of those, 98.5% (640/650) will be truly effective. When you see a statistically significant result for any particular drug (for which the prior probability is 80%), you can conclude that there is a 98.5% chance that it really lowers blood pressure and a 1.5% chance that it is really ineffective. In other words, you expect the FDR to equal 1.5%.

Drug C. Prior probability = 1%
This drug was randomly selected from the drug company's inventory of compounds. Nothing you know about this drug suggests that it affects blood pressure. Your best guess is that about 1% of such drugs will lower blood pressure.

	DECISION: REJECT NULL HYPOTHESIS	DECISION: DO NOT REJECT NULL HYPOTHESIS	TOTAL
Drug is ineffective	10	190	200
Drug works	640	160	800
Total	650	350	1,000

Table 18.6. Distribution of results with 80% power, 5% significance level, and prior probability = 80%.

	DECISION: REJECT NULL HYPOTHESIS	DECISION: DO NOT REJECT NULL HYPOTHESIS	TOTAL
Drug is ineffective	50	940	990
Drug works	8	2	10
Total	58	942	1,000

Table 18.7. Distribution of results with 80% power, 5% significance level, and prior probability = 1%.

What will happen if you screen 1,000 such drugs? The answer is shown in Table 18.7.

If you test 1,000 drugs like this one, you expect to see 58 statistically significant results. Of those, you expect that 14% (8/58) will be truly effective and that 86% (50/58) will be ineffective. When you see a statistically significant result for any particular drug (for which the prior probability is only 1%), you can conclude that there is a 14% chance that it will really lower blood pressure and an 86% chance that it is really ineffective. In other words, you expect the FDR to equal 86%.

Summary of drug screening example

The three examples had three different prior probabilities, and resulted in three different FDR values, ranging from 1.5% to 86%. These examples demonstrate that in some situations, a statistically significant result almost certainly means there is a true effect. In other situations, a statistically significant result is almost certainly a quirk of random sampling. It all depends on the situation, as assessed by the prior probability.

Bayes

These calculations use what is called a *Bayesian approach*, named after Thomas Bayes, who first published work on this problem in the mid-18th century. (Bayes was previously mentioned in Chapter 2.)

There are some situations in which the prior probabilities are well defined, for example, in analyses of genetic linkage. The prior probability that two genetic loci are linked is known, so Bayesian statistics are routinely used in analysis of genetic linkage. There is nothing controversial about using Bayesian inference when the prior probabilities are known precisely.

In many situations, including the drug examples above, the prior probabilities are little more than a subjective feeling. Many, but not all, statisticians think it is OK to convert these feelings to numbers (e.g., "99% sure" or "70% sure"), which are then treated as prior probabilities. Of course the calculated results (the FDRs) are no more accurate than are your estimates of the prior probabilities.

Q & A

How is it possible for an effect to be statistically significant but not scientifically significant (important)?

> You reach the conclusion that an effect is statistically significant when the P value is less than 0.05. With large sample sizes, this can happen even when the effect is tiny and irrelevant. The small P value tells us that the effect would not often occur by chance but says nothing about whether the effect is large enough to care about.

When is an effect large enough to care about, to be scientifically significant (important)?

> It depends on what you are measuring and why you are measuring it. This question can only be answered by someone familiar with your particular field of science. It is a scientific question, not a statistical one.

Does the context of the experiment (the prior probability) come into play when deciding if a result is statistically significant or not?

> Only if you take into account prior probability when deciding on a value for α. Once you have chosen α, the decision about when to call a result "statistically significant" depends only on the P value and not on the context of the experiment.

Does the context of the experiment (the prior probability) come into play when calculating the false discovery rate?

> Yes. See Tables 18.5 through 18.7.

Does your choice of a value for α influence the calculated value of the FDR?

> Yes. Your decision for a value of α determines the chance that a result will end up in the first or second column of Table 18.1.

When a P value is less than 0.05 and thus the comparison is deemed to be statistically significant, can you be 95% sure that the effect is real?

> No! It depends on the situation, on the prior probability.

CHAPTER SUMMARY

- P values depend on the size of the difference or association, as well as sample size. You'll see a small P value either with a large effect observed in tiny samples, or a tiny effect observed in large samples.
- A conclusion of statistical significance does not mean the difference is large enough to be interesting, does not mean the results are intriguing enough to be worthy of further investigation, and does not mean that the finding is scientifically or clinically significant.
- Don't mix up the significance level α (the threshold P value defining statistical significance) with the P value. You must choose α as part of the experimental design. The P value, in contrast, is computed from the data.
- Also don't mix up the significance level α with the false discovery rate (FDR), which is the chance that a statistically significant finding is actually due to a coincidence of random sampling.
- The false discovery rate depends on the context of the study. In other words, it depends on the prior probability that your hypothesis is true (based on prior data and theory).
- Even if you don't do formal Bayesian calculations, you should consider prior knowledge and theory when interpreting data.

TERMS INTRODUCED IN THIS CHAPTER

- Bayesian approach (p. 160)
- False discovery rate (FDR) (p. 154)
- Prior probability (p. 156)

CHAPTER 19

❖❖❖

Interpreting a Result That Is Not Statistically Significant

Extraordinary claims require extraordinary proof.

CARL SAGAN

When you see a result that is not statistically significant, don't stop thinking. "Not statistically significant" means only that the P value is larger than a preset threshold. Thus, a difference (or correlation, or association) as large as what you observed would not be unusual as a result of random sampling if the null hypothesis were true. This does not prove that the null hypothesis is true. This chapter explains how to use confidence intervals to help interpret those findings that are not statistically significant.

"NOT SIGNIFICANTLY DIFFERENT" DOES NOT MEAN "NO DIFFERENCE"

A large P value means that a difference (or correlation, or association) as large as what you observed would happen frequently as a result of random sampling. But this does not necessarily mean that the null hypothesis of no difference is true or that the difference you observed is definitely the result of random sampling.

Vickers (2010) tells a great story that illustrates this point:

> The other day I shot baskets with [the famous basketball player] Michael Jordan (remember that I am a statistician and never make things up). He shot 7 straight free throws; I hit 3 and missed 4 and then (being a statistician) rushed to the sideline, grabbed my laptop, and calculated a *P* value of .07 by Fisher's exact test. Now, you wouldn't take this *P* value to suggest that there is *no* difference between my basketball skills and those of Michael Jordan, you'd say that our experiment hadn't *proved* a difference.

A high P value does not prove the null hypothesis. Deciding not to reject the null hypothesis is not the same as believing that the null hypothesis is definitely true. The absence of evidence is not evidence of absence (Altman & Bland, 1995).

	CONTROLS	HYPERTENSION
Number of subjects	17	18
Mean receptor number (receptors/platelet)	263	257
SD	87	59

Table 19.1. Number of α_2-adrenergic receptors on the platelets of controls and people with hypertension.

EXAMPLE: α_2-ADRENERGIC RECEPTORS ON PLATELETS

Epinephrine, acting through α_2-adrenergic receptors, makes blood platelets stickier and thus helps blood clot. We counted these receptors and compared people with normal and high blood pressure (Motulsky, O'Connor, & Insel, 1983). The idea was that the adrenergic signaling system might be abnormal in people with high blood pressure (hypertension). We were most interested in the effects on the heart, blood vessels, kidney, and brain but obviously couldn't access those tissues in people, so we counted receptors on platelets instead. Table 19.1 shows the results.

The results were analyzed using an unpaired t test (see Chapter 30). The average number of receptors per platelet was almost the same in both groups, so of course the P value was high, 0.81. If the two populations were Gaussian with identical means, you'd expect to see a difference as large or larger than that observed in this study in 81% of studies of this size.

Clearly, these data provide no evidence that the mean receptor number differs in the two groups. When my colleagues and I published this study 30 years ago, we stated that the results were not statistically significant and stopped there, implying that the high P value proves that the null hypothesis is true. But that was not a complete way to present the data. We should have interpreted the CI.

The 95% CI for the difference between group means extends from –45 to 57 receptors per platelet. To put this in perspective, you need to know that the average number of α_2-adrenergic receptors/platelet is about 260. We can therefore rewrite the confidence interval as extending from –45/260 to 57/260, which is from –17.3% to 21.9%, or approximately plus or minus 20%.

It is only possible to properly interpret the confidence interval in a scientific context. Here are two alternative ways to think about these results:

- A 20% change in receptor number could have a huge physiological impact. With such a wide CI, the data are inconclusive, because they are consistent with no difference, substantially more receptors on platelets from people with hypertension, or substantially fewer receptors on platelets of people with hypertension.
- The CI convincingly shows that the true difference is unlikely to be more than 20% in either direction. This experiment counts receptors on a convenient tissue (blood cells) as a marker for other organs, and we know the number of receptors per platelet varies a lot from individual to individual. For these reasons, we'd only be intrigued by the results (and want to pursue

	ADVERSE OUTCOME	TOTAL	RISK	RELATIVE RISK
Routine ultrasound	383	7,685	5.0%	1.020
Only when indicated	373	7,596	4.9%	
Total	756	15,281		

Table 19.2. Relationship between fetal ultrasounds and outcome.
The risks in Column 4 are computed by dividing the number of adverse outcomes by the total number of pregnancies. The relative risk is computed by dividing one risk by the other (see Chapter 27 for more details).

this line of research) if the receptor number in the two groups differed by at least 50%. Here, the 95% CI extended about 20% in each direction. Therefore, we can reach a solid negative conclusion that either there is no change in receptor number in individuals with hypertension or that any such change is physiologically trivial and not worth pursuing.

The difference between these two perspectives is a matter of scientific judgment. Would a difference of 20% in receptor number be scientifically relevant? The answer depends on scientific (physiological) thinking. Statistical calculations have nothing to do with it. Statistical calculations are only a small part of interpreting data.

EXAMPLE: FETAL ULTRASOUNDS

Ewigman et al. (1993) investigated whether the routine use of prenatal ultrasounds would improve perinatal outcomes. The researchers randomly divided a large pool of pregnant women into two groups. One group received routine ultrasound exams (sonograms) twice during their pregnancy. The other group was administered sonograms only if there was a clinical reason to do so. The physicians caring for the women knew the results of the sonograms and cared for the women accordingly. The investigators looked at several outcomes. Table 19.2 shows the total number of adverse events, defined as fetal or neonatal deaths (mortality) or moderate to severe morbidity.

The null hypothesis is that the risk of adverse outcomes is identical in the two groups. In other words, the null hypothesis is that routine use of ultrasounds neither prevents nor causes perinatal mortality or morbidity, so the relative risk equals 1.00. Chapter 27 will explain the concept of relative risk in more detail.

Table 19.2 shows that the relative risk is 1.02. That isn't far from the null hypothesis value of 1.00. The two-tailed P value is 0.86.

Interpreting the results requires knowing the 95% CI for the relative risk, which a computer program can calculate. For this example, the 95% CI ranges from 0.88 to 1.17.

Our data are certainly consistent with the null hypothesis, because the CI includes 1.0. This does not mean that the null hypothesis is true. Our CI tells us that the data are also consistent (within 95% confidence) with relative risks ranging from 0.88 to 1.17.

Here are three approaches to interpreting the results:

- The CI is centered on 1.0 (no difference) and is quite narrow. These data convincingly show that routine use of ultrasound is neither helpful nor harmful.
- The CI is narrow, but not all that narrow. It certainly makes clinical sense that the extra information provided by an ultrasound will help obstetricians manage the pregnancy and might decrease the chance of a major problem. The CI goes down to 0.88, a risk reduction of 12%. If I were pregnant, I'd certainly want to use a risk-free technique that reduces the risk of a sick or dead baby by as much as 12% (from 5.0% to 4.4%)! The data certainly don't prove that a routine ultrasound is beneficial, but the study leaves open the possibility that routine ultrasound might reduce the rate of awful events by as much as 12%.
- The CI goes as high as 1.17. That is a 17% relative increase in problems (from 5.0% to 5.8%). Without data from a much bigger study, these data do not convince me that ultrasounds are helpful and make me worry that they might be harmful.

Statistics can't help resolve the differences among these three approaches. It all depends on how you interpret the relative risk of 0.88 and 1.17, how worried you are about the possible risks of an ultrasound, and how you combine the data in this study with data from other studies (I have no expertise in this field and have not looked at other studies).

In interpreting the results of this example, you also need to think about benefits and risks that don't show up as a reduction of adverse outcomes. The ultrasound picture helps reassure parents that their baby is developing normally and gives them a picture to bond with and show relatives. This can be valuable regardless of whether it reduces the chance of adverse outcomes. Although statistical analyses focus on one outcome at a time, you must consider all the outcomes when evaluating the results.

HOW TO GET NARROWER CIs

Both examples presented above demonstrate the importance of interpreting the CI in the scientific context of the experiment. Different people will appropriately have different opinions about how large a difference (or relative risk) is scientifically or clinically important and will therefore interpret a not statistically significant result differently.

If the CI is wide enough to include values you consider clinically or scientifically important, then the study is inconclusive. In some cases, you might be able to narrow the confidence intervals by improving the methodology and thereby reducing the SD. But in most cases, increasing the sample size is the only approach to narrowing the CI in a repeat study. This rule of thumb can help: if sample size is increased by a factor of four, the CI is expected to narrow by a factor of two. More generally, the width of a CI is inversely proportional to the square root of sample size.

WHAT IF THE P VALUE IS REALLY HIGH?

If you ran many experiments in which the null hypothesis was really true, you'd expect the P values to be uniformly distributed between 0.0 and 1.0. Half of the P values would have values greater than 0.5, and 10% would have values greater than 0.9. But what do you conclude when the P value is really high?

In 1866, Mendel published his famous paper on heredity in pea plants (Mendel, 1866). This was the first explanation of heredity and recessive traits and really founded the field of genetics. Mendel proposed a model of recessive inheritance, designed an experiment with peas to test the model, and showed that the data fit the model very well. Extremely well! Fisher (1936) reviewed these data and then pooled all of Mendel's published data to calculate a P value that answered the question: Assuming that Mendel's genetic theory is correct and every plant was classified correctly, what is the probability that the deviation between expected and observed would be as great or greater than actually observed?

The answer (the P value) is 0.99993. In other words, if all the data were honestly reported, there is only a 0.007% chance that the data would match the theory this well. You'd expect more random discrepancies between theory and data. If 100,000 scientists were to repeat the work, only 7 would be expected to gather observed data that match the theory so perfectly. Although it is possible

Q & A

that Mendel was extremely lucky, Fisher concluded that Mendel's data are simply too good to be real.

If a P value is greater than 0.05, can you conclude that you have disproven the null hypothesis?
> No.

If you conclude that a result is not statistically significant, it is possible that you are making a Type II error as a result of missing a real effect. What factors influence the chance of this happening?
> The probability of a Type II error depends on the significance level (α) you have chosen, the sample size, the size of the true effect, and the prior probability of that effect.

By how much do you need to increase the sample size to make a confidence interval half as wide?
> A general rule of thumb is that increasing the sample size by a factor of four will cut the expected width of the confidence interval by a factor of two. (Note that two is the square root of four.)

What if I want to make the confidence interval one-quarter as wide as it is?

Increasing the sample size by a factor of 16 will be expected to reduce the width of the confidence interval by a factor of 4. (Note that 4 is the square root of 16.)

CHAPTER SUMMARY

- If a statistical test computes a large P value, you should conclude that the findings would not have been unusual if the null hypothesis were true.
- You should *not* conclude that the null hypothesis of no difference (or association, etc.) has been proven.
- When interpreting a high P value, the first thing to do is look at the size of the effect.
- Also look at confidence interval of the effect.
- If the confidence interval includes effect sizes that you would consider to be scientifically important, then the study is equivocal.

CHAPTER 20

❖❖❖

Statistical Power

There are two kinds of statistics, the kind you look up, and the kind you make up.

REX STOUT

The power of an experimental design answers this question: If the true effect is of a specified size and the experiment is repeated many times, what fraction of the results will be statistically significant? The concept of power can help when deciding how large a sample size to use (see Chapter 26) and can help interpret results that are not statistically significant.

WHAT IS STATISTICAL POWER?

The definition of a P value begins with "If the null hypothesis is true . . . "

But what if the null hypothesis is false and the treatment really does affect the outcome? Even so, your data may reach the conclusion that the effect is not statistically significant. Just by chance, your data may yield a P value greater than 0.05 (or whatever significance level you chose). The following question therefore becomes relevant: If there really were an effect (or difference, or relative risk, or correlation, etc., which we will collectively call *effect*) of a specified value in the overall population, what is the chance of obtaining an effect that is statistically significant in one particular sample?

The answer, called the *power* of the experiment, depends on the sample size, the amount of scatter, and the size of the effect you hypothesize exists. Given these values, power is the fraction of experiments in which you would expect to find a statistically significant result.

Table 20.1 (which repeats Table 18.2) shows the results of many hypothetical statistical analyses, each analyzed to reach a conclusion of "statistically significant" or "not statistically significant." Assuming the null hypothesis is not true, power is the fraction of experiments that reach a statistically significant conclusion. So power equals C/(C + D). Beta (β) is 1.0 minus power, so it equals D/(C + D).

Note the similarity in the definitions of α and β. If the null hypothesis is true, then α is the chance of making the wrong decision (rejecting the null hypothesis). If the null hypothesis is false (with a specified alternative hypothesis), β is the chance of making the wrong decision (not rejecting the null hypothesis).

	DECISION: REJECT NULL HYPOTHESIS	DECISION: DO NOT REJECT NULL HYPOTHESIS	TOTAL
Null hypothesis is true	A	B	A + B
Null hypothesis is false	C	D	C + D

Table 20.1. Definition of power.

This table shows the results of A + B + C + D experiments. In some cases, the null hypothesis is true (top row), whereas in other cases it is not (bottom row). The first column tabulates results that are statistically significant (leading you to reject the null hypothesis), whereas the second column tabulates results that are not statistically significant. Power is the fraction of experiments that achieve a statistically significant result when the null hypothesis is false. Thus, power is defined as C/(C + D).

AN ANALOGY TO UNDERSTAND STATISTICAL POWER

Here is a silly analogy helps illustrate the concept of statistical power (Hartung, 2005). You send your child into the basement to find a tool. He comes back and says, "It isn't there." What do you conclude? Is the tool there or not?

There is no way to be sure, so the answer must be a probability. The question you really want to answer is, What is the probability that the tool is in the basement? But that question can't really be answered without knowing the prior probability and using Bayesian thinking (see Chapter 18). Instead, let's ask a different question: If the tool really is in the basement, what is the chance your child would have found it? The answer, of course, is: it depends. To estimate the probability, you'd want to know three things:

- How long did he spend looking? If he looked for a long time, he is more likely to have found the tool than if he looked for a short time. The time spent looking for the tool is analogous to sample size. An experiment with a large sample size has high power to find an effect, while an experiment with a small sample size has less power.
- How big is the tool? It is easier to find a snow shovel than the tiny screwdriver used to fix eyeglasses. The size of the tool is analogous to the size of the effect you are looking for. An experiment has more power to find a big effect than a small one.
- How messy is the basement? If the basement is a real mess, he was less likely to find the tool than if it is carefully organized. The messiness is analogous to experimental scatter. An experiment has more power when the data are very tight (little variation), and less power when the data are very scattered.

If the child spent a long time looking for a large tool in an organized basement, there is a high chance that he would have found the tool if it were there. So you could be quite confident of his conclusion that the tool isn't there. Similarly, an experiment has high power when you have a large sample size, are looking for a large effect, and have data with little scatter (small standard deviation). In this

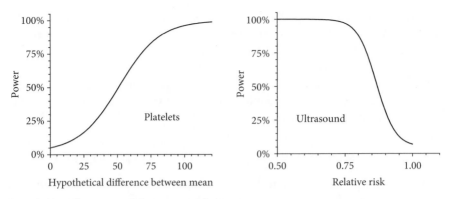

Figure 20.1. The power of the two experiments.
The graph on the left summarizes the experiment in Table 19.1, which compared a_2-receptor number in controls and people with hypertension. The graph on the right summarizes the data in Table 19.2, evaluating the advantages of a fetal ultrasound. For any hypothetical difference between means (X-axis), this graph shows the power of the study to find a statistically significant difference (P < 0.05). Note that an increasing effect for the platelet example (larger difference in receptor number) goes from left to right, whereas an increasing effect for the ultrasound example (smaller relative risk) goes from right to left.

situation, there is a high chance that you would have obtained a statistically significant effect if the effect existed.

If the child spent a short time looking for a small tool in a messy basement, his conclusion that the tool isn't there doesn't really mean very much. Even if the tool were there, he probably would not have found it. Similarly, an experiment has little power when you use a small sample size, are looking for a small effect, and the data have lots of scatter. In this situation, there is a high chance of obtaining a conclusion of "not statistically significant" even if the effect exists.

POWER OF THE TWO EXAMPLE STUDIES

The power of an experimental design can be computed to detect various hypothetical differences. The left side of Figure 20.1 shows the power of the platelet study (described in Chapter 19) to detect various hypothetical differences between means. The power was calculated using the program GraphPad StatMate from the sample size of the study, the SDs of the two groups, and the definition of significance (P < 0.05).

If there really is no difference between population means, there is a 5% chance of obtaining a statistically significant result. That is the definition of statistical significance (using the traditional 5% significance level). So the curve intersects the Y-axis at 5%. If there is a difference between population means, the power of this study depends on the size of that difference. If the difference is tiny (left side of the graph), then the power is low. If the difference is large, then so is the power, which approaches 100% with very large differences.

The right side of Figure 20.1 shows the power of the ultrasound study. If the relative risk is really 1.0 (no effect), there is a 5% chance of obtaining a statistically significant result, because that is the definition of statistical significance. So the curve shows 5% power with a relative risk of 1.00, and more power as the effect gets more pronounced (lower relative risk, showing that an ultrasound is beneficial).

The shapes in Figure 20.1 are universal. What varies between studies is the horizontal location of the curve.

POST HOC POWER ANALYSIS IS NOT HELPFUL

Some software programs try to make things easy and remove the need to think in a scientific context. These programs compute the power of an experiment to detect the effect size (or difference, relative risk, etc.) already observed in that particular experiment. The result is sometimes called *observed power*, and the procedure is sometimes called a *post hoc power analysis* or *retrospective power analysis*.

This approach is not helpful (Hoenig & Heisey, 2001; Lenth, 2001; Levine & Ensom, 2001). If your study reached a conclusion that the difference is not statistically significant, then by definition, its power to detect the effect actually observed is very low. You learn nothing new by calculating the observed power.

In contrast, it does make sense to ask how much power an experiment had to detect an effect of some specified size. That specified effect size must be chosen based on your scientific goals and not based on the effect actually observed in a particular experiment. It can also make sense to use the results of one experiment to help do power calculations that can be used to design a future experiment.

Q & A

If I want to use a program to compute power, what questions will it ask?
To calculate power, you must enter the values for alpha, the expected standard deviation, the planned sample size, and the size of the hypothetical difference (or effect) you are hoping to detect.

Why do I have to specify the effect size for which I am looking? I want to detect any size effect.
All studies will have a small power to detect tiny effects and a large power to detect enormous effects. You can't calculate power without specifying the effect size for which you are looking.

When does it make sense to do calculations involving power?
In two situations. When planning a study, you need to decide how large a sample size to use. Those calculations will require you to specify how much power you want to detect some hypothetical effect size. After completing a study that has results that are not statistically significant, it can make sense to ask how much power that study had to detect some specified effect size.

Can one do all sample size/power calculations using standard equations?
No. In many cases it is necessary to run computer simulations to compute the power of a proposed experimental design.

CHAPTER SUMMARY

- The *power* of an experimental design is the probability that you will obtain a statistically significant result assuming a certain effect size in the population.
- The power of an experiment depends on sample size, variability, the choice of α to define statistical significance, and the hypothetical effect size.
- For any combination of sample size, variability, and α, there will be a high power to detect a huge effect and a small power to detect a tiny effect.
- Power should be computed based on the desired effect size or the minimum effect size that would be scientifically worth detecting. Once the study is complete, it is not helpful to compute the power that study had to detect the effect size that was actually observed.

TERMS INTRODUCED IN THIS CHAPTER

- Observed power (p. 172)
- Post hoc power analysis (p. 172)
- Power (p. 169)
- Retrospective power analysis (p. 172)

✦✦✦

CHAPTER 21

Testing for Equivalence
or Noninferiority

The problem is not what you don't know, but what you know
that ain't so.

WILL ROGERS

In many scientific and clinical investigations, the goal is not to find
out whether one treatment causes a substantially different effect than
another treatment. Instead, the goal is to find out whether the effects of
a new treatment are equivalent (or not inferior) to those of a standard
treatment. This chapter explains how the usual approach of statistical
hypothesis testing is not useful. A conclusion that the difference between
two treatments is statistically significant does not answer the question
about equivalence, and neither does a conclusion that the difference is
not statistically significant.

EQUIVALENCE MUST BE DEFINED SCIENTIFICALLY,
NOT STATISTICALLY

Imagine that you've created a generic drug and want to show that it works just as
well as a standard drug. There are many aspects of drug action that can be compared, and we'll focus on one question: Is the peak blood plasma concentration
of the two formulations of the drug equivalent? In other words, is the new drug
absorbed as well as the standard drug? Patients are given first one drug formulation and then (after a washout period) the other, randomizing the order. The peak
concentration of each drug formulation is measured in blood samples taken from
each subject.

When you compare the two drugs, you will always see some difference in
peak plasma concentration. It doesn't make sense to ask whether the two produce precisely the same outcome. When asking about *equivalence*, the question is whether the outcomes are close enough to be clinically or scientifically
indistinguishable.

How close is close enough? "Close enough" must be defined as a range of
treatment effects that are considered scientifically or clinically trivial. This requires

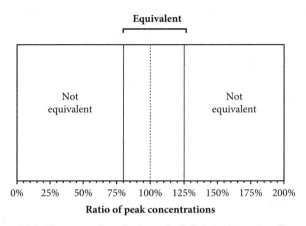

Figure 21.1. The zone of equivalence is defined using scientific criteria.

The FDA has said that for comparing two formulations of a drug, the results are deemed equivalent when the ratio of peak drug concentrations is between 80 and 125%.

thinking about the scientific or clinical context of your experiment. Statistical calculations and conclusions about statistical significance are completely irrelevant in defining this range, called the *equivalence zone* or *equivalence margin*.

When testing a generic drug, the U.S. Food and Drug Administration (FDA) defines two drug formulations to be equivalent when the ratio of their peak concentrations in blood plasma, and the entire 90% CI of that ratio, is between 0.80 and 1.25 (U.S. Food and Drug Administration [FDA], 2012). This definition is based on clinical thinking about drug action. In other circumstances, equivalence might be defined differently.

Figure 21.1 shows this definition. Make sure you understand Figure 21.1, because it is the basis for Figures 21.2 and 21.3.

Note that the zone of equivalence does not appear to be symmetrical around 100%, but in fact it is. It is fairly arbitrary whether you calculate the ratio of peak levels as the peak level of the new formulation divided by the peak level of the standard formulation, or whether you define it as the peak level of the standard formulation divided by the peak level of the new formulation. The reciprocal of 80% is 125% (1/0.8 = 1.25), and the reciprocal of 125% is 80% (1/1.25 = 0.80). So the zone of equivalence is, in a practical sense, symmetrical.

IF THE MEAN IS WITHIN THE EQUIVALENCE ZONE

Figure 21.2 shows data from three drugs for which the mean ratio of peak concentrations is within the zone of equivalence. The fact that the mean value is within the zone does not prove equivalence. Two drugs are defined to be bioequivalent

Figure 21.2. Three drug formulations for which the ratio of peak concentrations are within the equivalent zone.

For Drug A, the CI extends outside of the equivalent zone, so the results are inconclusive. The results are consistent with the two drugs being equivalent or not equivalent. In Drugs B and C, the 95% CIs lie completely within the equivalent zone, so the data demonstrate that the two drugs are equivalent to the standard drug.

Figure 21.3. Results of three drugs for which the mean ratio of peak concentrations is in the not-equivalent zone.

With Drugs D and E, the CI includes both equivalent and not-equivalent zones, so the data are not conclusive. The 95% CI for Drug F is entirely outside the equivalence zone, proving that it is not equivalent to the standard drug.

(by the FDA) only when the entire 95% CI for the ratio of peak concentrations lies within the equivalence zone.

This is the case for Drugs B and C. Those 95% CIs lie entirely within the equivalence zone, so the data demonstrate that Drugs B and C are equivalent to the standard drug with which they are being compared.

In contrast, the 95% CI for Drug A is partly in the equivalence zone and partly outside of it. The data are inconclusive.

IF THE MEAN IS OUTSIDE OF THE EQUIVALENCE ZONE

Figure 21.3 shows data from three drugs for which the mean ratio of peak concentrations is in the not-equivalent zone. This does not prove the drugs are not equivalent. You must look at the entire 95% CI to make such a determination.

The 95% CIs for Drugs D and E are partly within the equivalent zone and partly outside it. The data are not conclusive.

The 95% CI for Drug F lies totally outside the equivalence zone. These data prove that Drug F is not equivalent to the standard drug.

THE USUAL APPROACH OF STATISTICAL HYPOTHESIS TESTING IS NOT HELPFUL

A common mistake is to simply compare the peak blood concentration of the two drugs using a paired t test (see Chapter 31). It seems as though the P value ought to be informative, but this approach is not helpful. If the results are not statistically significant, it is a mistake to conclude that the blood concentrations of the two drugs are equivalent. Similarly, if the results are statistically significant, it is a mistake to conclude that the two drugs are not equivalent.

A paired t test would test the null hypothesis that the two drugs have the same mean peak levels, meaning that the ratio of peak concentrations would equal 100%. All three figures have a vertical dotted line marking that point.

You can look at the data shown in Figures 21.2 and 21.3 and instantly know whether the difference between peak levels is statistically significant. A result is statistically significant with a P value less than 0.05 when the 95% CI does not include the value that defines the null hypothesis. Data from Drug C (Figure 21.2) and Drugs E and F (Figure 21.3) meet this criterion, because their 95% CIs do not include 100%. The difference between the peak concentrations of each of these drugs and the standard drug is statistically significant. However, as explained above, Drug C is equivalent to the standard drug, Drug F is not equivalent, and the data for Drug E are inconclusive.

A result is not statistically significant (P > 0.05) when the 95% CI includes the value defined by the null hypothesis. The data for Drugs A, B, and D meet this criterion, because the CIs include 100%. The difference between the peak concentration of each of these drugs and the standard drug is not statistically significant. But this tells us nothing about equivalence. The data for Drug B demonstrate equivalence, whereas the data for Drugs A and D are inconclusive about whether the drugs are equivalent to the standard drug.

BENDING OVER BACKWARD TO ADAPT HYPOTHESIS TESTING TO EQUIVALENCE

It is possible to apply the ideas of statistical hypothesis testing to testing for equivalence, but it is tricky (Wellek, 2002). The approach is to pose two distinct null hypotheses and define statistical significance using two distinct one-tailed tests.

Figure 21.4. Applying the idea of statistical hypothesis testing to equivalence testing.

A conclusion of equivalence requires a statistically significant finding from two different tests of two different null hypotheses, shown by the vertical lines. Each null hypothesis is tested with a one-sided alternative hypothesis, shown by the arrows. Two drugs are considered equivalent when the ratio of peak concentrations is significantly greater than 80% and also significantly less than 125%.

Let's continue the example of comparing peak drug concentrations for two drugs. The two drugs are shown to be equivalent when both of the following conditions are true (Figure 21.4).

- The mean value of the ratio is *greater* than 0.80 (the lower limit that defines equivalence), and this increase is statistically significant.
- The mean value of the ratio is *less* than 1.25 (the upper limit that defines equivalence), and this decrease is statistically significant.

Each hypothesis is flipped to form a one-sided null hypothesis (that the mean ratio is less than 0.80, and that the mean ratio is greater than 1.25). If both one-tailed P values are smaller than a preset threshold, you reject both null hypotheses and therefore conclude that the mean ratio must be between 0.80 and 1.25. Thus, the results demonstrate equivalence.

Juggling two null hypotheses and two P values (each one-tailed) is not for the statistical novice. The results are the same as those obtained using the CI approach described above. The CI approach is much easier to understand.

NONINFERIORITY TRIALS

Equivalence trials attempt to prove that a new treatment or drug works about the same as the standard treatment. *Noninferiority* trials attempt to prove that a new treatment is not worse than the standard treatment.

To prove equivalence, all parts of the CI must be *within* the equivalence zone. To prove noninferiority, all parts of the CI must be to the right of the left (lower) border of the equivalence zone. The entire CI, therefore, is in a range that either shows the new drug is superior or shows that the new drug is slightly inferior but still in the zone defined to be practically equivalent.

YOU MUST BE SURE THE STANDARD TREATMENT WORKS

Snapinn (2000) and Kaul and Diamond (2006) reviewed the many issues one must consider when interpreting data that purport to demonstrate equivalence or noninferiority.

The most important issue is this: you must be 100% sure the standard drug works. A conclusion that a new treatment is equivalent (or not inferior) to a standard treatment is only useful when you are absolutely, positively sure that the standard treatment actually works better than a placebo. If the data from other studies leave any doubts about whether the standard treatment works, then it really doesn't make any sense to ask whether a new treatment is equivalent (or not inferior) to it.

Q & A

Why isn't a conclusion that a difference is not statistically significant enough to prove equivalence?
The P value, and thus the conclusion about whether an effect is statistically significant or not, is based entirely on analyzing the data. A conclusion about equivalence has to take the context into account. What is equivalent for one variable in one situation is not equivalent for another variable in another context.

Is it possible for a difference to be statistically significant but for the data to prove equivalence?
Surprisingly, yes. The conclusion that the difference is statistically significant just means the data convince you that the true difference is not zero. It doesn't tell you that the difference is large enough to care about. It is possible for the entire confidence interval to include values that you consider to be equivalent. Look at Drug C in Figure 21.2.

How is testing for noninferiority different than testing for superiority?
Although it might initially appear so, the double negatives are not all that confusing. When testing for noninferiority (in the example presented in this chapter), you are asking if the data prove that a drug is not worse than a standard drug. You will conclude that Drug A is "not worse" than Drug B when the two drugs are equivalent or when Drug A is better.

CHAPTER SUMMARY

• In many scientific and clinical investigations, the goal is not to find out whether one treatment causes a substantially different effect than another treatment. Instead, the goal is to find out whether the effects of a new treatment are equivalent to (or not inferior to) that of a standard treatment.

- The usual approach of statistical hypothesis testing does not test for equivalence. Determining whether or not a difference between two treatments is statistically significant tells you nothing about whether the two treatments are equivalent.
- There is no point testing whether a new treatment is equivalent to a standard one unless you are really sure that the standard treatment works.
- Equivalence trials attempt to prove that a new treatment or drug works about the same as the standard treatment. Noninferiority trials attempt to prove that a new treatment is not worse than the standard treatment.

TERMS INTRODUCED IN THIS CHPATER

- Equivalence (p. 174)
- Equivalence margin (p. 175)
- Equivalence zone (p. 175)
- Noninferiority (p. 178)

Challenges in Statistics

CHAPTER 22

❖❖❖

Multiple Comparisons Concepts

> If you torture your data long enough, they will tell you
> whatever you want to hear.
>
> MILLS (1993)

Coping with multiple comparisons is one of the biggest challenges in data analysis. If you calculate many P values, some are likely to be small just by random chance. Therefore, it is impossible to interpret small P values without knowing how many comparisons were made. This chapter explains three approaches to coping with multiple comparisons. Chapter 23 will show that the problem of multiple comparisons is pervasive, and Chapter 40 will explain special strategies for dealing with multiple comparisons after ANOVA.

THE PROBLEM OF MULTIPLE COMPARISONS

The problem of multiple comparisons is easy to understand

If you make two independent comparisons, what is the chance that one or both comparisons will result in a statistically significant conclusion just by chance? It is easier to answer the opposite question. Assuming both null hypotheses are true, what is the chance that both comparisons will not be statistically significant? The answer is the chance that the first comparison will not be statistically significant (0.95) times the chance that the second one will not be statistically significant (also 0.95), which equals 0.9025, or about 90%. That leaves about a 10% chance of obtaining at least one statistically significant conclusion by chance.

It is easy to generalize that logic to more comparisons. With K independent comparisons (where K is some positive integer), the chance that none will be statistically significant is 0.95^K, so the chance that one or more comparisons will be statistically significant is $1.0 - 0.95^K$. Figure 22.1 plots this probability for various numbers of independent comparisons.

Consider the unlucky number 13. If you perform 13 independent comparisons (with the null hypothesis true in all cases), the chance is about 50% that one or more of these P values will be less than 0.05 and will thus lead to a conclusion that the effect is statistically significant. In other words, with 13 independent comparisons, there is about a 50:50 chance of obtaining at least one false positive finding of statistical significance.

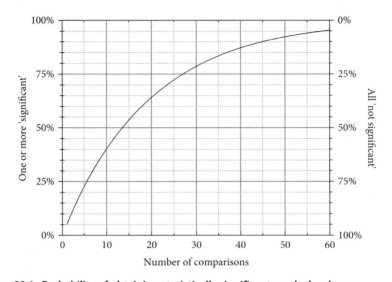

Figure 22.1. Probability of obtaining statistically significant results by chance.
The X-axis shows various numbers of statistical comparisons, each assumed to be independent of the others. The left Y-axis shows the chance of obtaining one or more statistically significant results (P < 0.05) by chance. The value on the left Y axis is computed as $1.0 - 0.95^X$, where X is the number of comparisons.

With more than 13 comparisons, it is more likely than not that one or more conclusions will be statistically significant just by chance. With 100 independent null hypotheses that are all true, the chance of obtaining at least one statistically significant P value is 99%.

A dramatic demonstration of the problem with multiple comparisons

Bennett and colleagues (2011) dramatically demonstrated the problem of multiple comparisons. They used functional magnetic resonance imaging (fMRI) to map blood flow in thousands of areas of a brain. Their experimental subject was shown a photograph of a person and asked to identify which emotion the person in the photograph was experiencing. The investigators measured blood flow to thousands of areas of the brain before and after presenting the photograph to their experimental subject. In two particular areas of the brain, blood flow increased substantially, and those differences were statistically significant (P < 0.001).

Sounds compelling. The investigators have identified regions of the brain involved with recognizing emotions. Right? Nope! The investigators could prove beyond any doubt that both findings were false positives resulting from noise in the instruments, and had nothing to do with changes in blood flow. How could they be sure? Their experimental subject was a dead salmon! When they properly corrected for multiple comparisons to account for the thousands of brain regions at which they looked, neither finding was statistically significant.

CORRECTING FOR MULTIPLE COMPARISONS IS NOT ALWAYS NEEDED

Before considering two general approaches to correcting for multiple comparisons, let's pause to consider three scenarios in which corrections for multiple comparisons are not needed.

Corrections for multiple comparisons are not needed if the people reading the data take into account the number of comparisons

Some statisticians recommend that investigators never correct for multiple comparisons but instead report uncorrected P values and CIs with a clear explanation that no mathematical correction was made for multiple comparisons (Rothman, 1990). The people reading the results must then informally account for multiple comparisons. If all the null hypotheses are true, you'd expect 5% of the comparisons to have uncorrected P values of less than 0.05. Compare this number with the actual number of small P values.

This approach requires that all comparisons (or at least the *number* of comparisons) be reported. If the investigators only show the small P values, it is impossible to interpret the results.

Corrections for multiple comparisons are not essential when you have clearly defined one outcome as primary and the others as secondary

Many clinical trials clearly define, as part of the study protocol, that one outcome is primary. This is the key outcome on which the conclusion of the study is based. The study may do other secondary comparisons, but those are clearly labeled as secondary. Correction for multiple comparisons is often not used with a set of secondary comparisons.

Corrections for multiple comparisons may not be needed if you make only a few planned comparisons

Even if a study collects lots of data, you may want to focus on only a few scientifically sensible comparisons, rather than every possible comparison. The term *planned comparison* is used to describe this situation. These comparisons must be planned as part of the experimental design. It is cheating to look at the data and then decide which comparisons you wish to make. When you make only a few preplanned comparisons, many statisticians think it is OK to not correct for multiple comparisons.

Corrections for multiple comparisons are not essential when the results are complementary

Ridker and colleagues (2008) asked whether lowering LDL cholesterol would prevent heart disease in patients who did not have high LDL concentrations, did not have a prior history of heart disease, but did have an abnormal blood test suggesting the presence of some inflammatory disease. The study included almost

18,000 people. Half received a statin drug to lower LDL cholesterol and half received a placebo.

The investigators' primary goal (planned as part of the protocol) was to compare the number of end points that occurred in the two groups, including deaths from a heart attack or stroke, nonfatal heart attacks or strokes, and hospitalization for chest pain. These events happened about half as often to people treated with the drug compared with people taking the placebo. The drug worked.

The investigators also analyzed each of the end points separately. Those people taking the drug (compared with those taking the placebo) had fewer deaths, fewer heart attacks, fewer strokes, and fewer hospitalizations for chest pain.

The data from various demographic groups were then analyzed separately. Separate analyses were done for men and women, old and young, smokers and nonsmokers, people with hypertension and those without, people with a family history of heart disease and those without, and so on. In each of 25 subgroups, patients receiving the drug experienced fewer primary end points than those taking the placebo, and all of these effects were statistically significant.

The investigators made no correction for multiple comparisons for all these separate analyses of outcomes and subgroups, because these were planned as secondary analyses. The reader does not need to try to informally correct for multiple comparisons, because the results are so consistent. The multiple comparisons each ask the same basic question, and all the comparisons lead to the same conclusion—people taking the drug had fewer cardiovascular events than those taking the placebo. In contrast, correction for multiple comparisons would be essential if the results showed that the drug worked in a few subsets of patients but not other subsets.

THE TRADITIONAL APPROACH TO CORRECTING FOR MULTIPLE COMPARISONS

The Familywise Error Rate

When each comparison is made individually without any correction for multiple comparisons, the traditional 5% significance level applies to each individual comparison. This is therefore known as *the per-comparison error rate*, and it is the chance that random sampling would lead *this particular comparison* to an incorrect conclusion that the difference is statistically significant when this particular null hypothesis is true.

When you correct for multiple comparisons, the significance level is redefined to be the chance of obtaining *one or more* statistically significant conclusions if *all* of the null hypotheses in the family are actually true. The idea is to make a stricter threshold for defining significance. If α is set to the usual value of 5% and all the null hypotheses are true, then the goal is to have a 95% chance of obtaining zero statistically significant results and a 5% chance of obtaining one or more statistically significant results. That 5% chance applies to the entire family of comparisons performed in the experiment, so it is called a *familywise error rate* or the *per-experiment error rate*.

What is a family of comparisons?

What exactly is a family of related comparisons? Usually, a family consists of all the comparisons in one experiment or all the comparisons in one major part of an experiment. That definition leaves lots of room for ambiguity. When reading about results corrected for multiple comparisons, ask about how the investigators defined the family of comparisons.

The Bonferroni correction

The simplest approach to achieving a familywise error rate is to divide the value of α (often 5%) by the number of comparisons. Then define a particular comparison as statistically significant only when its P value is less than that ratio. This is called a *Bonferroni correction*.

Imagine that an experiment makes 20 comparisons. If all 20 null hypotheses are true and there are no corrections for multiple comparisons, about 5% of these comparisons are expected to be statistically significant (using the usual definition of α). Table 22.1 and Figure 22.1 show that there is about a 65% chance of obtaining one (or more) statistically significant result.

If the Bonferroni correction is used, a result is only declared to be statistically significant when its P value is less than 0.05/20, or 0.0025. This ensures that if all the null hypotheses are true, there is about a 95% chance of seeing no statistically significant results among all 20 comparisons and only a 5% chance of seeing one (or more) statistically significant results. The 5% significance level applies to the entire family of comparisons rather than to each of the 20 individual comparisons.

Note a potential point of confusion. The value of α (usually 0.05) applies to the entire family of comparisons, but a particular comparison is declared to be statistically significant only when its P value is less than α/K (where K is the number of comparisons).

Example of a Bonferroni correction

Hunter and colleagues (1993) investigated whether vitamin supplementation could reduce the risk of breast cancer. The investigators sent dietary questionnaires

PROBABILITY OF OBSERVING SPECIFIED NUMBER OF SIGNFICANT COMPARISONS		
NUMBER OF SIGNIFICANT COMPARISONS	NO CORRECTION	BONFERRONI
Zero	35.8%	95.1%
One	37.7%	4.8%
Two or more	26.4%	0.1%

Table 22.1. How many significant results will you find in 20 comparisons?

This table assumes you are making 20 comparisons, that all 20 null hypotheses are true, and that α is set to its conventional value of 0.05. If there is no correction for multiple comparisons, there is only a 36% chance of observing no statistically significant findings. With the Bonferroni correction, this probability goes up to 95%.

to over 100,000 nurses in 1980. From the questionnaires, they determined the participants' intake of vitamins A, C, and E and divided the women into quintiles for each vitamin (i.e., the first quintile contains the 20% of the women who consumed the smallest amount). They then followed these women for eight years to determine the incidence rate of breast cancer. Using a test called the chi-square test for trend, the investigators calculated a P value to test the null hypothesis that there is no linear trend between vitamin-intake quintile and the incidence of breast cancer. There would be a linear trend if increasing vitamin intake was associated with increasing (or decreasing) incidence of breast cancer. There would not be a linear trend if (for example) the lowest and highest quintiles had a low incidence of breast cancer compared with the three middle quintiles. The authors determined a different P value for each vitamin. For Vitamin C, P = 0.60; for Vitamin E, P = 0.07; and for Vitamin A, P = 0.001.

Interpreting each P value is easy: if the null hypothesis is true, the P value is the chance that random selection of subjects will result in as large (or larger) a linear trend as was observed in this study. If the null hypothesis is true, there is a 5% chance of randomly selecting subjects such that the trend is statistically significant.

If no correction is made for multiple comparisons, there is a 14% chance of observing one or more significant P values, even if all three null hypotheses are true. The Bonferroni method sets a stricter significance threshold by dividing the significance level (0.05) by the number of comparisons (3), so a difference is declared statistically significant only when its P value is less than 0.050/3, or 0.017. According to this criterion, the relationship between Vitamin A intake and the incidence of breast cancer is statistically significant, but the intakes of Vitamins C and E are not significantly related to the incidence of breast cancer.

The terminology can be confusing. The significance level is still 5%, so α still equals 0.05. But now the significance level applies to the family of comparisons. The lower threshold (0.017) is used to decide whether each particular comparison is statistically significant, but α (now the familywise error rate) remains 0.05.

Example of Bonferroni corrections: Genome-wide association studies

Genome-wide association studies (GWAS) use the Bonferroni correction to cope with a huge number of comparisons. These studies look for associations between diseases (or traits) and genetic variations (usually single nucleotide polymorphisms). They compare the prevalence of many (up to 1 million) genetic variations between many (often tens of thousands) cases and controls, essentially combining about 1 million case-control studies (to be explained in Chapter 28). To correct for multiple comparisons, the threshold is set using the Bonferroni correction. Divide α (0.05) by the number of comparisons (1,000,000) to calculate the threshold, which is 0.00000005 (5×10^{-8}).

A tiny P value (less than is 0.00000005) in a GWAS is evidence that the prevalence of a particular genetic variation differs in the two groups you are studying. One of the groups is composed of people with a disease and the other group is composed of controls, so one explanation is that the disease is associated with that genetic variant. But there is another possible explanation. If the patients and controls are not well matched and have different ancestry (say the patients are largely of Italian ancestry and the controls are largely Jewish), you'd expect genetic differences between the two groups that have nothing to do with the disease being studied (see "The challenge of case-control studies" in Chapter 28).

CORRECTING FOR MULTIPLE COMPARISONS WITH THE FALSE DISCOVERY RATE

The false discovery rate approach is an alternative approach to multiple comparisons that is especially useful when the number of simultaneous comparisons is large (Benjamini & Hochberg, 1995).

Lingo: FDR

This approach does not use the term *statistically significant* but instead uses the term *discovery*. A finding is deemed to be a discovery when its P value is lower than a certain threshold. A discovery is false when the null hypothesis is actually true for that comparison. The false discovery rate is the answer to these two equivalent questions (this definition actually defines the positive FDR, or pFDR, but the distinction between the pFDR and the FDR is subtle and won't be explained in this book):

- If a comparison is classified as a discovery, what is the chance that the null hypothesis is true?
- Of all discoveries, what fraction is expected to be false?

Controlling the FDR

When analyzing a set of P values, you can set the FDR to a desired value, abbreviated Q. If you set Q to 10%, then your goal is for at least 90% of the discoveries to be true and no more than 10% to be false discoveries (for which the null hypothesis is actually true). Of course, you can't know which are which.

A method developed by Benjamini and Hochberg (1995) sets the threshold values for deciding when a P value is low enough to be deemed a discovery. The method actually sets a different threshold value for each comparison. The threshold is tiny for the comparison with the smallest P value and much larger for the comparison with the largest P value. To understand why this makes sense, imagine that you computed 100 P values and all the null hypotheses were true. You'd expect the P values to be randomly distributed between 0.0 and 1.0. It would not be at all surprising for the smallest P value to equal 0.01. But it would be surprising (if all null hypotheses were true) for the median P value to be 0.01. You'd expect that value to be about 0.5. So it makes perfect sense to rank the P values from low to high and use that ranking when choosing the threshold that defines a discovery.

Here is a brief explanation of how those thresholds are determined. If all the null hypotheses were true, you'd expect the P values to be randomly scattered between 0 and 1. Half would be less than 0.50, 10% would be less than 0.10, and so on. Let's imagine that you are making 100 comparisons and you have set Q (the desired FDR) to 5%. If all the null hypotheses were true, you'd expect that the smallest P value would be about 1/100, or 1%. Multiply that value by Q. So you declare the smallest P value to be a discovery if its P value is less than 0.0005. You'd expect the second-smallest P value to be about 2/100, or 0.02. So you'd call that comparison a discovery if its P value is less than 0.0010. The threshold for the third-smallest P value is 0.0015, and so on. The comparison with the largest P value is called a discovery only if its value is less than 0.05. This description is a bit simplified but does provide the general idea behind the method.

Other ways to use the FDR

The section above explained one approach: choose a desired FDR and use that to decide which results count as a discovery. In this approach, there is one FDR for the whole set of comparisons.

An alternative approach is to first decide on a threshold for defining discovery. For example, in a gene-chip assay, you might choose the 5% of genes whose expression changed the most. Given that definition, you would then compute the FDR. Again, there is one FDR for the entire set of comparisons.

A third alternative approach is to compute a distinct FDR for each comparison. For each comparison, define a threshold for discovery so that particular comparison just barely satisfies the definition. Using that definition, compute the overall FDR for all the comparisons. This value is called a q value. Repeat for each comparison. If you make 1,000 comparisons, you'll end up with 1,000 distinct q values.

COMPARING THE TWO METHODS OF CORRECTING FOR MULTIPLE COMPARISONS

Table 22.2 shows the results of many comparisons. You can't create this table with actual data, because the entries in the rows assume that you are Mother Nature and therefore know whether each null hypothesis is actually true. In fact, you never know that, so this table is conceptual.

The top row represents the results of comparisons for which the null hypothesis is in fact true—the treatment really doesn't work. The second row shows the results of comparisons for which there truly is a difference. The first column tabulates comparisons for which the P value was low enough to be deemed statistically significant (or a discovery, in the lingo of the FDR method discussed earlier). The second column tabulates comparisons for which the P value was high enough to be deemed not statistically significant (or not a discovery).

It would be nice if all comparisons ended up in Cells B or C, leaving A and D empty. This will rarely be the case. Even if the null hypothesis is true, random

	DECISION: STATISTICALLY SIGNIFICANT, OR DISCOVERY	DECISION: NOT STATISTICALLY SIGNIFICANT, OR NOT A DISCOVERY	TOTAL
Null hypothesis: True	A (false positive)	B	A + B
Null hypothesis: False	C	D (false negative)	C + D
Total	A + C	B + D	A + B + C + D

Table 22.2. This table (identical to Table 18.2) shows the results of many statistical analyses, each analyzed to reach a decision to reject or not reject the null hypothesis.

The top row tabulates results for experiments for which the null hypothesis is really true. The second row tabulates experiments for which the null hypothesis is not true. When you analyze data, you don't know whether the null hypothesis is true or not, so you could never create this table from an actual series of experiments. A, B, C, and D are integers (not proportions) that count the number of analyses.

sampling will ensure that some comparisons will mistakenly yield a statistically significant conclusion and contribute to Cell A. And even if the null hypothesis is false, random sampling will ensure that some results will be not statistically significant and will contribute to Cell D.

A, B, C, and D each represent a number of comparisons, so the sum of A + B + C + D equals the total number of comparisons you are making.

If you make no correction for multiple comparisons

What happens if you make no correction for multiple comparisons and set α to its conventional value of 5%? Of all experiments done when the null hypothesis is true, you would expect 5% to be statistically significant just by chance. In other words, you would expect the ratio A/(A + B) to equal 5%. This 5% value applies to each comparison separately so is called a per-comparison error rate. In any particular set of comparisons, that ratio might be greater than 5% or less than 5%. But on average, if you make many comparisons, that is the value you'd expect.

If you use the Bonferroni method to correct for multiple comparisons

If you use the Bonferroni method and set α to its conventional value of 5%, then there is a 5% chance of designating one or more comparisons (for which the null hypothesis is really true) as statistically significant. In other words, there is a 5% chance that the value A in Table 22.2 is greater than 0, and a 95% chance that A equals 0.

If you use the false discovery method to correct for multiple comparisons

In Table 22.2, the total number of discoveries equals A + C. If you set Q (the desired FDR) to 5%, then you would expect the ratio A/(A + C) to be no higher than 5% and the ratio C/(A + C) to exceed 95%.

Table 22.3 compares the three methods for dealing with multiple comparisons.

APPROACH	WHAT YOU CONTROL	FROM TABLE 22.1
No correction for multiple comparisons	α = if all null hypotheses are true, the fraction of all experiments for which the conclusion is statistically significant	$\alpha = A/(A + B)$
Bonferroni	α = if all null hypotheses are true, the chance of obtaining one or more statistically significant conclusions	α = probability that A > 0
False discovery rate	Q = the fraction of all discoveries for which the null hypothesis really is true	$Q = A/(A + C)$

Table 22.3. Three approaches to handling multiple comparisons.

Q & A

If you make 10 independent comparisons and all null hypotheses are true, what is the chance that none will be statistically significant?

If you use the usual 5% significance level, the probability that each test will be not statistically significant is 0.95. The chance that all 10 will be not significant is 0.95^{10}, or 59.5%.

Is the definition of a family of comparisons always clear?

No, it is a somewhat arbitrary definition, and different people seeing the same data could apply that definition differently.

If you make only a few planned comparisons, why don't you have to correct for those comparisons?

It makes sense to correct for all comparisons that were made, whether planned or not. But some texts say that if you only plan a few comparisons, you should get rewarded by not having to correct for multiple comparisons. This policy doesn't really make sense to me.

When using the Bonferroni correction, does the variable α refer to the overall familywide significance level (usually 5%) or the threshold used to decide when a particular comparison is statistically significant (usually 0.05/K, where K is the number of comparisons).

α refers to the overall familywide significance level (usually 5%).

Will you know when you make a false discovery?

No. You never know for sure if the null hypothesis is really true or not, so you won't know when a discovery is false. Similarly, you won't know when you make Type I or Type II errors with statistical hypothesis testing.

Are there other ways to deal with multiple comparisons?

Yes. One way is to fit a multilevel hierarchical model, but this requires special expertise and is not commonly used, at least in biology (Gelman, 2012).

CHAPTER SUMMARY

- The multiple comparisons problem is clear. If you make lots of comparisons (and make no special correction for the multiple comparisons), you are likely to find some statistically significant results just by chance.
- Coping with multiple comparisons is one of the biggest challenges in data analysis.

- One way to deal with multiple comparisons is to analyze the data as usual, but to fully report the number of comparisons that were made and then let the reader account for the number of comparisons.
- If you make 13 independent comparisons with all null hypotheses true, there is about a 50:50 chance that one or more P values will be less than 0.05.
- The most common approach to multiple comparisons is to define the significance level to apply to an entire family of comparisons, rather than to each individual comparison.
- A newer approach to multiple comparisons is to control or define the false discovery rate.

TERMS INTRODUCED IN THIS CHAPTER

- Bonferroni correction (p. 187)
- Discovery (p. 189)
- Family of comparisons (p. 187)
- Familywise error rate (p. 186)
- Genome-wide association studies (GWAS) (p. 188)
- Multiple comparisons (p. 183)
- Per-comparison error rate (p. 186)
- Per-experiment error rate (p. 186)
- Planned comparisons (p. 185)

❖

CHAPTER 23

The Ubiquity of Multiple Comparisons

If the fishing expedition catches a boot, the fishermen should throw it back, and not claim that they were fishing for boots.

JAMES L. MILLS

Chapter 22 explained the problem of multiple comparisons. This chapter demonstrates how pervasive this problem is in many kinds of data analysis.

OVERVIEW

Berry (2007, p. 155) concisely summarized the importance and ubiquity of multiple comparisons:

> Most scientists are oblivious to the problems of multiplicities. Yet they are everywhere. In one or more of its forms, multiplicities are present in every statistical application. They may be out in the open or hidden. And even if they are out in the open, recognizing them is but the first step in a difficult process of inference. Problems of multiplicities are the most difficult that we statisticians face. They threaten the validity of every statistical conclusion.

This chapter points out some of the ways in which the problem of multiple comparisons affects the interpretation of many statistical results.

MULTIPLE COMPARISONS IN MANY CONTEXTS

Multiple subgroups

Analyzing data divided into multiple subgroups is a common form of multiple comparisons. A simulated study by Lee and colleagues (1980) points out potential problems. They pretended to compare survival following two "treatments" for coronary artery disease. They studied a group of real patients with coronary artery disease randomly divided them into two groups, A and B. In a real study, they would give the two groups different treatments and compare survival after treatment. In this simulated study, they treated all subjects identically but analyzed the data as if the two random groups were actually given two distinct treatments. As expected, the survival of the two groups was indistinguishable.

They then divided the patients in each group into six subgroups depending on whether they had disease in one, two, or three coronary arteries and whether the heart ventricle contracted normally. Because these are variables that are expected to affect survival of the patients, it made sense to evaluate the response to "treatment" separately in each of the six subgroups. Whereas they found no substantial difference in five of the subgroups, they found a striking result among patients with three-vessel disease who also had impaired ventricular contraction. Of these patients, those assigned to Treatment B had much better survival than those assigned to Treatment A. The difference between the two survival curves was statistically significant, with a P value less than 0.025.

If this were an actual study comparing two alternative treatments, it would be tempting to conclude that Treatment B is superior for the sickest patients, and to recommend Treatment B to those patients in the future. But in this study, the random assignment to Treatment A or Treatment B did not alter how the patients were actually treated. Because the two sets of patients were treated identically, we can be absolutely, positively sure that the survival difference was a coincidence.

It is not surprising that the authors found one low P value out of six comparisons. Figure 22.1 shows that there is a 26% chance that at least one of six independent comparisons will have a P value of less than 0.05, even if all null hypotheses are true.

If the difference between treatments had not been statistically significant in any of the six subgroups used in this example, the investigators probably would have kept looking. Perhaps they would have gone on to separately compare Treatments A and B only in patients who had previously had a myocardial infarction (heart attack), only in those with abnormal electrocardiogram findings, or only those who had previous coronary artery surgery. Since you don't know how many comparisons were possible, you cannot rigorously account for the multiple comparisons.

For analyses of subgroups to be useful, it is essential that the study design specify all subgroups analyses, including methods to be used to account for multiple comparisons.

Multiple sample sizes

If you perform an experiment and the results aren't quite statistically significant, it is tempting to run the experiment a few more times (or add a few more subjects) and then analyze the data again with the larger sample size. If the results still aren't significant, then you might do the experiment a few more times (or add more subjects) and reanalyze once again. The problem is you can't really interpret the results when you informally adjust sample size as you go.

If the null hypothesis of no difference is in fact true, the chance of obtaining a statistically significant result using that informal sequential approach is far higher than 5%. In fact, if you carry on that approach long enough, then every single experiment will eventually reach a statistically significant conclusion, even if the null hypothesis is true. Of course, "long enough" might be very long indeed, exceeding your budget or even your life span.

The problem with this approach is that you continue collecting data only when the result is not statistically significant and stop when the result is statistically significant. If the experiment were continued after reaching significance, adding more data might then result in a conclusion that the results are not statistically significant. But you'd never know this, because you would have stopped once significance was reached. If you keep running the experiment when you don't like the results but stop the experiment when you do like the results, you can't interpret the results at face value.

Statisticians have developed rigorous ways to allow you to dynamically adjust sample size. Look up *sequential data analysis*. These methods use more stringent criteria to define significance in order to make up for the multiple comparisons. Without these special methods, you can't interpret the results unless the sample size is set in advance.

Multiple end points and misreporting which end point is "primary"

When conducting clinical studies, investigators often collect data on multiple clinical outcomes. When planning the study, investigators should specify the outcome they care most about, called the *primary outcome*. That is the outcome that they most care about. The other outcomes are called *secondary outcomes*. By now you should realize why it is so important that the primary outcome be carefully defined in advance. If many outcomes are measured, you should expect some to show statistically significant differences by chance. The analysis of secondary outcomes can strengthen the scientific argument of a paper and lead to new hypotheses to study. But statistical analyses must be focused on the primary outcome chosen in advance.

Every once in a while, it can make sense to change the definition of the primary outcome between study design and publication in response to a new understanding of the disease being studied. But this kind of change is appropriate only when fully documented in the paper and when the decision is made by people who have not yet seen any of the study data (Evans, 2007). It is cheating to first look at the data for all the outcomes, identify the outcome that has the biggest effect or the smallest P value, and then label that outcome as "primary."

Chan and colleagues (2004) showed that some investigators do misreport their choice of primary outcome. The researchers identified all protocols for randomized clinical trials approved in two cities over two years. A decade later, they looked at how the results of the studies were reported. In half the studies, they found that the primary outcome specified in the published paper was not the primary outcome defined in the original protocol. The investigators had switched which outcome they deemed to be primary.

Vedula and colleagues (2009) found the same problem when they reviewed 20 protocols for clinical trials to test a particular drug. Only 12 of these trials were published (read about publication bias in Chapter 43). The protocols for these

12 trials included 21 primary outcomes. Only half were included (and labeled properly) in the published paper. Of the 28 primary outcomes reported in the published papers, 12 were not defined as primary outcomes in the study protocol. Vera-Badillo and colleagues (2013) found similar problems with the reporting of randomized clinical trials of various treatments of breast cancer. They identified 30 studies that had posted the experimental protocol in advance. In seven of those studies, the primary end point reported in the publication differed from the primary end point planned in advance. They also found that when the results for the primary end point were not statistically significant, over half of the studies used "spin" to misrepresent the results, and many didn't even mention the primary outcome in the concluding sentence of the abstract.

You can't draw proper inferences from clinical studies unless you can be sure that the investigators chose the primary outcome as part of the study plan and stuck with that decision when writing up the results.

Multiple geographic areas

Five children in a particular school developed leukemia last year. Is that a coincidence? Or does the clustering of cases suggest the presence of an environmental toxin that caused the disease? These questions are very difficult to answer (Thun & Sinks, 2004). It is tempting to estimate the answer to a different question, What is the probability that five children in this particular school would all get leukemia this year? You could calculate (or at least estimate) the answer to that question if you knew the overall incidence rates of leukemia among children and the number of children enrolled in the school. The probability would be tiny. Everyone intuitively knows that and so is alarmed by the cluster of cases.

But you've asked the wrong question once you've already observed the cluster of cases. The school only came to your attention this year because of the cluster of cases, so you must consider all other schools and other years. The right question is, What is the probability that five children in any school would develop leukemia in the same year? Clearly the answer to this question is much higher than the answer to the previous one. To calculate a precise answer, you would have to decide which geographical areas to include, which years to include, and which diagnoses to include.

About 1,000 cancer clusters are reported to U.S. health officials each year. About three-quarters of these are not really clusters of similar kinds of cancers, but that still leaves several hundred cancer clusters each year. These are fully investigated to look for known toxins and to be alert to other findings that might suggest a common cause. But most *disease clusters* turn out to be coincidences. It is surprising to find a cluster of one particular disease in one particular place at any one particular time. But chance alone will result in many clusters of various diseases in various places at various times.

Multiple predictions or hypotheses

In 2000, the Intergovernmental Panel on Climate Change made predictions about future climate conditions. Pielke (2008) asked what seemed like a straightforward question: How accurate were those predictions over the next seven years? That's not long enough to seriously assess predictions of global warming, but it is a necessary first step. Answering this question proved impossible (Tierney, 2008). The problems were that the report contained numerous predictions and didn't specify which sources of climate data should be used. Did the predictions come true? The answer depends on the choice of which prediction to test and which data set to use for testing—"a feast for cherry pickers" (Pielke, 2008).

You can only evaluate the accuracy of a prediction or diagnosis when the timing, methodology, and data source are precisely defined. If there are multiple ways to assess the accuracy of predictions, you may get contradictory results.

It is also essential to record your predictions and not rely on memory alone. Psychologists have repeatedly demonstrated a phenomenon they call *hindsight bias* (Goodwin, 2010). People do a really bad job of remembering their predictions after the event occurs. Here's an example. College students were asked to state what they thought the chances were of a "not guilty" verdict in the O. J. Simpson murder trial (Bryant & Brockway, 1997). The trial was heavily reported in the media, and almost everyone had an opinion. A week after Simpson was acquitted by the jury, the students were asked to recall their prediction of the probability of acquittal. These recollected probabilities were much higher than the recorded predictions, especially among people who originally predicted a small chance of acquittal. This effect is also called the *I-knew-it-all-along effect*. People tend to change their memory of their own predictions to match what actually ended up happening.

Kerr (1998) defined the acronym *HARK*, which means *hypothesizing after the results are known*. This occurs when a scientist first looks at the data which includes many variables or subgroups, uses those data to pose an hypothesis, and then publishes the results so it appears that the hypothesis was stated before the data collection began. This can be viewed as another form of multiple comparisons, as complex data sets allow you to pose many hypotheses. If you search through the data to find a difference or association, that finding is likely to be a coincidence and so will not hold up in repeat experiments.

Multiple definitions of groups
When comparing two groups, the groups must be defined as part of the study design. If the groups are defined by the data, many comparisons are being made implicitly and the results cannot be interpreted.

Austin and Goldwasser (2008) demonstrated this problem. They looked at the incidence of hospitalization for heart failure in Ontario, Canada, in 12 groups of patients defined by their astrological sign (determined by their birthday). People born under the sign of Pisces happened to have the highest incidence of heart failure. After seeing this association, the investigators did a simple statistics test to compare the incidence of heart failure among people born under Pisces with the incidence of heart failure among all others (people born under all other 11 signs combined into one group). Taken at face value, this comparison showed that the people born under Pisces have a statistically significant higher incidence of heart failure than do people born under the other 11 signs (the P value was 0.026).

The problem is that the investigators didn't really test a single hypothesis; they implicitly tested 12. They only focused on Pisces *after* looking at the incidence of heart failure for people born under all 12 astrological signs. So it isn't fair to compare that one group against the others without considering the other 11 implicit comparisons.

Let's do the math. If you do 12 independent comparisons, what is the chance that every P value will be greater than 0.026? Since the probabilities of independent comparisons are independent of each other, we can multiply them. The probability of getting a P value greater than 0.026 in one comparison is 1.0 − 0.026, or 0.974. The probability of two P values both being greater than 0.026 in two independent comparisons is 0.974 × 0.974 = 0.949. The chance that all 12 independent comparisons would have P values greater than 0.026 is 0.974^{12} = 0.729. Subtract that number from 1.00 to find the probability that at least one P value of 12 will be less than 0.026. The answer is 0.271. In other words, there is a 27.1% chance that one or more of the 12 comparisons will have a P value less than or equal to 0.026 just by chance, even if all 12 null hypotheses are true. Once you realize this, the association between astrological sign and heart failure does not seem impressive.

Multiple ways to select variables in regression

Multiple regression (as well as logistic and proportional hazards regression) models predict an outcome as a function of multiple independent (input) variables. Some programs offer automated methods to choose which independent variables should be included in the model and which should be omitted. Chapters 37 and 38 will discuss these techniques.

Freedman (1983) did simulations to show how these methods can lead to misleading results. His paper is reprinted as an appendix in a text by Good and Hardin (2006). He simulated a study with 100 subjects using data from 50 independent variables recorded for each subject. The simulations were performed so all the variation was random. The simulated outcome had no relationship at all to any of the simulated inputs. In other words, the data were all noise and no signal.

A multiple regression model (see Chapter 37) was used to find the equation that best predicted the outcome from all the inputs, and a P value was computed for each input variable. Each P value tested a null hypothesis that a particular input variable had no impact on predicting the outcome. The results were what you would expect: the P values were randomly spread between 0 and 1. An overall P value, testing the null hypothesis that the overall model was not helpful in predicting the outcome, was high (0.53).

No surprises so far. Random variables went in, and a random assortment of P values came out. Anyone looking at the conclusion (and not knowing the data were randomly generated) would conclude that the 50 input variables have no ability to predict the outcome.

Freedman then selected the 15 input variables that happened to have the lowest P values (less than 0.25) and refit the multiple regression model using only those 15 input variables, rather than all 50 as before. Now the overall P value was tiny (0.0005). Of the 15 variables now in the model, 6 had an associated P value smaller than 0.05.

Anyone who saw only this analysis and didn't know the 15 input variables were selected from a larger group of 50 would conclude that the outcome variable is related to, so can be predicted from, the 15 input variables,. Because these are simulated data, we know that this conclusion would be wrong. There is in fact no relationship between the input variables and the outcome variable.

By including so many variables and selecting a subset that happened to be predictive, the investigators were performing multiple comparisons. Freedman knew this, and that is the entire point of his paper. But the use of this kind of variable selection happens often in analyzing large studies. When multiple regression analyses include lots of opportunity for variable selection, it is easy to be misled by the results, especially if the investigator doesn't explain exactly what he has done.

Multiple ways to preprocess the data

As mentioned in Chapter 7, it is very common for scientists not to analyze the direct experimental measurements, but instead to first process the data by transforming, normalizing, smoothing, adjusting (for other variables), or removing outliers (see Chapter 25). These kinds of manipulations can make lots of sense when done in a careful, preplanned way. However, you engage in multiple comparisons if you repeat the statistical analyses on data processed in multiple ways and then report analyses on the form that gives you the results for which you are hoping.

Multiple ways to analyze the data

This story told by Vickers (2006) is only a slight exaggeration:

> STATISTICIAN: Oh, so you have already calculated the P value?
> SURGEON: YES, I USED MULTINOMIAL LOGISTIC REGRESSION.
> STATISTICIAN: REALLY? HOW DID YOU COME UP WITH THAT?
> SURGEON: Well, I tried each analysis on the SPSS drop-down menus, and that was the one that gave the smallest P value.

If you analyze your data many ways—perhaps first with a t test, then with a nonparametric Mann-Whitney test, and then with a two-way ANOVA (adjusting for another variable)—you are performing multiple comparisons.

BE SKEPTICAL OF RESULTS OBTAINED BY DATA TORTURE OR P-HACKING

The results of statistical tests can only be interpreted at face value if the experimental design and analysis methods were decided in advance.

This chapter lists the many ways that investigators can try to wring statistical significance out of data. Mills (1993) calls this *data torture*. Simmons and colleagues (2012) call this *p-hacking* and lament *researcher degrees of freedom* (Simmons, Nelson, & Simonsohn, 2011).

Simmons and colleagues (2011) simulated a set of experiments that demonstrate how allowing researcher degrees of freedom can increase the chance of a false positive result. The simulation is for a simple situation in which the investigator compares two groups of subjects (n = 20 per group, each group being a mixture of men and women). The investigator first tests whether the mean value of a variable differs significantly between the two groups. If not, he tests a second variable, and if that doesn't "work," he computes the mean of the two variables and tests whether that mean differs between groups. If that doesn't yield a "statistically significant" result, he collects data from 10 more subjects in each group to increase the sample size from 20 to 30 per group and reanalyzes the data. If even that doesn't produce a "statistically significant" result, he runs a more sophisticated analysis that accounts for differences between men and women. This set of steps doesn't seem very different from what lots of scientists do routinely.

The simulated data were all sampled from a single Gaussian distribution, so we can be absolutely sure that there is no difference between the two populations and that the null hypothesis is true. In this situation, using the standard definition of statistical significance, the investigator should reach an incorrect conclusion that the difference is statistically significant in 5% of the simulations. In fact, they found a "statistically significant" (P < 0.05) result in 31% of the simulated experiments! The use of researcher degrees of freedom as described above (also called p-hacking) increased the chance of a false positive (Type I) error six-fold, from 5% to 31%. It would be easy to add even more researcher degrees of freedom by looking only at subgroups, removing outliers, analyzing logarithms, switching to nonparametric tests, and so on. These methods would increase the false positive rate even further.

If investigators use even a few researcher degrees of freedom (i.e., allow themselves to p-hack), the chance of finding a bogus (false positive) "statistically significant" result increases. Statistical results can only be interpreted at face value if the analyses were planned in advance and the investigators don't try alternative analyses once the data are collected. When the number of possible analyses is not defined in advance (and is almost unlimited), results simply cannot be trusted.

HOW TO COPE WITH MULTIPLE COMPARISONS

This chapter shows you many ways in which multiple comparisons can impede progress in science. The best way to deal with multiple comparisons is to plan studies with focused hypotheses and defined methods to account for multiple comparisons in analyses. When evaluating results that include multiple comparisons, your best defense is the Bonferroni method discussed in Chapter 22.

When reviewing completed studies, you can correct for multiple comparisons if you know how many comparisons were made. Simply divide the usual threshold that defines statistical significance (often 0.05) by the number of comparisons. So if the study is making 100 comparisons, divide 0.05 by 100 and only declare a result to be statistically significant (at the 5% significance level with correction for multiple comparisons) when the P value is less than 0.0005.

CHAPTER SUMMARY

- The problem of multiple comparisons shows up in many situations:
  - Multiple end points
  - Multiple time points
  - Multiple subgroups
  - Multiple geographical areas
  - Multiple predictions
  - Multiple geographical groups
  - Multiple ways to select variables in multiple regression
  - Multiple methods of preprocessing data
  - Multiple statistical analyses
- You can only interpret statistical results when you know how many comparisons were made.
- All analyses should be planned, and all planned analyses should be conducted and reported. These simple guidelines are often violated.

TERMS INTRODUCED IN THIS CHAPTER

- Data torture (p. 201)
- Disease cluster (p. 197)
- Hindsight bias (p. 198)
- Hypothesizing after the results are known (HARK) (p. 198)
- I-knew-it-all-along effect (p. 198)
- P-hacking (p. 201)
- Primary outcome (p. 196)
- Researcher degrees of freedom (p. 201)
- Secondary outcome (p. 196)

CHAPTER 24

⬧⬧⬧

Normality Tests

You need the subjunctive to explain statistics.

DAVID COLQUHOUN

Many statistical tests (including t tests, ANOVA, and regression) assume that the data were sampled from a Gaussian distribution (Chapter 10). When you analyze data or review analyses that others have done, you need to ask whether that assumption is reasonable. This sounds like a simple question, but it really isn't.

THE GAUSSIAN DISTRIBUTION IS AN UNREACHABLE IDEAL

In almost all cases, we can be 100% sure that the data were not sampled from an ideal Gaussian distribution. Why? Because an ideal Gaussian distribution extends infinitely in both directions to include both superlarge values and extremely negative values. In most scientific situations, there are constraints on the possible values. Blood pressures, concentrations, weights, and many other variables cannot have negative values, so they can never be sampled from perfect Gaussian distributions. Other variables can be negative but have physical or physiological limits that don't allow extremely large values (or extremely low negative values). These variables also cannot follow a perfect Gaussian distribution. Micceri (1989) looked at 440 different variables that psychologists measure and concluded that every one deviated substantially from a Gaussian distribution. I suspect the same is true for most scientific fields.

Because almost no variables you measure follow an ideal Gaussian distribution, why use tests that rely on the Gaussian assumption? Plenty of studies with simulated data have shown that the statistical tests based on the Gaussian distribution are useful when data are sampled from a distribution that only approximates a Gaussian distribution. These tests are fairly robust to violations of the Gaussian assumption, especially if the sample sizes are large and equal. So the question that matters is not whether the data were sampled from an ideal Gaussian population, but rather whether the distribution from which they were sampled is close enough to the Gaussian ideal that the results of the statistical tests are still useful.

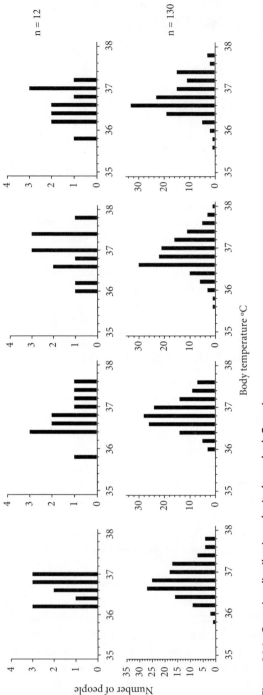

Figure 24.1. Gaussian distributions don't always look Gaussian.

These graphs show frequency distributions of values randomly chosen from a Gaussian distribution. The top four graphs are for samples with 12 values. The mean is 36.77 and the SD is 0.40. The bottom four graphs have a mean of 36.82 and an SD of 0.41, and n = 130. The variation from graph to graph (sample to sample) is simply the result of random variation when sampling from a Gaussian distribution. You rarely see smooth bell-shaped curves unless the sample size is enormous.

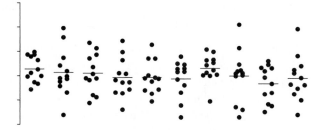

Figure 24.2. Samples from Gaussian distributions don't always look Gaussian.

All 10 samples were randomly sampled from Gaussian distributions. It is too easy to be fooled by random variation and think that the data are far from Gaussian.

WHAT A GAUSSIAN DISTRIBUTION REALLY LOOKS LIKE

The ideal Gaussian frequency distribution was shown in Figure 10.1. With huge data sets, this is what you expect a Gaussian distribution to look like. But what about smaller data sets?

Figures 24.1 shows simulated data. Each of the eight frequency distributions shows the distribution of values randomly chosen from a Gaussian distribution. The top four distributions are four different samples with 12 values in each sample, and the bottom four graphs show samples with 130 values. Because of random sampling variation, none of these frequency distributions really looks completely bell shaped and symmetrical.

The bell-shaped Gaussian distribution is an ideal distribution of the population. Unless the samples are huge, actual frequency distributions tend to be less symmetrical and more jagged. This is just the nature of random sampling.

Figure 24.2 makes the same point by showing individual values rather than frequency distributions. Each sample was drawn from a Gaussian distribution. To many people, however, some of these distributions don't look Gaussian. Some people prefer to plot the cumulative frequency distributions, perhaps transforming the axes so that a Gaussian distribution becomes linear (called a QQ plot). But replotting the data doesn't change the issue. Unless the sample size is huge, random sampling can create distributions that don't really look Gaussian. It takes a lot of experience to interpret such plots.

TESTING FOR NORMALITY

Statistical tests can be used to quantify how much a data set deviates from the expectations of a Gaussian distribution. Such tests are called *normality tests*.

How normality tests work

The first step in running a normality test is to quantify how far a set of values differs from the predictions of a Gaussian distribution. One of the more popular normality tests is the *D'Agostino–Pearson omnibus K2 normality test.* This test first computes two values that quantify how far the distribution deviates from the Gaussian ideal:

- *Skewness* quantifies symmetry. A symmetrical distribution (such as the Gaussian) has a skewness of 0. An asymmetrical distribution with a heavy right tail (more large values) has a positive distribution, and one with a heavy left tail (more small numbers) has a negative skew.
- *Kurtosis* quantifies how peaked the distribution is. A Gaussian distribution is the ideal and has a kurtosis of 0.0. Distributions with a sharper peak and thinner tails have a positive kurtosis. Distributions with a rounded peak and wide shoulders have a negative kurtosis.

The D'Agostino–Pearson omnibus K2 normality test combines the skewness and kurtosis into a single value that describes how far a distribution is from Gaussian. Other normality tests (such as the Shapiro–Wilk test, the Kolmogorov–Smirnov test, and the Darling–Anderson test) use other approaches to quantify the discrepancy between the observed distribution and the Gaussian distribution.

The meaning of a P value from a normality test

All normality tests compute a P value that answers the following question: If you randomly sample from a Gaussian population, what is the probability of obtaining a sample that deviates from a Gaussian distribution as much as (or more so than) this sample does?

Interpreting a high P value

If the P value from a normality test is large, all you can say is that the data are not inconsistent with a Gaussian distribution. Ugh! Statistics requires the use of double negatives.

A normality test cannot prove the data were sampled from a Gaussian distribution. All the normality test can do is demonstrate that the deviation from the Gaussian ideal is not more than you'd expect to see from chance alone.

Interpreting a small P value

If the P value from the normality test is small, you reject that null hypothesis and so accept the alternative hypothesis that the data are not sampled from a Gaussian population. But this is less useful than it sounds. As explained at the beginning of this chapter, few variables measured in scientific studies are completely Gaussian. So with large sample sizes, normality tests will almost always report a small P value, even if the distribution only mildly deviates from a Gaussian distribution.

ALTERNATIVES TO ASSUMING
A GAUSSIAN DISTRIBUTION

If you don't wish to assume that your data were sampled from a Gaussian distribution, you have several choices:

- Identify another distribution from which the data were sampled. In many cases, you can then transform your values to create a Gaussian distribution.

Most commonly, if the data come from a lognormal distribution (see Chapter 11), you can transform all values to their logarithms.

- Ignore small departures from the Gaussian ideal. Statistical tests tend to be quite robust to mild violations of the Gaussian assumption.
- Identify and remove outliers (see Chapter 25).
- Switch to a nonparametric test that doesn't assume a Gaussian distribution (see Chapter 41).

Don't make the mistake of jumping directly to the fourth option, using a nonparametric test. It is very hard to decide when to use a statistical test based on a Gaussian distribution and when to use a nonparametric test. This decision really *is* difficult, requiring thinking and perspective and consistency. For that reason, the decision should not be automated.

COMMON MISTAKES: NORMALITY TESTS

Mistake: Using a normality test to automate the decision of when to use a nonparametric test

Nonparametric tests (explained in Chapter 41) do not assume a Gaussian distribution, so it seems logical to use the results of a normality test to decide whether or not to use a nonparametric test. In fact, the decision of when to use a nonparametric test is not straightforward, and reasonable people can disagree about when to use them. The results of a normality test are not very helpful in making that decision. The problem is that a normality test is designed to detect evidence that the distribution deviates from a Gaussian distribution, but it does not quantify whether that deviation is large enough to invalidate the usual tests that assume a Gaussian distribution. I'll delve into that subject in more detail in Chapter 41.

Mistake: Using a normality test with tiny samples

Normality tests try to deternube the distribution (population) from which the data were sampled. That requires data. With tiny samples (three or four values) it really is impossible to make useful inferences about the distribution of the population.

Q & A

Does it make sense to ask whether a particular data set is Gaussian?

No. A common misconception is to think that the normality test asks whether a particular data set is Gaussian. But the term *Gaussian* refers to an entire population. It only makes sense to ask about the population from which your data were sampled. Normality tests ask whether the data are consistent with the assumption of sampling from a Gaussian distribution.

Is it ever possible to know for sure whether a particular data set was sampled from a Gaussian distribution?

Not unless the data were simulated. Otherwise, you can never know for sure the distribution from which the data were sampled.

Do normality tests determine whether the data are parametric or nonparametric?

No! The terms *parametric* and *nonparametric* refer to statistical tests, not data distributions.

Should a normality test be run as part of every experiment?

Not necessarily. You want to know whether a certain kind of data are consistent with sampling from a Gaussian distribution. The best way to find out is to run a special experiment just to ask about the distribution of data collected using a particular method. This experiment would need to generate plenty of data points but would not have to make any comparisons or ask any scientific questions. If analysis of many data points convinces you that a particular experimental protocol generates data that are consistent with a Gaussian distribution, there is no point in testing smaller data sets from individual runs of that experiment.

Will every normality test give the same result?

No. The different tests quantify deviations from normality in different ways, so they will return different results.

How many values are required to run a normality test?

It depends on which test you choose. Some work with as few as five values.

Does a normality test decide whether the data are far enough from Gaussian to require a nonparametric statistical test?

No. It is hard to define what "far enough" means, and the normality tests were not designed with this aim in mind.

How useful are normality tests?

Not very. It would be very rare for any set of scientific data to truly be sampled from an ideal Gaussian distribution. With small samples, the normality tests don't have much power to detect non-Gaussian distributions. With large samples, the tests have too much power, since they can detect minor deviations from the ideal that wouldn't affect the validity of a t test or ANOVA.

CHAPTER SUMMARY

- The Gaussian distribution is an ideal that is rarely achieved. Few, if any, scientific variables completely follow a Gaussian distribution.
- Data sampled from Gaussian distributions don't look as Gaussian as many expect. Random sampling is more random than many appreciate.
- Normality tests are used to test for deviations from the Gaussian ideal.
- A small P value from a normality test only tells you that the deviations from the Gaussian ideal are more than you'd expect to see by chance. It tells you nothing about whether the deviations from the Gaussian ideal are large enough to affect conclusions from statistical tests that assume sampling from a Gaussian distribution.

TERMS INTRODUCED IN THIS CHAPTER

- D'Agostino–Pearson omnibus K2 normality test (p. 206)
- Kurtosis (p. 206)
- Normality test (p. 205)
- Skewness (p. 206)

CHAPTER 25

✦✦✦

Outliers

There are liars, outliers, and out-and-out liars.

ROBERT DAWSON

An outlier is a value that is so far from the others that it appears to have come from a different population. More informally, an outlier is a data point that is too extreme to fit your preconceptions. The presence of outliers can invalidate many statistical analyses. This chapter explains the challenges of identifying outliers.

HOW DO OUTLIERS ARISE?

Outliers—also called anomalous, spurious, rogue, wild, or contaminated observations—can occur for several reasons.

- Invalid data entry. The outlier may simply be the result of transposed digits or a shifted decimal point. If you suspect an outlier, the first thing to do is to make sure the data were entered correctly and that any calculations (changing units, normalizing, etc.) were done accurately.
- Biological diversity. If each value comes from a different person or animal, the outlier may be a correct value. It is an outlier because that individual may truly be different from the others. This may be the most exciting finding in your data!
- Random chance. In any distribution, some values are far from the others by chance.
- Experimental mistakes. Most experiments have many steps, and it is possible that a mistake was made.
- Invalid assumption. If you assume data are sampled from a Gaussian distribution, you may conclude that a large value is an outlier. But if the distribution is, in fact, lognormal, then large values are common and are not outliers. See Chapter 11.

THE NEED FOR OUTLIER TESTS

The presence of an outlier can spoil many analyses, either creating the appearance of differences (or associations, correlations, etc.) or blocking discovery of real differences (or associations, correlations, etc.).

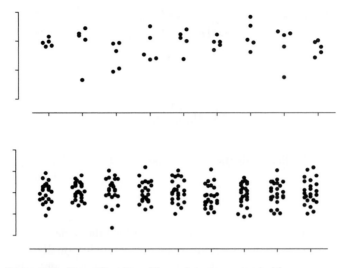

Figure 25.1. No outliers here. These data were sampled from Gaussian distributions.

All of these data sets were computer generated and sampled from a Gaussian distribution. But when you look at them, some points just seem too far from the rest to be part of the same distribution. They seem like real outliers—but they are not. The human brain is very good at seeing patterns and exceptions from patterns, but it is poor at recognizing random scatter.

It would seem that the presence of outliers would be obvious. If this were the case, we could deal with outliers informally. But identifying outliers is much harder than it seems.

Figure 25.1 shows the problem with attempting to identify outliers informally. It shows 18 data sets all sampled from a Gaussian distribution. Half of the samples have five values and half have 24 values. When you look at the graph, some points just seem to be too far from the rest. It seems obvious they are outliers. But in fact, all of these values were sampled from the same Gaussian distribution.

One problem with ad hoc removal of outliers is our tendency to see too many outliers. Another problem is that the experimenter is almost always biased. Even if you try to be fair and objective, your decision about which outliers to remove will probably be influenced by the results you want to see.

FIVE QUESTIONS TO ASK BEFORE
TESTING FOR OUTLIERS

Before using an outlier test, ask yourself these five questions:

- Was there a mistake in data entry? If so, fix it. An outlier may simply be the result of transposed digits or a shifted decimal point.

- Is the extreme value really a code for missing values? In some programs, you simply leave a blank spot when a value is missing. In other programs, you might enter a value like 999. If you entered the wrong code for a missing value (or configured the program incorrectly), that code might appear to be an outlier.
- Did you notice a problem during the experiment? If so, don't bother with outlier tests. Eliminate values if you noticed a problem with a value during the experiment.
- Could the extreme values be a result of biological variability? If each value comes from a different person or animal, the outlier may be a correct value. It is an outlier not because of an experimental mistake, but because that individual may be different from the others. This may be the most exciting finding in your data!
- Is it possible the distribution is not Gaussian? If so, it may be possible to transform the values to make them Gaussian. Most outlier tests assume that the data (except the potential outliers) come from a Gaussian distribution.

OUTLIER TESTS

The question an outlier test answers

If you've answered no to all five questions above, two possibilities remain:

- The extreme value came from the same distribution as the other values and just happened to be larger (or smaller) than the rest. In this case, the value should not be treated specially.
- The extreme value was the result of a mistake. This could be something like bad pipetting, a voltage spike, or holes in filters. Or it could be a mistake in recording a value. These kinds of mistakes can happen and are not always noticed during data collection. Because including an erroneous value in your analyses will give invalid results, you should remove it. In other words, the value comes from a different population than the other values and is misleading.

The problem, of course, is that you can never be sure which of these possibilities is correct. Mistake or chance? No mathematical calculation can tell you for sure whether the outlier came from the same or a different population than the others. An outlier test, however, can answer this question: If the values really were all sampled from a Gaussian distribution, what is the chance one value would be as far from the others as the extreme value you observed?

Interpreting a small P value

If this P value is small, conclude that the outlier is not from the same distribution as the other values. Assuming you answered no to all five questions listed above, you have justification to exclude it from your analyses.

Interpreting a high P value

If the P value is high, you have no evidence that the extreme value came from a different distribution than the rest. This does not prove that the value came from the same distribution as the others. All you can say is that there is no strong evidence that the value came from a different distribution.

How do outlier tests work?

Statisticians have devised several methods for detecting outliers. All of the methods first quantify how far the outlier is from the other values. This distance can be the difference between the extreme value and the mean of all values, the difference between the extreme value and the mean of the remaining values, or the difference between the extreme value and the next closest value. This value is then standardized by dividing by some measure of variability, such as the SD of all values, the SD of the remaining values, the distance to the closest value, or the range of the data. To determine if the extreme value can be considered a statistically significant outlier, the calculated ratio is compared with a table of critical values.

One of the most popular outlier tests is the *Grubbs outlier test* (also called the *extreme studentized deviate test*). This test divides the difference between the extreme value and the mean of all values by the SD of all values.

IS IT LEGITIMATE TO REMOVE OUTLIERS?

Some people feel that removing outliers is cheating. It can be viewed that way when outliers are removed in an ad hoc manner, especially when you remove only those outliers that get in the way of obtaining results you like. But leaving outliers in the data you analyze can also be considered cheating, because it can lead to invalid results.

It is not cheating when the decision of whether to remove an outlier is based on rules and methods established before the data were collected and these rules (and the number of outliers removed) are reported when the data are published.

When your experiment has a value flagged as an outlier, it is possible that a coincidence occurred, the kind of coincidence that happens in 5% (or whatever level you pick) of experiments even if the entire scatter is Gaussian. It is also possible that the value is a "bad" point. Which possibility is more likely? The answer depends on your experimental system.

- If your experimental system generates one or more bad points in a small percentage of experiments, eliminate the value as an outlier. It is more likely to be the result of an experimental mistake than to come from the same distribution as the other points.
- If your system is very pure and controlled, such that bad values almost never occur, keep the value. It is more likely that the value comes from the same distribution as the other values than that it represents an experimental mistake.

AN ALTERNATIVE: ROBUST STATISTICAL TESTS

Rather than eliminate outliers, an alternative strategy is to use statistical methods that are designed to ensure that outliers have little effect on the results. Methods of data analysis that are not much affected by the presence of outliers are called *robust*. You don't need to decide when to eliminate an outlier, because the method is designed to accommodate them. Outliers just automatically fade away.

The simplest robust statistic is the median. If one value is very high or very low, the value of the median won't change, whereas the value of the mean will change a lot. The median is robust; the mean is not.

To learn more about robust statistics, start with the book by Huber (2003).

COMMON MISTAKES: OUTLIER TESTS

Mistake: Not realizing that outliers are common in data sampled from a lognormal distribution

Most outlier tests are based on the assumption that all data—except the potential outlier(s)—are sampled from a Gaussian distribution. The results are misleading if the data were sampled from some other distribution.

Figure 25.2 shows that lognormal distributions can be especially misleading. These simulated values were all generated from a lognormal distribution. An outlier test, based on the assumption that most of the values were sampled from a Gaussian distribution, identified an outlier in three of the four data sets. But these values are not outliers. Extremely large values are expected in a lognormal distribution.

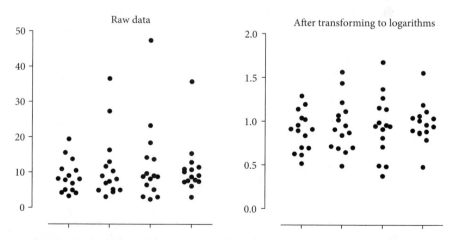

Figure 25.2. No outliers here. These data were sampled from lognormal distributions.
(Left) The four data sets were randomly sampled by a computer from a lognormal distribution. A Grubbs outlier test found a significant (P < 0.05) outlier in three of the four data sets. (Right) Graph of the same values after being transformed to their logarithms. No outliers were found. Most outlier tests are simply inappropriate when values are not sampled from a Gaussian distribution.

If you don't realize that the data came from a lognormal distribution, an outlier test would be very misleading. Excluding these high values as outliers would be a mistake and would lead to incorrect results. If you recognize that the data are lognormal, it is easy to analyze them properly. Transform all the values to their logarithms (see the right side of Figure 25.2), and those large values no longer look like outliers.

Mistake: Using a test designed to detect a single outlier when there are several outliers

Detecting multiple outliers is much harder than identifying a single outlier. The problem is that the presence of the second outlier can *mask* the first one so that neither is identified. Special tests are therefore needed. It is tempting to first run an outlier test, then remove the extreme value, and then run the outlier test again on the remaining values, but this method does a bad job of identifying multiple outliers.

Mistake: Eliminating outliers only when you don't get the results you want

If you are going to remove outliers from your data, it must be done in a systematic way, using criteria set up in advance. It is not OK to first analyze the data and then only look to remove outliers when you don't get the results you wanted. That is a sure way to get invalid results.

Mistake: Truly eliminating outliers from your notebook

It can make sense to exclude outliers from graphs and analyses when you do so in a systematic way that you report. But the outliers should still be recorded in your notebook or computer files so that you have a full record of your work.

Q & A

What does it mean to remove or eliminate an outlier?
> When an outlier is eliminated, the analyses are performed as if that value were never collected. If the outliers are graphed, they are clearly marked. Removing an outlier from an analysis does not mean the outlier should be erased from the lab notebook. Instead, the value of the outlier and the reason it was excluded should be recorded.

Can outlier tests be used with linear or nonlinear regression?
> Yes. In fact, I published a method for doing so (Motulsky & Brown, 2006).

How should outlier removal be reported?
> A scientific paper should state how many values were removed as outliers, the criteria used to identify the outliers, and whether those criteria were chosen as part of the experimental design. When possible, report the results computed in two ways, with the outliers included and with the outliers excluded.

Can reasonable scientists disagree about how to deal with outliers?
> Yes!

Should outlier tests automatically be applied to all data sets?
> No! Before running any outlier test, ask yourself the five questions listed at the beginning of this chapter.

CHAPTER SUMMARY

- Outliers are values that lie very far from the other values in the data set.
- The presence of a true outlier can lead to misleading results.
- If you try to remove outliers manually, you are likely to be fooled. Random sampling from a Gaussian distribution creates values that often appear to be outliers but are not.
- Before using an outlier test, make sure the extreme value wasn't simply a mistake in data entry.
- Don't try to remove outliers if it is likely that the outlier reflects true biological variability.
- Beware of multiple outliers. Most outlier tests are designed to detect one outlier and do a bad job of detecting multiple outliers. In fact, the presence of a second outlier can cause the outlier test to miss the first outlier.
- Beware of lognormal distributions. Very high values are expected in lognormal distributions, but these can easily be mistaken for outliers.
- Rather than removing outliers, consider using robust statistical methods that are not much influenced by the presence of outliers.

TERMS INTRODUCED IN THIS CHAPTER

- Extreme studentized deviate test (p. 212)
- Grubbs outlier test (p. 212)
- Masking (p. 214)
- Outlier (p. 204)
- Robust statistical tests (p. 213)

❖❖❖

CHAPTER 26

Choosing a Sample Size

The aim of science is to seek the simplest explanation of
complex facts. Seek simplicity and distrust it.

A. N. WHITEHEAD

Many experiments and clinical trials are run with a sample size that is
too small, so even substantial treatment effects may go undetected.
When planning a study, therefore, you must choose an appropriate
sample size. This chapter explains how to understand a sample size
calculation and how to compute sample size for simple experimental
designs. Sample size calculations for multiple and logistic regression
are briefly covered in Chapters 37 and 38, respectively. Sample size for
nonparametric tests is discussed in Chapter 41.

SAMPLE SIZE AND STATISTICAL HYPOTHESIS TESTING

Interpreting a sample size statement

Reports of clinical trials often include a statement that reads something like this:
"We chose to study 313 subjects in each group in order to have 80% power to
detect a 33% reduction in the recurrence rate from a baseline rate of 30% with a
significance level of 0.05 (two tailed)."

To understand this statement, you have to understand all four questions
it answers.

Question 1: What significance level?

Recall that the significance level (α) of a study is the answer to the question, If the
null hypothesis is actually true, what is the chance your result will be statistically
significant? This study defined $\alpha = 0.05$, so P values less than 0.05 are defined to
be statistically significant. If the investigators had chosen a stricter threshold (say,
$\alpha = 0.01$), they would have needed a larger sample size.

Question 2: What power?

Statistical power was discussed in Chapter 20. The power of an experimental
design is the answer to the question, If there truly is an effect of a specified size,
what is the chance that your result will be statistically significant?

If you want lots of statistical power, then you'll need a huge sample size. If
moderate power is adequate, then the sample size can be smaller. Some sample

size programs ask you to enter β instead of power. β equals 1.0 – power. For this example, β = 20%, more commonly entered into sample size programs as 0.20.

The example study chose a power of 80%. That means that if the effect the investigators wanted to see really exists, there is an 80% chance that the study would have resulted in a statistically significant result, leaving a 20% chance of missing the real difference.

Question 3: What effect size?

Every sample size calculation requires you to decide the effect size for which you are looking, abbreviated Δ. Depending on the experimental design, the effect size might be a difference between means, a risk ratio, a correlation coefficient, an odds ratio, or something else. Here, the investigators were looking for a 33% reduction in recurrence rate (from 30% down to 20%, a relative risk reduction of 33%, which is an absolute reduction of 10%). Entering this value is an essential part of sample size determination, because it takes more subjects to find a smaller effect than a larger effect. Effect sizes can be expressed in many ways, so be careful to understand what the program you are using is asking for.

Question 4: What is the standard deviation?

When comparing means, you must estimate the expected value of the standard deviation. If the data have lots of variability, a large sample size is needed. If the data are very consistent (tiny SD), a smaller sample size will suffice. This makes sense. The signal for which you are looking is the difference between means, but it is obscured by the "noise" of variation within each group, quantified by the SD. If there is a lot of noise compared to the signal, you'll need larger samples.

The example above compares proportions, so you don't need to enter the standard deviation directly. Instead, you must estimate how far the proportion will be from 50%. If you expect the outcome to occur about half the time, the sample size must be larger than if you expect the proportion to be closer to 0% or 100%. Here, the investigators estimated that the baseline recurrence rate (which they are hoping to reduce) is 30%.

The sample size statement restated

Now we can restate the sample size statement in simpler language: the investigators assumed that the recurrence rate is normally 30% and asked whether the new treatment reduced the recurrence rate by one-third (or more), down to 20% (or less). They chose to use 313 subjects in each group so that if the true effect was a one-third reduction in recurrence rate, they would have an 80% chance of obtaining a statistically significant result (defined as P < 0.05).

A CALCULATION OR A NEGOTIATION?

Although it always sounds like the investigator calculated sample size from α, β, and Δ, often the process is more of a negotiation. The investigators may have first said they were looking for a 10% chance in recurrence rates and then were horrified at the enormous number of subjects that would have been required to produce such a result. They then may have changed their goal until n seemed "reasonable." They may also have fussed with the values of α and β. Such an approach is not cheating, but rather is a smart way to decide on sample size. Parker and Berman (2003) point out that the goal often isn't to calculate the number of subjects needed, but rather to ask, If I use n subjects, what information can I learn?

In some cases, the calculations may convince you that it is not possible to learn what you want to know with the number of subjects you are able to use. This realization can be very helpful. It is far better to cancel such an experiment in the planning stage than to waste time and money on a futile experiment that won't have sufficient power. If the experiment involves any clinical risk or expenditure of public money, performing such a study can even be considered unethical.

SAMPLE SIZE AND THE MARGIN OF ERROR OF CONFIDENCE INTERVALS

Everything else being equal, a larger sample size produces a narrower confidence interval (CI). If you can state the maximum acceptable width of the CI, you can calculate how many subjects you need. You can use the same calculations, programs, and tables as those used for computing sample size needed to achieve statistical significance. Instead of entering the smallest difference you wish to detect, enter the largest acceptable margin of error, which equals half the width of the entire confidence interval.

When calculating sample size to achieve a set margin of error, it is customary to do the calculations so that if the assumptions are correct and you perform many experiments using the calculated sample size, the average margin of error will equal the value you specified. About 50% of the time, the margin of error will be narrower than that specified value, and 50% of the time, the margin of error will be wider. Although the term *power* is often not used, these calculations are the same as those used to calculate sample size for 50% power. If you chose, say, 80% power instead, then you'd need a larger sample size and the margin of error would be expected to be narrower than the specified value 80% of the time (Metcalfe, 2011).

SAMPLE SIZE TRADE-OFFS

Sample size versus power

Everything else being equal, if you increase the sample size you will increase the power of the study. The left half of Figure 26.1 shows this trade-off when comparing two means (fixing the standard deviations within the group, the value of α, and the effect size for which you are looking).

A standard for sample size calculations is to specify 80% power. Here are two approximate rules of thumb for changing to different values of power:

- If you only need 50% power, a sample size about half as large as required for 80% power will suffice.
- A sample size about twice as large is needed to go from 80% to 99% power.

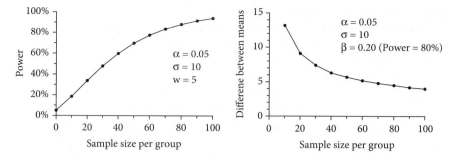

Figure 26.1. Graphing the trade-offs in choosing sample size.
(Left) Power increases as you increase sample size. These calculations were for an unpaired t test with the significance level set to 5%, the standard deviation within each group set to 10, and the effect size set to 5. (Right) Here we hold the power equal to 80% and plot the decrease in the effect size that we can detect as sample size increases.

Sample size versus effect size

The required sample size is inversely proportional to the square of the effect size. If you want to detect an effect size that is half as big, you need to quadruple (multiply by four) the sample size. If you want to detect an effect size that is one-third what you computed, you need to multiply the sample size by nine. The right half of Figure 26.1 shows this trade-off for one particular example.

Equal versus unequal sample sizes

It is standard to use the same sample size in both groups. However, it may make sense to plan studies with unequal sample sizes if one treatment is much more expensive, difficult, or risky than the other, or when it is difficult to find subjects for one group (e.g., patients with a rare disease). The trade-off is that if you reduce the number of subjects in one group, you must increase the number of subjects in the other group even more, which means the total number of subjects must increase. You can't reduce the number of subjects in the smaller group to less than half of the number required had the sample sizes been equal (without losing power), and it won't help to increase the size of the larger group to more than four or five times the number required had the sample sizes been equal (van Belle, 2008).

OTHER APPROACHES TO CHOOSING SAMPLE SIZE

Adaptive trials

The idea behind *adaptive trials* is simple: interim analyses, performed while the trial is proceeding, are used to decide the course of the study. This approach to designing large clinical trials is gaining acceptance (Kuehn, 2006).

    The simplest adaptive designs allow researchers to analyze the data only once at the middle of a trial and then decide to stop the trial if patients in one group are having substantially better outcomes than those in the other. This approach is also called *sequential analysis*. Newer adaptive techniques are used to decide when to stop adding new patients to the study, when to stop the study, and what proportion of new subjects should be assigned to each alternative treatment. These decisions must be based on unambiguous protocols established before collecting data to ensure that the statistical results are meaningful.

In some cases, the adaptive design can lead investigators to end a study early when it is clear that one treatment is much better than the other, or when it becomes clear that the difference between the treatments is at best trivial. In other cases, the adaptive design can prolong a study until it reaches a crisp conclusion.

The methods used to adapt a study to interim findings are not straightforward, have some drawbacks (Fleming, 2006), and are not standard.

Simulations

Simulations provide a versatile approach to optimizing sample size for any kind of experimental design. Run multiple sets of computer simulations, each with various assumptions about variability, effect size, and sample sizes. Of course, you'll include a random factor so each simulation has different data. For each set of simulations, tabulate whatever outcome makes the most sense—often the fraction of the simulated results that have a P value less than 0.05 (the power). Then you can review all the simulations and see what advantages increasing sample size gives you. Knowing the cost and effort of the experiment, you can then choose a sample size that makes sense.

HOW TO COMPUTE SAMPLE SIZE IN SIMPLE SITUATIONS

Sample sizes can be computed for all kinds of study designs. Here I explain the calculations for a few simple experimental designs.

How to compute sample size to compare two means

1. Define the smallest difference between means you want to detect, or (equivalently) the largest margin of error you wish to see in the confidence interval for the difference between means. Call this value w.
2. Estimate the standard deviation, s, you expect to see. Here we assume that the standard deviation is the same for both groups. If you expect different standard deviations, you'll need to use a program that can handle that situation.
3. Decide on what power you need. Many investigators set power to 80%, but this decision really should depend on the context of the work. If you are computing sample size based on the desired width of a confidence interval, the power is the chance that the margin of error of the confidence interval generated in the experiment will be smaller than w.
4. Decide on what significance level you need. Many investigators set the significance level to 5% (or 0.05), as this is traditional. If you are computing sample size based on the desired width of a confidence interval, compute 100% minus the desired confidence level. If you want to create a 95% confidence interval, set the significance level to 5%, or 0.05.

Power	SIGNIFICANCE LEVEL OR α (TWO-TAILED)			
	0.1	0.05	0.01	0.001
50%	2.7	3.8	6.6	10.8
80%	6.2	7.8	11.7	17.1
90%	8.6	10.5	14.9	20.9
95%	10.8	13.0	17.8	24.4
99%	15.8	18.4	24.0	31.5

Table 26.1. Multiplier (C) used when computing required sample size.

Choose a row corresponding to the power you desire and a column corresponding to your required significance level (the two-tailed P value that defines statistical significance). Then use the tabulated value as C in Equations 26.1–26.4.

Each cell in this table was computed using the Excel formula =(NORMSINV(Alpha/2)+NORMSINV(1-Power/100))^2.

     If you want to create a 99% confidence interval, set the significance level to 1%, or 0.01.

5.   Use Table 26.1 to look up a value C that is computed from the power and significance level you chose. For 80% power and a 5% significance level, C = 7.8.

6.   Calculate the sample size you need in each group (n):

$$n = 2C\left(\frac{s}{w}\right)^2 \tag{26.1}$$

Example: Comparing two means

How many participants are needed to compare the means of two groups when the SD of blood pressure measurements is about 10 mmHg and we want to be able to detect a difference of 5 mmHg? We'll choose a significance level of 0.05 and compute sample size to obtain 80% power.

    We can also state the problem in terms of confidence intervals. We expect the SD to be 10 and want the margin of error of the 95% confidence interval for the difference between the two means to be five (or narrower), with 80% power.

    Set s = 10 and w = 5. From Table 26.1, C = 7.8. Equation 26.1 calculates sample size (n).

$$n = 2 \cdot 7.8\left(\frac{10}{5}\right)^2 = 62.4 \approx 63$$

Let's review exactly what this means:

- Assume that the true difference between means is 5.00 mmHg.
- Now imagine that you perform many experiments, with n = 63 per group in each experiment.
- Because of random sampling, you won't find that the difference between means equals exactly 5.00 mmHg in every experiment. Instead, you'll find

that the difference between means varies, and will be greater than 5.00 mmHg in about half the experiments and less than 5.00 mmHg in the other half.

- In 80% (the power) of those experiments, the P value will be less than 0.05 (two-tailed), so the results will be deemed statistically significant. In the remaining 20% of the experiments, the difference between means will be deemed not statistically significant, so you will have made a Type II (β) error.
- This can be summarized as follows: a sample size of 63 in each group has 80% power to detect a difference between means of 5.00 mmHg with a significance level (α) of 0.05 (two-tailed). Equivalently, a sample size of 63 in each group has an 80% chance of generating a 95% confidence interval (for the difference between the two means) with a margin of error less than or equal to 5.00 mm Hg.

Example: Comparing weights of men and women

You "hypothesize" that men on average weigh more than women, but want to prove it with statistics. How many men and women do you need to weigh be *really* sure, defined by setting the risk of both Type I and Type II error to 1%? For this hypothetical experiment, we can look up the SD and mean of weights. It turns out that the difference between the mean weight of men and the mean weight of women is very close to the standard deviation of each (Environmental Protection Agency, 2011, page 8–36 for 30-year olds).

From Table 26.1, set power to 99%, α to 0.01, and look up C which equals 24. In Equation 26.1 the ratio of s/w equals 1.0, so the sample size needed per group is twice C or 48. This example is a helpful way to appreciate the need for large samples. Look around. It is obvious that men, on average, weigh more than women. But to prove that (with only 1% chance of error in either direction) requires nearly 100 subjects in all. If you use the more conventional choices where power is set to 80% and α to 0.05, then you'll need 16 in each group.

How to compute sample size to compare two proportions

1. Define the smallest difference between proportions you want to detect, or (equivalently) the largest margin of error you wish to see in the confidence interval for the difference between proportions. Call this variable w. If you set w to a small value, you'll need a larger sample size than if you set it to a large value.

2. Estimate the variable p_{av}, which is the anticipated average of the two proportions. If estimating p_{av} is impossible, calculate a worst-case (largest) sample size by setting p_{av} equal to 0.5. Any other value will require a smaller sample size.

3. Decide on what power you need. Many investigators set power to 80%, but this decision really should depend on the context of the work. If you are computing sample size based on the desired width of a confidence interval, the power is the chance that the margin of error of the confidence interval generated in the experiment will be narrower than w.

4. Decide on what significance level you need. Many investigators set the significance level to 5% (or 0.05), as this is traditional. If you are computing sample size based on the desired width of a confidence interval, compute 100% minus the desired confidence level. If you want to create a 95% confidence interval, set the significance level to 5%, or 0.05. If you want to create a 99% confidence interval, set the significance level to 1%, or 0.01.
5. Use Table 26.1 to look up a value C that is computed from the power and significance level you chose. For 80% power and a 5% significance level, C = 15.7.
6. Calculate the sample size you need in each group (n):

$$n = \frac{2C \cdot p_{av}(1 - p_{av})}{w^2} \qquad\qquad (26.2)$$

Example: Comparing two proportions

How many values are needed to determine the difference in incidence rates of a disease in two populations when you expect the incidence to be about 0.10, you want to be able to detect a difference in incidence rates as small as 0.02, you define statistical significance as $\alpha = 0.05$, and you wish to have 90% power?

1. Set w to 0.02.
2. You are hoping to detect a difference between 0.10 and 0.08, so set p_{av} equal to their average, 0.09.
3. Look up C on Table 26.1 for 90% power and $\alpha = 0.05$. C = 10.5.
4. Compute the sample size using Equation 26.2. The required sample size is 4,300 per group.

Let's review exactly what this means:

• Assume that the true difference between proportions is 0.02.
• Now imagine that you perform many experiments, with n = 4,300 per group in each experiment.
• Because of random sampling, you expect to see a difference larger than 0.02 half the time and smaller than 0.02 the other half of the time.
• In 90% of those experiments, you expect to find a statistically significant result, with P < 0.05 (two-tailed). In the remaining 10% of the experiments, you expect to observe a difference that is not statistically significant, so you will have made a Type II error.

How to determine sample size to compare one group with a hypothetical value

Sometimes you wish to compute a sample size for only one group. Say you want to set the sample size so the confidence interval will be of a specified width. Or perhaps you want to run a one-sample t test to compare the observed sample mean with a theoretical population mean, or a binomial test to compare an observed sample proportion with a theoretical population proportion.

Calculating a sample size for one group uses the same principles as explained above for calculating a sample size to compare two groups, except that the needed sample size is cut in half. The sample size needed in your one group is half the sample size you'd need in each group, if you were comparing two groups. When comparing two groups, there are two sources of random variation, so you need twice the sample size in each group. Thus, the equations equivalent to 26.1 and 26.2 are:

$$n = C\left(\frac{s}{w}\right)^2 \tag{26.3}$$

$$n = \frac{C \cdot p(1-p)}{w^2} \tag{26.4}$$

With one group, there is only one proportion, so p_{av} in Equation 26.4 has been changed to plain p, the expected proportion.

For example, the SD of blood pressure is about 10 mmHg (s), and you wish to define the mean blood pressure with a margin of error of 5 mmHg (w). How large does your sample need to be? Let's set the confidence level to 95%, which is equivalent to setting α equal to 0.05. When doing these kinds of calculations for confidence intervals, it is traditional to set power to 50%. Looking in Table 26.1, C = 3.8. Using Equation 26.3 above, n = 15.2, which we'll round up to 16. If you want 80% power, then C = 7.8, so n = 31.2, which we'll round up to 32.

COMMON MISTAKES: SAMPLE SIZE

Mistake: Computing sample size based on published effect sizes without thinking

All sample size calculations require you to specify how large of an effect you are looking for. It takes a huge sample size to reliably detect a tiny effect but a smaller sample size to detect a huge effect. However, specifying the effect size is not easy. Many people prefer to use results from a prior study (or a pilot study) and then calculate a sample size large enough to detect an effect that large. This approach seems to make sense, but it presents two problems.

One problem is that you might care about an effect, even if it is smaller than the published (or pilot) effect size. You should compute the sample size required to detect the smallest effect worth detecting, not the sample size required to detect an effect size someone else has published (or you determined in a pilot study).

The second problem is that the published difference or effect is likely to inflate the true effect (Ioannidis, 2008; Zollner & Pritchard, 2007). If many studies are performed, the average of the effects detected in these studies should be close to the true effect. Some studies will happen to find larger effects, and some studies will happen to find smaller effects. The problem is that the studies that find small effects are much less likely to get published than the studies that find large effects. Therefore, published results tend to overestimate the actual effect size.

What happens if you use the published effect size to compute the necessary sample size for a repeat experiment? You'll have a large enough sample size to have 80% power (or whatever power you specify) to detect that published effect size. But if the real effect size is smaller, the power of the study to detect that real effect will be less than the power you chose.

Mistake: Computing sample size based on standard effect sizes without thinking

In order to compute sample size, you need to specify an effect size. To avoid the difficulty of defining the minimum effect size they want to detect, many are tempted to pick a standard effect size instead. They use standard values for the significance level and power, so why not use a *standard effect size* too?

In a book that is almost considered the bible of sample size calculations in the social sciences, Cohen (1988) defines some standard effect sizes. He limits these recommendations to the behavioral sciences (his area of expertise) and warns that all general recommendations are more useful in some circumstances than others. Here are his guidelines for an unpaired t test:

- A "small" difference between means is equal to one-fifth the SD.
- A "medium" effect size is equal to one-half the SD.
- A "large" effect is equal to 0.8 times the SD.

If you are having trouble deciding the effect size for which you are looking (and therefore are stuck and can't determine a sample size), you can decide whether you are looking for a small, medium, or large effect and then use Cohen's definitions. Lenth (2001) argues that you should avoid these "canned" effect sizes, and I agree. You must decide how large a difference you care to detect based on understanding the experimental system you are using and the scientific questions you are asking. Cohen's recommendations seem a way to avoid thinking about the goals of the experiment. If you choose standard definitions of α (0.05), power (80%), and effect size (see above), then there is no need for any calculations. If you are comparing two groups (with a t test), you will need a sample size of 26 in each group to detect a large effect, 65 in each group to detect a medium effect, and 400 in each group to detect a small effect. Choosing a standard effect size is really the same as picking a standard sample size.

Mistake: Failing to plan for dropouts

Sample size calculations tell you the number of values you need to enter into your analyses. If any subjects drop out of your experiment, you'll need to start with more than the computed number. For example, if you expect about 20% of the subjects to drop out, you should start with 20% more subjects than calculated. You should also take the following into consideration as you are designing your study:

- Dropouts are especially an issue when the study is lengthy or when complying with the experimental protocol is inconvenient.
- The number of dropouts may vary among the different treatment arms of the study.
- If dropouts are expected to be random, then you can solve the problem by starting with additional subjects. But what if the people who drop out of the study are those who feel especially great (and don't want to spend time with doctors) or those who are feeling awful (so don't want to put up with extra tests)? More generally, what if the reason for dropping out is related to the outcome being assessed? In these cases, the results might be biased, because the research subjects who completed the trial may not be representative of the population you care about, and more may drop out of one treatment arm of the study than another. Starting the study with additional subjects won't solve the problem.

Mistake: Using standard values for significance level and power without thinking

Sample size calculations require you to specify the significance level you wish to use, as well as the power. Many simply choose the standard values, a significance level (α) of 5% (P < 0.05 is significant) and a power of 80%. It is better to choose values based on the relative consequences of Type I and Type II errors.

Mistake: Ad hoc sequential sample size determination

The following approach will lead to invalid results: Collect and analyze some data. If the results are not statistically significant, collect some more data and reanalyze. Keep collecting more data until you obtain a statistically significant result or until you run out of money, time, or curiosity. This point is also discussed in Chapter 23.

Figure 26.2 is a simulation that demonstrates this point. The simulated data were sampled from populations with Gaussian distributions and identical means and standard deviations. An unpaired t test (see Chapter 30) was computed with n = 5 in each group, and the resulting P value (0.406) is plotted at the left of the graph. Note that the Y-axis is logarithmic to focus attention on the small P values. Then the t test was repeated with one more value added to each group. That P value was 0.825. Then one more value was added to each group (n = 7), and the P value was computed again (0.808). This was continued up to n = 75.

With these simulated data, we know for sure that the null hypothesis is true. If you ran t tests on many different samples of data, you'd expect to see a statistically significant result 5% of the time. But here, each sample included all the values in the prior sample, so the P values are not independent. Rather, they mimic

Figure 26.2. Beware of ad hoc sample size decisions.

This figure shows a simulation of a common mistake in data analysis. The (simulated) investigator started with n = 5 in each group and calculated a t test to determine a P value. The P value (plotted on the logarithmic Y-axis) was (much) greater than 0.05, so the investigator increased sample size to six in each group, and then seven, eight, nine, and so on. All the values analyzed were sampled from the same Gaussian distribution, so the null hypothesis is known to be true. The figure shows what would have happened had the study continued to include 75 in each group. However, the investigator actually would have stopped at 21 in each group, since the result happened to hit statistical significance then. Unless the analyses are done with special methods, P values simply cannot be interpreted when the sample size isn't determined in advance. Note that the values in this figure were calculated via simulation with random numbers. Another simulation would have looked different, but the main point would persist: the cumulative P value bounces around a lot, especially with small sample sizes.

what some scientists do: run the experiment a bunch of times, check the P value, increase sample size if the P value isn't as low as desired, calculate the P value again, and so on.

If you could extend the sample size out to infinity, you'd always eventually see a P value less than 0.05, even if the null hypothesis were true. In practice, the investigator would eventually run out of money, time, or patience. Even so, the chance of hitting a P value less than 0.05 under the null hypothesis is way higher than 5%.

Someone viewing these results would have been very misled if the investigator had stopped collecting data the first time the P value dropped below 0.05 (at n = 21) and had simply reported that the P value was 0.0286 and the sample size was 21 in each group. It is impossible to interpret results if an ad hoc method is used to choose sample size.

Q & A

Does 80% power mean that 80% of the subjects would be improved by the treatment and 20% would not?

No. Power refers to the probability that a proposed study would reach a conclusion that a given difference (or effect size) is statistically significant. Power has nothing to do with the fraction of patients who benefit from a treatment.

How much power do I need?

Sample size calculations require you to choose how much power the study will have. If you want more power, you'll need a larger sample size. Often power is set to a standard value of 80%. Ideally, the value should be chosen to match the experimental setting, the goals of the experiment, and the consequences of making a Type II error (see Chapter 16).

What are Type I and Type II errors?

See Table 16.2. A Type I error occurs when the null hypothesis is true but your experiment gives a statistically significant result. A Type II error occurs when the alternative hypothesis is true but your experiment yields a result that is not statistically significant.

My program asked me to enter β. What's that?

β is defined as 100% minus power. A power of 80% means that if the effect size is what you predicted, your experiment has an 80% chance of producing a statistically significant result and thus a 20% chance of obtaining a not significant result. In other words, there is a 20% chance that you'll make a Type II error (missing a true effect of a specified size). β, therefore, equals 20%, or 0.20.

Why don't the equations that compute sample size to achieve a specified margin of error require you to enter a value for power?

The equations you'll find in other books for calculating sample size to obtain a specified margin of error (half-width of the CI) often don't require you to choose power. That is because they assume 50% power. If the assumptions are all true and you use the computed sample size, there is a 50% chance that the computed margin of error will be less than the desired value and a 50% chance that it will be larger.

Can a study ever have 100% power?

No. Or at least not with the usual kinds of statistical testing.

Why does power decrease if you set a smaller value for α (without changing sample size)?

When you decide to make α smaller, you set a stricter criterion for finding a significant difference. The advantage of this decision is that it decreases the chance of making a Type I error. The disadvantage is that it will now be harder to declare a difference significant, even if the difference is real. By making α smaller, you increase the chance that a real difference will be declared not significant, and thus decrease statistical power .

Why does power increase if you choose a larger n?

If you collect more evidence, conclusions will be more certain. Collecting data from a larger sample decreases the standard error and thus increases statistical power.

What is a power analysis?

Some investigators refer to sample size calculations as a *power analysis*.

What does it mean when an investigator uses the phrase "powered to detect . . . "?

That is just slang for choosing a large enough sample size for the specified effect size, significance level, and power.

How fuzzy are sample size calculations?

Sample size calculations depend on one fuzzy estimate (the expected SD) and three fairly arbitrary decisions (α, power, and the smallest difference worth detecting). So the calculations are pretty fuzzy. Different knowledgeable people facing the same situation may come up with different estimates and decisions and so would compute different sample sizes. Sample size statements in journals tend to make sample size determination sound far more unambiguous than it is.

How can I do a sample size calculation if I have no idea what I expect the SD to be?

You can't. You'll need to perform a pilot study to find out.

How large a sample size do I need for that pilot study?

Julious (2005) suggests that 12 is a good number.

Is it useful to compute power of a completed study?

When a study reaches a conclusion that an effect is not statistically significant, it can be useful to ask what power that experimental design had to detect various effects (but it is rarely helpful to compute the power to detect the effect actually observed). See Chapter 19.

CHAPTER SUMMARY

- Sample size calculations require that you specify the significance level, the effect size for which you are looking, the scatter (SD) of the data or the expected proportion, and the desired power.
- You need larger samples when you are looking for a small effect, when the SD is large, and when you desire lots of statistical power.
- It is often more useful to ask what you can learn from a proposed sample size than to compute a sample size from the four variables listed in the previous bullet.
- It is important to decide the magnitude of the effect for which you are looking. Using standard effect sizes doesn't really help.
- Instead of determining sample size to detect a specified effect size with a specified power, you can compute the sample size needed to make the margin of error of a confidence interval have a desired size.
- *Ad hoc sequential sample size determination* is commonly done but not recommended. You can't really interpret P values when the sample size was determined that way.
- Adaptive trials do not require you to choose sample size at the beginning of the study. Instead, these methods pick a sample size based on data collected during the study based on algorithms chosen as part of the study protocol.

TERMS INTRODUCED IN THIS CHAPTER

- Adaptive trials (p. 219)
- Ad hoc sequential sample size determination (p. 226)
- Power analysis (p. 228)
- Sequential analysis (p. 226)
- Standard effect sizes (p. 225)

PART F

Statistical Tests

CHAPTER 27

❖❖❖

Comparing Proportions

No one believes an hypothesis except its originator, but everyone believes an experiment except the experimenter.

W. I. B. BEVERIDGE

This chapter and the next one explain how to interpret results that compare two proportions. This chapter explains analyses of cross-sectional, prospective, and experimental studies in which the results are summarized as the difference or ratio of two incidence or prevalence rates.

EXAMPLE: APIXABAN FOR TREATMENT OF THROMBOEMBOLISM

Study design

Agnelli and colleagues (2012) tested whether apixaban would be beneficial in extended treatment of thromboembolism. They studied patients who had a venous thromboembolism—a clot in the leg vein—which can be dangerous if it breaks loose and goes to the lung. After all the patients completed the standard antico-agulation (blood thinner) therapy, the investigators randomly divided them into three groups that were randomly assigned to receive placebo or one of two doses of apixaban, which inhibits clotting factor Xa. We'll ignore the patients who got the higher dose and only discuss the lower dose here.

The investigators followed the patients for a year. The primary outcome they assessed was whether there was a recurrence of the thromboembolism. They measured other outcomes and analyzed these results too.

This study is called a randomized, double-blind prospective study.

It is a *randomized* study because the assignment of subjects to receive a drug or placebo was determined randomly. Patients or physicians could not request one treatment or the other.

It is *double blind* because neither patient nor investigator knew who was getting the drug and who was getting the placebo. This is sometimes called a *double-masked* study. Until the study was complete, the information about which patient got which drug was coded, and the code was not available to any of the participating subjects or investigators (except in case of a medical emergency).

TREATMENT	RECURRENT THROMBOEMBOLISM	NO RECURRENCE	TOTAL
Placebo	73	756	829
Apixaban (2.5 mg twice a day)	14	826	840
Total	87	1,582	1,669

Table 27.1. Data from the Agnelli et al. (2012) apixaban study. Each value is an actual number of patients.

P value (Fisher's test)	< 0.0001
P value summary	****
One- or two-sided	Two-sided
Statistically significant? ($\alpha = 0.05$)	Yes
Relative risk	5.283
95% confidence interval	3.006 to 9.286
Difference between fractions	0.07139
95% confidence interval of difference	0.05006 to 0.09273
Number needed to treat (NNT)	14.0
95% confidence interval	10.8 to 20.0

Table 27.2. Statistical results of the Agnelli et al. (2012) apixaban study.

It is *prospective* because the subjects were followed forward over time. Chapter 28 explains retrospective case-control studies, which look back in time.

Results of the study

The results are shown in Table 27.1, arranged in a customary manner with the rows representing alternative treatments and the columns representing alternative outcomes. Each subject was assigned to one row based on treatment and ended up in one column based on outcome. Therefore, each cell in the table is the actual number of subjects that were given one particular treatment and had one particular outcome.

Table 27.1 is called a *contingency table*. These tables show how the outcome is contingent on the treatment or exposure. By convention, a contingency table always shows the exact number of subjects (or some other experimental unit) with a certain treatment or exposure and a certain outcome. Tables of percentages or normalized rates are not called contingency tables.

The goal of the analyses is to generalize about the general population of patients who have had a thromboembolism by computing confidence intervals and a P value. The results (computed by GraphPad Prism) are shown in Table 27.2. Lots of programs can do these calculations, so this chapter only discusses interpretation of the results, not the details of the calculations.

Confidence intervals of the proportions

The disease progressed in 8.8% of the patients receiving the placebo (73/829), with the 95% CI ranging from 7.0% to 10.9% (Chapter 4 explained confidence intervals of proportions). In contrast, the disease progressed in 1.7% of the people receiving apixaban, with the 95% CI ranging from 0.9% to 2.8%. These confidence intervals present the findings clearly. The patients who received apixaban were a lot less likely to have a recurrent thromboembolism.

The attributable risk

One way to summarize the results is to calculate the difference between the two proportions (often expressed as percentages). Disease progressed in 8.8% of placebo-treated patients and in 1.7% of patients treated with apixaban, a difference of 7.1%. This difference between two incidence rates is called *attributable risk*.

The 95% CI of the attributable risk ranges from 5.0% to 9.3%. If we assume our subjects are representative of the larger population of adults with a thromboembolism, we are 95% confident that treatment with apixaban will reduce the incidence of disease progression by somewhere between 5.0% and 9.3%. Note that these calculations deal with the actual difference in incidence rates (subtraction), not the relative change (division).

Number needed to treat (NNT)

Laupacis, Sackett, and Roberts (1988) have suggested reporting the reciprocal of the attributable risk, a value they call the *number needed to treat* (NNT). This value tells you how many patients would require treatment to reduce the expected number of cases of a defined end point by one. In this example, the reciprocal of 0.0714 is 14.0. This is easy to understand. For every 14 patients who receive the treatment, you'd expect to prevent a recurrent thromboembolism in one. The other 13 would receive no benefit from the treatment. Of course, you don't know before treating a patient whether he or she will benefit.

The CI of the NNT is obtained by taking the reciprocal of each end of the CI of the attributable risk, so it extends from 10.8 to 20.0. In other words, you'd expect somewhere between 1 of 11 and 1 of 20 patients to benefit from the treatment.

There are two advantages to reporting results as the NNT. One advantage is that this approach avoids the need to think about small fractions. The other advantage is that it puts the results in a clinically relevant context. Is it worth treating 13 patients who won't benefit to help the one patient who will? Answering that question requires that you compare the benefit to that one patient (huge in this case) with the risks, costs, and inconvenience to the other 12. You also need to account for the availability of alternative treatments. Making that decision requires clinical judgment. Expressing the result as the NNT makes it easier to think about the trade-offs.

When the treatment or exposure causes harm, the term *NNT* doesn't fit, so this value is renamed the *number needed to harm*.

The relative risk

It is often more intuitive to think of the ratio of two proportions rather than their difference. This ratio is termed the *risk ratio* or *relative risk*. Disease progressed in 8.8% of placebo-treated patients and in 1.7% of patients treated with apixaban. The ratio is 8.8/1.7, or 5.2. In other words, subjects treated with the placebo were 5.2 times as likely as patients treated with apixaban to have a recurrent thromboembolism.

That seems a bit backwards. Let's flip the ratio to 1.7/8.8, which is 0.19. Patients receiving the drug had 19% the risk of a recurrent thromboembolism than did patients receiving the placebo.

The 95% CI of the relative risk extends from 0.11 to 0.33. The interpretation is now familiar. If we assume our subjects are representative of the larger population of adults with thromboembolism, we are 95% sure that treatment with apixaban will reduce the relative incidence of disease progression by a factor between 0.11 to 0.33 times the risk.

Don't get confused by the two uses of percentages. In this example, the drug lowered the *absolute* risk of thromboembolism by 7.1% (the difference between 8.8% and 1.7%) and reduced the *relative* risk by 81% (which is 100% minus the ratio 1.7/8.8), down to a risk that is 19% that of placebo-treated patients.

In this example, the term *risk* is appropriate because it refers to disease recurrence. In other contexts, one alternative outcome may not be worse than the other, and the relative risk is therefore more appropriately termed the *relative probability* or *relative rate*.

P value

All P values test a null hypothesis, which in this case is that apixaban does not alter the risk of a recurrent thromboembolism. The P value answers the question, If the null hypothesis were true, what is the chance that random sampling of subjects would result in incidence rates as different (or more different) from what we observed?

The P value depends on sample size and on how far the relative risk is from 1.0. Two statistical methods can be used to compute the P value. Most would agree that the best test to use is *Fisher's exact test*. With large sample sizes (larger than used here), Fisher's test is mathematically unwieldy, so a chi-square test is used instead. With large samples, the two compute nearly identical P values.

The P value (calculated by a computer using either test) is tiny, less than 0.0001. Interpreting the P value is straightforward: if the null hypothesis is true, there is less than a 0.01% chance of randomly picking subjects with such a large (or larger) difference in incidence rates.

Other results

The results discussed above convincingly show that the lower dose (of two tested) of apixaban prevents recurrence of thromboembolism. Although statistical tests focus on one result at a time, you must integrate various results before reaching an overall conclusion. Here the investigators also showed the following:

- The results with the higher dose (5 mg twice a day) were very similar to the results summarized above for 2.5 mg twice a day.
- Patients receiving apixaban (an anticoagulant) did not have more bleeding events than those treated with the placebo.
- The reduction in the number of recurrent thromboembolisms was similar in old and young, and in men and women.

ASSUMPTIONS

The ability to interpret the results of a prospective or experimental study depends on the following assumptions.

Assumption: Random (or representative) sample

The 95% CI is based on the assumption that your sample was randomly selected from the population. In many cases, this assumption is not true. You can still interpret the CI as long as you assume that your sample is representative of the population.

The patients in the example were certainly not randomly selected, but it is reasonable to think that they are representative of adults completing anticoagulant therapy following a venous thromboembolism.

Beware of the two uses of the term *random*. The subjects in the study were not randomly selected from the population of all people with venous thromboembolism. But each subject was randomly assigned to receive the drug or placebo.

Assumption: Independent observations

The 95% CI is only valid when all subjects are sampled from the same population and each has been selected independently of the others. Selecting one member of the population should not change the chance of selecting anyone else.

This assumption would be violated, for example, if the study included several people from one family.

Assumption: No difference between the two groups except treatment

In this study, subjects were randomly assigned to receive the drug or placebo, so there is no reason to think that the two groups differed. But it is possible that, just by chance, the two groups differed in important ways. The authors presented data showing that the two groups were indistinguishable in terms of age, weight, sex, kidney function, and other relevant variables.

COMPARING OBSERVED AND EXPECTED PROPORTIONS

Mendel's peas

Mendel pioneered the field of genetics. The chi-square test (developed after he died) can be used to compare actual data with the predictions of Mendel's model. Table 27.3 shows one of his experiments (from Cramer, 1999). In this experiment, he recorded whether peas were round or angular and also whether they were yellow or green. The yellow and round traits were dominant. Table 27.3 also shows the expected proportions of the four phenotypes and the expected number in an experiment with 556 peas. These expected numbers are not integers, but that is OK, because they are the average expectations if you did many experiments.

Is the discrepancy between the observed and expected distributions greater than we would expect to see by chance? The test that answers this is called the

PHENOTYPE	OBSERVED NUMBER OF SEEDS IN THIS EXPERIMENT	EXPECTED PROPORTION	EXPECTED NUMBER IN THIS EXPERIMENT
Round and yellow	315	9/16	312.75
Round and green	108	3/16	104.25
Angular and yellow	101	3/16	104.25
Angular and green	32	1/16	34.75
Total	556	16/16	556.00

Table 27.3. Mendel's peas.
This shows one of Mendel's experiments (adapted from Cramer, 1999). The first column shows the actual distribution of traits in the seeds he collected. The last column shows the distribution expected by genetic theory.

chi-square test. It computes a P value that answers the question, If the theory that generated the expected distribution is correct, what is the chance that random sampling would lead to a deviation from expected as large or larger as that observed in this experiment?

The P value is 0.93 (computed here using GraphPad Prism, but lots of programs can do the calculations). With such a high P value, there is no reason to doubt that the data follow the expected distribution. Note that the large P value does not prove the theory is correct, only that the deviations from that theory are small and consistent with random variation. (Many of Mendel's P values were high, which makes some wonder whether the data were fudged, as was briefly discussed in Chapter 19.)

How it works: Chi-square goodness-of-fit test

When running the chi-square test to compare observed and expected distributions, it is essential that you enter observed values that are the actual numbers of subjects in each category. Don't try to run the chi-square test with percentages or any kind of normalized value. The expected values are the number of subjects you expect to see in each category. These values do not need to be integers. Each expected value is the average number you'd expect in that category if the experiment were repeated many times. In any one experiment, the observed value must be an integer. But the expected value, averaged over many experiments, can be a fraction. The sum of all the observed values must equal the sum of all the expected values.

To quantify the discrepancy between the observed and expected distributions, the chi-square test does the following: For each category, it computes the difference between the observed and expected values and divides that difference by the expected value. It then adds up the square of all those ratios. The result is a chi-square value, pronounced *ki* (rhymes with *eye*) *square* and written as χ^2. Here is the value as an equation:

$$X^2 = \sum \frac{(\text{Observed} - \text{Expected})^2}{\text{Expected}}$$

For the Mendel example, $\chi^2 = 0.470$. The relationship between χ^2 and P value depends on how many categories there are. The number of degrees of freedom is defined to equal the number of categories minus one. The example data have four categories and so have 3 df. This makes sense. Once you know the total number of peas and the number in three of the phenotype categories, you can figure out the number of peas in the remaining category. Knowing that $\chi^2 = 0.470$ and df = 3, the P value can be determined by computer or using a table.

The test is based on some approximations, which are only reasonably accurate when all the expected values are fairly large. If any expected value is less than five, the results are dubious. This matters less when there are lots of categories (rows) and matters the most when there are only two categories (in which case the expected values should be 10 or higher).

Binomial test

The chi-square test described above is an approximation. When there are only two categories, the *binomial test* computes the exact P value, without any approximation or worry about sample size.

The coin-flipping example of Chapter 15 used the binomial test. To compute the binomial test, you must enter the total number of observations, the fraction that had one of the two outcomes, and the fraction expected (under the null hypothesis) of having that outcome. In the coin-flipping example, the expected proportion was 50%, but this is not always the case.

COMMON MISTAKES: COMPARING PROPORTIONS

Mistake: Mixing up the two uses of the chi-square test

The chi-square test is used in two distinct ways, and it is easy to mix them up.

The first part of the chapter explained how to compare two proportions. One way to compute a P value is with a chi-square test (although the Fisher's exact test is preferred). The data provide four observed values. The expected values are computed from those observed data (assuming the null hypothesis is true). The chi-square test then combines the discrepancies between observed and expected and computes a P value. To analyze the data using a program that performs the chi-square test, you would enter only the four values of the contingency table (see Table 27.1). The program would compute the expected values automatically.

The second part of this chapter explained how the chi-square test can be used to compare an observed distribution with a distribution expected by theory. To do this analysis using a computer program, you must enter both an observed and an expected count for each category. The expected counts must come from theory or external data and cannot be derived from the data being analyzed.

Mistake: Not entering actual counts into a contingency table

When analyzing a contingency table or when comparing observed and expected distributions, it is essential that you enter the actual number of observed counts.

The results will be incorrect if you enter fractions or percentages or normalized rates (e.g., deaths per 1,000 people).

Mistake: Analyzing observed and expected data (like Mendel's peas) in a program designed to analyze a contingency table

If you want to compare observed and expected distributions, it is essential that you use a program designed for that purpose. Don't mistakenly use a program designed to analyze a contingency table. If you enter observed data into the first column and expected counts into the second column, you have *not* created a contingency table. If you analyze that table with a program designed to analyze a contingency table, the results will be meaningless.

Q & A

What if there are more than three groups or more than three outcomes?

It won't be possible to compute an attributable risk or relative risk, but it is possible to compute a P value. Fisher's test is limited to use on tables with two rows and two columns, but the chi-square test can analyze any size contingency table. Some programs offer exact tests for this situation.

If there are more than three rows or columns, does it matter in what order they are placed?

The usual chi-square test pays no attention to the order of rows or columns. If your table has two columns and three or more rows in which the order matters (e.g., doses or ages), the *chi-square test for trend* questions whether there is a significant trend between row number and the distribution of the outcomes.

What is Yates's correction?

When you use a program to analyze contingency tables, you might be asked about Yates's correction. The chi-square test used to analyze a contingency table can be computed in two ways. Yates's correction increases the resulting P value to adjust for a bias in the usual chi-square test, but it overcorrects.

Are special analyses available for paired data in which each subject is measured before and after an intervention?

Yes. McNemar's test is explained in Chapter 31.

Is it better to express results as the difference between two proportions or the ratio of the two proportions?

It is best, of course, to express the results in multiple ways. If you have to summarize results as a single value, I think the NNT is often the most useful way to do so. Imagine a vaccine that halves the risk of a particular infection so that the relative risk is 0.5. Consider two situations. In Situation A, the risk in unexposed people is 2 in 10 million, so halving the risk brings it down to one in 10 million. The difference is 0.0000001, so NNT = 10,000,000. In Situation B, the risk in unexposed people is 20%, so halving the risk would bring it down to 10%. The difference in risks is 0.10, so the NNT = 10. The relative risk is the same in the two situations, and the risk difference involves tiny fractions that are hard to say and remember. But the NNT is clear. In the first situation, you'd treat 10 million people to prevent one case of disease. In the second situation, you'd treat 10 people to prevent one case of disease. Expressed as NNT, the results are easy to understand and explain.

CHAPTER SUMMARY

- A contingency table displays the results of a study with a categorical outcome. Rows represent different treatments or exposures. Columns represent different outcomes. Each value is an actual number of subjects.
- When there are two treatments and two outcomes, the results can be summarized as the ratio (the relative risk), the difference between two incidence rates (the attributable risk), or the reciprocal of that difference (the number needed to treat, or NNT).
- Evidence against the null hypothesis that the outcome is not related to the treatment is expressed as a P value, computed using either Fisher's exact test or the chi-square test.
- An observed distribution can be compared against a theoretical distribution using the chi-square test. The result is expressed as a P value testing the null hypothesis that the data were in fact sampled from a population that followed the theoretical distribution. When there are only two categories, the binomial test is preferred.

TERMS INTRODUCED IN THIS CHAPTER

- Attributable risk (p. 235)
- Binomial test (p. 239)
- Chi-square test (p. 238)
- Chi-square test for trend (p. 240)
- Contingency table (p. 234)
- Double-blind study (p. 233)
- Fisher's exact test (p. 236)
- Number needed to harm (p. 235)
- Number needed to treat (NNT) (p. 235)
- Prospective study (p. 233)
- Randomized study (p. 233)
- Relative risk (p. 235)
- Risk ratio (p. 235)

CHAPTER 28

❖❖❖

Case-Control Studies

> It is now proven beyond doubt that smoking is one of the
> leading causes of statistics.
>
> FLETCHER KNEBEL

This chapter explains how to interpret results from a case-control study (also called a retrospective study). In this kind of study, the investigators select two groups of subjects: cases with the disease or condition being studied and controls selected to be similar in many ways but not to have the condition. The investigators then look back in time to compare the exposure of the two groups to a possible risk factor (or treatment).

EXAMPLE: DOES A CHOLERA VACCINE WORK?

Study design

Cholera kills many people in Africa. Although vaccines were not very effective against cholera in the past, newer vaccines hold great promise for preventing this awful disease.

Lucas and colleagues (2005) investigated whether a cholera vaccine was effective. An ideal approach would be to recruit unvaccinated people, randomly assign them to be vaccinated or not, and follow both groups for many years to compare the incidence of cholera. Such a study would take many years to conduct, require a huge number of subjects, and require withholding a vaccine from many. Instead, the investigators performed a case-control study to compare whether people with cholera are less likely to have been vaccinated than those who have not gotten cholera.

This kind of study is called a *case-control study* because the investigators pick cases and controls to study. It is also called a *retrospective study*, because the investigators start with the disease and look back in time to try to learn about the cause.

Table 28.1 shows the results of Lucas et al.'s study. Note the difference between this contingency table and the one from the example in Chapter 27 (concerning apixaban). In the apixaban example, the investigators set the row totals by choosing how many subjects would receive each treatment and then observed

	CASES (CHOLERA)	CONTROLS
Received vaccine	10	94
No vaccine	33	78
Total	43	172

Table 28.1. Case-control study to investigate association of cholera with lack of vaccination. Each value is a number of patients.
Source: Lucas et al. (2005).

the outcome tabulated in the two columns. In this example, the investigators set the column totals by choosing how many cases and controls to study and then determined whether each person had been previously vaccinated.

Computing the relative risk from case-control data is meaningless

It would be a mistake to look at Table 28.1 and divide 10 (number of people with cholera who were vaccinated) by 104 (the total number of vaccinated people in the study) to calculate a 10% risk that vaccinated people will get cholera. This is not a helpful calculation and is not a correct statement. The authors chose to have four times as many controls as cases. If they had chosen to use only twice as many controls as cases, that same calculation would have divided 10 into 52 and concluded that there is a 20% risk that vaccinated people will get cholera. Computing risks, and the relative risk, directly from a case-control study is invalid.

The odds ratio

Results from a case-control study are summarized as an *odds ratio*. Among the cases, the odds of being vaccinated are 10:33, or 0.303. The difference between odds and probability was explained in Chapter 2. Among the controls, the odds of being vaccinated are 94:78, or 1.205. The odds ratio therefore equals 0.303/1.205, or 0.25. A computer program can compute the 95% CI, which extends from 0.12 to 0.54.

It isn't obvious, but if the disease is fairly rare (defined as affecting less than about 10% of the population studied), then the odds ratio calculated from a case-control study will be approximately equal to the true relative risk. For this example, the odds ratio is 0.25. If we assume that cholera is rare in the population studied, we can conclude that vaccinated individuals are about 25% as likely to get cholera as unvaccinated people.

In the case of vaccination studies, it makes sense to subtract the odds ratio from 1.0. Here that difference is 0.75, which means the vaccine is 75% effective in preventing cholera. Vaccinated individuals are 25% as likely to get cholera as unvaccinated people, so the vaccine is 75% effective. The CI for this is obtained by subtracting each end of the odds ratio CI from 1.0. The 95% CI for vaccine effectiveness extends from 46 to 88%. (The numbers reported in the original paper are slightly different than these, because the researchers analyzed the data using a fancier method called *conditional logistic regression* to account for the matching of controls with each case.)

Interpreting a P value

When you analyze these data with a computer program, you will be asked which test to use to compute the P value. Most would agree that Fisher's exact test is preferred over a chi-square test. An alternative is to use a chi-square test (much easier if you are doing the calculations manually, but why would you?).

The P value for the example is 0.0003 (computed by Fisher's exact test). The null hypothesis is that there is no association between the vaccination and whether or not someone got cholera. The P value is easy to interpret. If the vaccination rates were really the same among patients and controls, there is only a 0.03% chance that random sampling in an experiment of this size would end up showing such a strong association between disease and vaccination. This is a two-tailed P value, so it includes the possibilities of seeing either a higher or a lower incidence of cholera in people who were vaccinated.

EXAMPLE: ISOTRETINOIN
AND BOWEL DISEASE

Study design

Isotretinoin is a very effective drug for treating severe acne, but there have been concerns about its toxicity. Several case reports have suggested that use of this drug might cause inflammatory bowel disease (IBD, which encompasses both ulcerative colitis and Crohn's disease). But the case reports really can't be interpreted except as vague warnings. Of course, some people treated with isotretinoin get inflammatory bowel disease. The question is whether people who take the drug have a higher risk of getting the disease than do people who don't take the drug. Because IBD is fairly rare, it would be very hard to answer such a question in a prospective study, so Bernstein and colleagues (2009) performed a case-control study.

The cases consisted of every single case of IBD that was reported in a person less than 40 years old and living in Manitoba, Canada, between 1995 and 2007. For each case, the researchers used 10 controls matched by age, sex, postal code, and medical insurance coverage.

The odds ratio

They analyzed the data for each disease (ulcerative colitis and Crohn's disease) separately and also analyzed the two diseases combined. The combined data are shown in Table 28.2. Of the 1,960 cases of people with IBD, 25 (1.28%) had taken

	CASES (IBD)	CONTROLS (NO IBD)
Received isotretinoin	25	213
No isotretinoin	1,935	19,216
Total	1,960	19,419

Table 28.2. Case-control study to investigate association of inflammatory bowel disease (IBD) with isotretinoin.

Source: Bernstein et al. (2009).

isotretinoin prior to their diagnosis. Of the 19,419 controls, 213 (1.10%) had taken isotretinoin. The odds ratio is 1.16, with a 95% confidence interval ranging from 0.73 to 1.77. Since that interval contains 1.0, the investigators concluded there was no evidence of an association between isotretinoin use and IBD.

Even thought the samples are huge, the confidence interval is quite wide. This is because only a small fraction (about 1%) of either group took the drug.

The P value

The researchers presented the results as an odds ratio with confidence interval, but did not also present a P value or statements of statistical significance. I think this was a wise decision. The odds ratio with its CI summarizes the results very clearly.

Even though the P value wasn't presented, we can infer something about it. Since the 95% confidence interval contains 1.0 (the value that denotes no association), the P value must be greater than 0.05 (since $100\% - 95\% = 5\% = 0.05$). Since 1.0 is not very near either end of the interval, the P value must be quite a bit greater than 0.05.

Generalizing the results

This study used data from every single case of IBD reported in Manitoba over a 12-year period. The people included in the study are not a random sample. Are statistical calculations valid? Yes. But you would need to assume that these patients in Manitoba during the period of 1995–2007 are representative of future patients all over the world. That seems like a reasonable assumption, since IBD (as far as I know) is not a disease that manifests itself differently in various locations or a disease that changes from year to year.

EXAMPLE: GENOME-WIDE ASSOCIATION STUDIES

As discussed in Chapter 22, genome-wide association studies look for associations between a disease or condition and genetic variations (usually single nucleotide polymorphisms). Many cases (often tens of thousands) and controls are tested, and a separate analysis is done for each variation. Essentially, a GWAS simultaneously runs many (perhaps a million) case-control studies at once. Because so many hypotheses are tested at once, analysis of the data must account for multiple comparisons, as explained in Chapter 22.

THE CHALLENGE OF CASE-CONTROL STUDIES

The advantage of case-control studies is that they can be done relatively quickly with a relatively small sample size drawn from previously recorded data. The alternative, in most cases, would be huge, time-consuming studies. The challenge of case-control studies is to pick the right controls. The idea is to control for extraneous factors that might confuse the results but not control away the effects you are looking for.

Cholera example

In the cholera example, the authors picked controls by visiting homes near the patient. There are some problems with this approach:

- The controls were picked because they were the same gender as the subject (with a few exceptions). Therefore, the study could not determine whether the vaccine reduces the incidence of cholera differently in men and women.
- The controls were picked to have the same age as the subjects. Therefore, the study could not determine anything about whether the vaccine reduces the incidence of cholera more in people of one age group than another.
- The subjects obviously knew they had cholera and so may have recalled their vaccination more vividly than the controls.
- The interviewers knew whether they were talking to a control subject or a case (someone who had cholera). Although they tried to ask the questions consistently, they may have inadvertently used different emphasis with the cases.
- The subjects had suffered from cholera and may have been motivated to help the researchers learn more about the disease. In contrast, the controls may have been more focused on ending the interview as quickly as possible. Thus, cases and controls may not have given equally detailed or accurate information.
- The only way patients became part of the study was by coming to the Cholera Treatment Center. Thus, the study did not include people who had mild cases of the disease and did not seek medical attention. Not everyone who is sick seeks medical attention, and this study selected for people who go to doctors when ill. This selection criterion was not applied to the controls.
- The only way that controls became part of the study was to be home when the investigators visited. This method thus selected for people who stay home a lot. This selection criterion was not applied to the cases.

These kinds of problems are inherent in all case-control studies. The investigators in this study were very aware of these potential problems. To test whether these biases affected their results, they ran a second case-control study. Here, the cases were patients with bloody diarrhea that turned out to not be caused by cholera. The controls were selected in the same way as in the first study. The second study, which had almost the same set of biases as the first, did not detect an association between diarrhea and cholera vaccination. The odds ratio was 0.64, with a 95% CI extending from 0.34 to 1.18. Because the CI spans 1.0, the study shows no association between cholera vaccination and bloody diarrhea not caused by cholera. The negative results of the second study suggest that the association reported in the main study really was caused by vaccination, rather than by any of the biases listed above.

Because of these problems, some statisticians recommend being very skeptical of odds ratios between 0.33 and 3, and even odds ratios between 0.25 and 4, even if the P value is tiny and the CI narrow (Taubes, 1995). There are too many ways for subtle biases to affect the results. How do you know when to accept the results? Results are more likely to be true when the odds ratio (or its reciprocal) is large, when the results are consistent when repeated, and when the results make sense biologically.

Genome-wide association study example

A tiny P value in a GWAS is evidence that the prevalence of that particular genetic variation differs in the two groups you are studying. One of the groups is composed of people with a disease and the other group is composed of controls, so one explanation is that the disease is associated with, and perhaps even caused by, that genetic variation. However, it may also be the case that the association is due to different ancestry of patients and controls and so has nothing to do with the disease being studied. For example, there may be more people of Jewish ancestry in one group and more of Italian ancestry in the other. Geneticists call this *population stratification*.

The huge size of a GWAS (thousands or tens of thousands of cases and controls) makes it very difficult to avoid population stratification but makes it possible to measure and (attempt to) correct for it (Price et al., 2010).

ASSUMPTIONS IN CASE-CONTROL STUDIES

Assumption: Random (or representative) sample

The 95% CI is based on the assumption that your cases and controls were randomly selected from the populations of cases and controls. In many cases, this assumption is not true. You can still interpret the CI as long as you assume that your sample is representative of the population about which you are generalizing.

The patients in the cholera example were certainly not a random selection of all those with cholera, but it is reasonable to think that they are representative of adults with cholera in that particular city and perhaps all of Africa. The patients in the isotretinoin study were certainly not randomly selected. The investigators studied every single person in Manitoba who reported having an IBD during a 12-year period. The results can only be interpreted if you assume that these subjects are similar to those who will exist elsewhere in future years.

Assumption: Independent observations

The 95% CI is only valid when all subjects are sampled from the same population and each has been selected independently of the others. Selecting one member of the population should not change the chance of selecting anyone else.

This assumption would be violated if the study included several people from one family or if there were several unique strains of cholera.

Assumption: Accurate data

The 95% CI is only valid when the number of subjects in each category is tabulated correctly. This assumption would be violated in the cholera example if some of the patients actually didn't have cholera or if some of the controls did.

Assumption: Assessing an event you really care about

The 95% CI allows you to extrapolate from the sample to the population for the event that you tabulated. In many cases, the event measured in the study is a few steps removed from the outcome you assessed. This was not an issue either of in the two examples of this chapter.

Assumption: No difference between the two groups except for the disease

The validity of a case-control study depends on the assumption that the controls do not differ systematically from the cases in any way except for the absence of disease. As discussed above, it can be very difficult to be sure a case-control study has complied with this assumption.

EPIDEMIOLOGY LINGO

Incidence versus prevalence

Incidence is the rate of new cases of a disease. *Prevalence* is the fraction of the group that already has the disease.

Four kinds of studies tabulated in a contingency table

In a *cross-sectional study*, the investigator selects a single sample of subjects without regard to either the disease or the risk factor. The subjects are then divided into two groups based on previous exposure to the risk factor. Finally, the investigators compare the prevalence of the disease in the two groups.

In a *prospective study* (introduced in Chapter 27 and also called a *longitudinal study*), the investigators select two groups of subjects. One group has been exposed to a possible risk factor. The other group hasn't. The investigator then waits while the natural history of the disease progresses and compares the incidence rates in the two groups.

In an *experimental study*, the investigator selects a single sample of subjects, who are randomly divided into two groups. Each of the groups gets a different treatment (or no treatment), and the investigators compare the incidence of the disease. The apixaban example in Chapter 27 was an experimental study.

In the *case-control (or retrospective) study*, the investigators select two groups of subjects. One group has the disease or condition being studied. These are the cases. The other group is selected to be similar in many ways but not to have the condition. These are the controls. The investigators then look back in time to compare the exposure of the two groups to a possible risk factor or a treatment of uncertain efficacy. Both examples in this chapter are case-control studies.

Contingency table

Contingency tables (first introduced in Chapter 27) show how the outcome is contingent on the treatment or exposure. The rows represent exposure (or lack of exposure) to alternative treatments or possible risk factors. Each subject belongs to one row based on exposure or treatment. Columns denote alternative outcomes. Each subject belongs to one column based on outcome. Therefore, each cell in the table is the number of subjects that were in one particular exposure (or treatment) group and had one particular outcome. Data from all four of the study designs listed above can be shown on a contingency table.

Not all tables are contingency tables. Contingency tables always show the actual number of subjects (or some other experimental unit) in various categories. Thus, each number must be a positive integer or zero. Tables of fractions, proportions, percentages, averages, changes, or durations are not contingency tables. Nor are tables of rates, such as number of cases per 1,000 people. Table 27.3 compares observed and expected counts and therefore is not a contingency table.

COMMON MISTAKES: ODDS RATIOS

Mistake: Interpreting an odds ratio as a relative risk

If you interpret the odds ratio as a relative risk, you'll think the effect size is bigger than it is. If the odds ratio is greater than 1.0, it exceeds the relative risk. If it is a fraction (less than 1.0), then it is smaller than the relative risk. Only if a tiny fraction of the control group has the condition being studied is the odds ratio a reasonable approximation of the relative risk. How tiny is that? It is hard to say, as it depends on the odds ratio.

The following equation from Zhang and Yu (1998) shows how to compute the relative risk (RR) from the odds ratio (OR) and the fraction of the control group that has the condition (prevalence, P_0).

$$RR = \frac{OR}{(1 - P_0) + (P_0 \cdot OR)}$$

I have included this equation in this nonmathematical book, because it is a common mistake to interpret an odds ratio as if it were a relative risk even when the prevalence is not tiny.

Mistake: Entering normalized data

When entering data into a contingency table, you must enter the actual number of subjects or events counted in each group. The results will be misleading if you enter percentages, normalized values, or any kind of rate. This mistake will not affect the computation of the odds ratio but will affect the calculation of its confidence interval and the P value.

Mistake: Trying to compute an odds ratio when one of the four values is zero

If one of the four values is zero, then the computed odds ratio is likely to be undefined as a result of division by zero. In this case, many investigators and programs add 0.5 to each value before calculating the odds ratio.

CHAPTER SUMMARY

- Case-control studies start by identifying people with and without a disease or condition and then look back in time to compare exposure to a possible risk factor or to a treatment.
- The advantage of a case-control study compared to an experimental study is that it can be performed more quickly, can sometimes use existing data, and requires smaller sample sizes.
- The results of a case-control study are expressed as an odds ratio, along with its 95% confidence interval. If the disease is rare, the odds ratio can be interpreted as an approximate relative risk.
- Fisher's exact test (or the chi-square test) can compute a P value that tests the null hypothesis of no association between disease and exposure to the possible risk factor or treatment.

TERMS INTRODUCED IN THIS CHAPTER

- Case-control study (p. 242)
- Cross-sectional study (p. 248)
- Experimental study (p. 248)
- Incidence (p. 248)
- Longitudinal study (p. 248)
- Odds ratio (p. 243)
- Population stratification (p. 247)
- Prevalence (p. 248)
- Retrospective study (p. 248)

CHAPTER 29

❖❖❖

Comparing Survival Curves

A statistician is a person who draws a mathematically precise
line from an unwarranted assumption to a foregone conclusion.

UNKNOWN AUTHOR

Chapter 5 explained how survival curves were generated and how
to interpret the confidence interval of survival curves. This chapter
explains how to compare two survival curves.

EXAMPLE SURVIVAL DATA

In one study, patients with chronic active hepatitis were treated with prednisolone
or placebo, and their survival was compared (Kirk et al., 1980; raw data from
Altman & Bland, 1998). Some patients were still alive at the time the data were
collected, and these individuals are indicated in Table 29.1 with an asterisk. The
information that these patients lived at least up to the time of data collection is an
essential part of the data collected from the study. The values for these individuals
are entered into a computer program as censored values. Additionally, one patient
taking prednisolone was lost to follow-up at 56 months. His data are also said to
be censored and are entered into a survival analysis program the same as the other
censored values. He lived at least 56 months, but we don't know what happened
after that.

Figure 29.1 shows the Kaplan-Meier graph of the survival data, as previously
explained in Chapter 5.

ASSUMPTIONS WHEN COMPARING
SURVIVAL CURVES

The following list summarizes the assumptions, explained in detail in Chapter 5,
that must be accepted in order to interpret survival analyses.

- Random (or representative) sample
- Independent subjects
- Consistent entry criteria
- Consistent definition of the end point
- Clear definition of the starting point

251

PREDNISOLONE	CONTROL
2	2
6	3
12	4
54	7
56 (left study)	10
68	22
89	28
96	29
96	32
125*	37
128*	40
131*	41
140*	54
141*	61
143	63
145*	71
146	127*
148*	140*
162*	146*
168	158*
173*	167*
181*	182*

Table 29.1. Sample survival data from Kirk et al. (1980), with raw data from Altman and Bland (1998).

Patients with chronic active hepatitis were treated with prednisolone or placebo, and their survival was compared. The values are survival time in months. The asterisks denote patients still alive at the time the data were analyzed. These values are said to be *censored*.

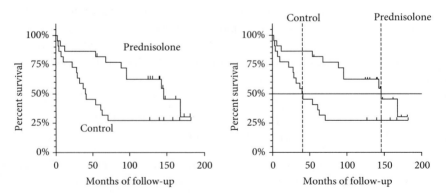

Figure 29.1. Kaplan-Meier survival curves created from the data in Table 29.1.

Each drop in the survival curve shows the time of one (or more) patient's death. The blips show the time of censoring. One subject was censored when he left the study. The data for the others were censored because they were alive when the data collection ended. The graph on the right shows how median survival times are determined.

- Time of censoring is unrelated to survival
- Average survival doesn't change during the study

When comparing survival curves, there are additional assumptions, as follows.

Assumption: Treatment groups were defined before data collection began

When you compare groups, the groups must be defined before data collection begins. It is not valid to divide a single group of patients (all treated the same way) into two groups based on whether they responded to treatment. It can be tempting, for example, to compare survival of subjects whose tumor got smaller as treatment progressed with those whose tumors stayed the same size, or to compare those whose lab tests improved with those whose lab tests didn't. Comparing survival of responders with nonresponders is invalid for two reasons.

The first reason is that a patient cannot be defined as a "responder" unless he or she survived long enough for you to measure the tumor. Any patient who died early in the study would certainly be defined as a member of the nonresponder group. In other words, survival influenced the assignment of a patient to a particular group. Therefore, you can't learn anything by comparing survival in the two groups.

The second reason is that the disease may be heterogeneous. The patients who responded may have had a different form of the disease than those who didn't respond. In this case, the responders may have survived longer even if they hadn't been treated.

You must define the groups you are comparing before starting the experimental phase of the study. Be very wary of studies that use data collected during the experimental phase of the study to divide patients into groups or to adjust the data.

Assumption: Groups are defined consistently as data are accrued

Some survival studies accrue patients over several years. In these studies, it is essential that the diagnostic groups be defined consistently.

Here is an example of when that assumption might be violated. Imagine you are comparing the survival of patients whose cancer has metastasized to the survival of patients whose cancer remains localized. Average survival is longer for the patients without metastases. Now imagine that a fancier scanner becomes available, making it possible to detect metastases earlier. What happens to the survival of patients in the two groups?

The group of patients without metastases will be smaller. The patients who are removed from the group will be those with small metastases that could not have been detected without the new technology. These patients will tend to die sooner than those patients without detectable metastases. By taking away these patients, the average survival of the patients remaining in the "no metastases" group will probably improve.

What about the other group? The group of patients with metastases will now be larger. The additional patients, however, will be those with small metastases. These patients will tend to live longer than patients with larger metastases. Thus, the average survival of all patients in the "with metastases" group will probably also improve.

Changing the diagnostic method paradoxically increases the average survival of both groups! Feinstein, Sosin, and Wells (1985) termed this paradox the *Will Rogers phenomenon* from a quote attributed to the humorist Will Rogers: "When the Okies left Oklahoma and moved to California, they raised the average intelligence in both states."

This phenomenon also causes a problem when making historical comparisons. The survival experience of people diagnosed with early-stage breast cancer is much better now than it was prior to widespread use of screening mammography. But does that mean that screening mammography saves lives? That is a hard question to answer, since the use of screening mammography has nearly doubled the number of diagnoses of early breast cancer while barely changing the number of diagnoses of advanced breast cancer. Nearly a third of the people who are diagnosed with breast cancer by mammography have abnormalities that probably would never have caused illness (Bleyer & Welch, 2012). Comparing the survival curves of people diagnosed with early breast cancer then and now is pointless, as the diagnostic criteria have changed so much.

Assumption: Proportional hazards

Hazard is defined as the slope of the survival curve—a measure of how rapidly subjects are dying. The *hazard ratio* compares two treatments. If the hazard ratio is 2.0, then the rate of deaths in one treatment group is twice the rate in the other group. Figure 29.2 shows ideal curves with a consistent hazard ratio. Both curves

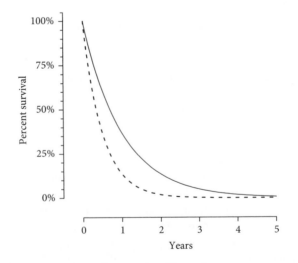

Figure 29.2. Definition of proportional hazards.
At every time point, the dotted curve is twice as steep as the solid curve. The hazard ratio is 2.0 (or 0.5, depending on how you compute it) at all time points.

start, of course, with 100% and end at 0%. At any time along the curve, the slope of one curve is twice the other. The hazards are proportional.

The assumption of *proportional hazards* is required for most comparisons of survival curves. This means that the hazard ratio is consistent over time, and any differences are caused by random sampling. This assumption would be true if, at all times, patients from one treatment group are dying at about half the rate of patients from the other group. It would not be true if the death rate in one group is much higher at early times but lower at late times. This situation is common when comparing surgery (high initial risk, lower later risk) with medical therapy (less initial risk, higher later risk).

If two survival curves cross, the assumption of proportional hazards is unlikely to be true. The exception would be when the curves only cross at late time points when few patients are still being followed.

COMPARING TWO SURVIVAL CURVES

Survival curves can be compared several ways

Hazard ratio

If the assumption of proportional hazards is accepted (which does not seem unreasonable given the sample data), the two survival curves can be summarized with the hazard ratio, which is essentially the same as the relative risk (see Chapter 27). For the example data, the hazard ratio is 0.42, with a 95% CI ranging from 0.19 to 0.92. In other words, the treated patients are dying at 42% of the rate of control patients, and we have 95% confidence that the true ratio is between 19% and 92%. More specifically, this means that in any particular day or week, a treated patient has 42% of the chance of dying as a control patient.

If the two survival curves are identical, the hazard ratio would equal 1.0. Because the 95% CI of the sample data does not include 1.0, we can be at least 95% certain that the two populations have different survival experiences.

Ratio of median survival times

Median survival time is defined as the time it takes until half the subjects have died. The right panel of Figure 29.1 shows the median survival of the sample data. The horizontal line is at 50% survival. The time at which each survival curve crosses that line is the median survival. The median survival of the control patients is 40.5 months, whereas the median survival of the patients treated with prednisolone is 146 months, 3.61 times longer.

It is possible to compute a 95% CI for the ratio of median survivals. This calculation is based not only on the assumption of proportional hazards (see prior section), but also on an additional assumption: that the chance of dying in a small time interval is the same early in the study and late in the study. In other words, the survival curve follows the same pattern as an exponential decay curve. This assumption does not seem unreasonable for the sample data. For these data, the ratio of median survivals is 3.61, with a 95% CI ranging from 3.14 to 4.07. In

other words, we are 95% sure that treatment with prednisolone triples to qua-
druples the median survival time.

Confidence bands

Figure 29.3 shows the example survival curves with their 95% confidence bands.
The curves overlap a bit, so plotting both curves together creates a very cluttered
graph. Instead, the two are placed side by side. Both curves start at 100%, of
course, so they overlap somewhat, but note that they don't overlap at all for many
months. This graph should be enough to convince you that the prednisolone
worked.

Note the difference between the confidence band, which is an area plotted on
a graph, and a confidence interval, which is a range of values.

P Value

The comparison of two survival curves can be supplemented with a P value. All
P values test a null hypothesis. When comparing two survival curves, the null
hypothesis is that the survival curves of the two populations are identical and any
observed discrepancy is the result of random sampling error. In other words, the
null hypothesis is that the treatment did not change survival overall and that any
difference observed was simply the result of chance.

The P value answers the question, If the null hypothesis is true, what is the
probability of randomly selecting subjects whose survival curves are as different
(or more different) than what was actually observed?

The calculation of the P value is best left to computer programs. The *log-
rank method*, also known as the *Mantel–Cox method* (and nearly identical to
the Mantel–Haenszel method), is used most frequently. The assumption of
proportional hazards (explained earlier in this chapter) is required to interpret this
P value.

For the sample data, the two-tailed P value is 0.031. If the treatment was
really ineffective, it is possible that the patients who were randomly selected to

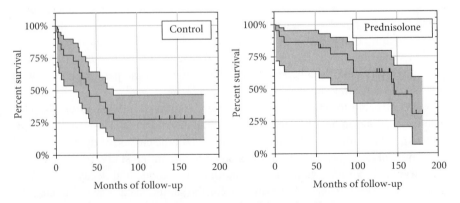

Figure 29.3. Survival curves showing the 95% confidence bands.

receive one treatment just happened to live longer than the patients who received the other treatment. The P value tells us that the chance of this happening is only 3.1%. Because that percentage is less than the traditional significance cutoff of 5%, we can say that the increase in survival with treatment by prednisolone is statistically significant.

An alternative method to compute the P value is known as the *Gehan–Breslow–Wilcoxon method*. Unlike the log-rank test, which gives equal weight to all time points, the Gehan–Breslow–Wilcoxon method gives more weight to deaths at early time points. This often makes a lot of sense, but the results can be misleading when a large fraction of patients are censored at early time points. The Gehan–Breslow-Wilcoxon test does not require a consistent hazard ratio, but it does require that one group consistently have a higher risk than the other.

For the sample data, the P value computed by the Gehan–Breslow–Wilcoxon method is 0.011.

WHY NOT JUST COMPARE MEAN OR MEDIAN SURVIVAL TIMES OR FIVE-YEAR SURVIVAL?

Why compare entire survival curves? Why not just compare the mean or median time until death?

Why it is rarely helpful to compare mean time until death

There are three reasons why it is rarely helpful to compare the mean age of death:

- You usually will want to tabulate the data before every subject has died. If anyone is still alive at the time you compile the data, it will not be helpful to compute the average age of death of only those who have died.
- If the data for any subject are censored, it will be impossible to calculate a meaningful mean time of death.
- Even if every subject has died (and none were censored), the mean time until death may still not be meaningful. It is likely that the distribution of ages is not Gaussian. If so, summarizing the data with the mean may not be informative. Furthermore, calculations like the 95% CI of the mean won't be helpful if the distribution is far from Gaussian.

Median time until death

Why not just compare the median time until death? This would avoid the problem with mean times, since the median survival time can be calculated once half the subjects have died. The median time until death is a useful way to summarize survival data, but focusing on a single number (the median survival) can be misleading.

Figure 29.4 shows two simulated survival curves. The first parts of the curves are identical, so they appear superimposed. They have almost the same median survival (7.5 years) . But after Year 7, the two curves diverge drastically. If you had a choice between the two treatments, you'd be nuts not to choose the one with

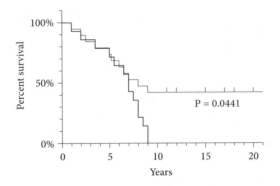

Figure 29.4. Two curves with identical five-year survival rates and identical median survival times.

The curves are very different, and it would be easy to pick which treatment you wanted. The log-rank test shows that the difference between the two survival curves is statistically significant, with P = 0.0232.

a survival curve that levels off with 42% of the patients still living after 20 years, rather than the treatment in which all subjects had died by Year 9.

Five-year survival

The success of cancer therapies is often expressed as a five-year survival. This tells you what fraction of the patients are still alive after five years. Five years is an arbitrary but standard length of time to tabulate survival. Figure 29.4 shows why focusing only on that one number can be misleading. The two curves in that figure have the same five-year survival (72%), but diverge a lot after five years. Even though the five-year survival is the same for both treatments, the ten-year survival is very different.

INTENTION TO TREAT

When comparing the survival of people randomly assigned to receive alternative treatments, you will often hit a snag. Some patients don't actually get the treatment they were randomized to receive. Others stop following the protocol.

It seems sensible to just exclude from analysis all data from people who didn't receive the full assigned treatment. This is called the *per protocol approach*, because it requires that you only compare survival of subjects who got the full treatment according to the study protocol. But this approach will lead to biased results if the noncompliance has any association with disease progression and treatment. Hollis and Campbell (1999) review an example of this. The study in question randomized patients with severe angina to receive a surgical or medical treatment. Some of the patients randomly assigned to the surgical group died soon after being randomized so never actually received the surgery. It would seem sensible to exclude them from the analysis, since they never received the treatment to which they were assigned. But that would exclude the earliest deaths from one assigned treatment (surgery) but not the other (medical). The results would then not be interpretable. The surgery group would have better survival, simply because people who died early were excluded from the analysis.

Dallal (2012) gives another example. Imagine a study of two alternative weight-loss treatments. One of the treatments is basically ineffective, but some people lose weight anyway. They stay in the study. Others don't lose weight and drop out of the study. If you only study those who remain in the study, you are selecting for those who lose weight, which would make a comparison between the two treatments invalid.

The preferred approach is called *intention to treat* (Hollis & Campbell, 1999; Montori & Guyatt, 2001). This approach is abbreviated ITT and summarized as "analyze as randomized." With ITT, data are analyzed based on the randomly assigned treatments even if:

- it later turns out that the patient didn't quite meet the entry criteria of the study;
- the treatment was not given; or
- they stopped the treatment for any reason.

Some studies analyze their data using both the per protocol and the intention-to-treat approaches. If the results are similar, then you know that the subjects who didn't get the assigned treatment (or didn't get a complete treatment) didn't affect the results much.

The ITT method applies to any analysis of clinical data, not just survival analysis. Some investigators use the intention-to-treat principle for analyzing efficacy but use the per protocol method for analyzing toxicity (e.g., Agnelli et al., 2012).

Q & A

Is the log-rank test the same as the Mantel–Haenszel test?

Almost. The two differ only in how they deal with multiple deaths at exactly the same time point. The results will be very similar.

Is the Gehan–Breslow-Wilcoxon test also the same?

No. The Gehan–Breslow-Wilcoxon method gives more weight to deaths at early time points, which makes a lot of sense. But the results can be misleading when a large fraction of patients are censored at early time points. In contrast, the log-rank test gives equal weight to all time points. The Gehan–Breslow-Wilcoxon test does not require a consistent hazard ratio, but it does require that one group consistently has a higher risk than the other. Of course, you should choose the test as part of the experimental design.

What if the two survival curves cross?

If two survival curves truly cross, then one group has a higher risk at early time points and the other group has a higher risk at later time points. If the two curves are based on plenty of data and the crossing point is near the middle of the time span, then the data violate the assumptions of both the log-rank and the Gehan–Breslow–Wilcoxon test, and you'd need to use more specialized methods that are beyond the scope of this book.

When survival curves overlap only at later time points, it may be a matter of chance and mean nothing. At those late time points, fewer patients are followed, and the two curves may cross as the result of chance.

CHAPTER SUMMARY

- Survival analysis is used to analyze data for which the outcome is elapsed time until a one-time event (often death) occurs.
- Comparisons of survival curves can only be interpreted if you accept a list of assumptions.
- The hazard ratio is one way to summarize the comparison of two survival curves. If the hazard ratio is 2.0, then subjects given one treatment have died at twice the rate of subjects given the other treatment. Hazard ratios should be reported with a 95% confidence interval.
- Another way to summarize the comparison of two survival curves is to compute the ratio of the median survival times, along with a confidence interval for that ratio.
- A P value tests the null hypothesis that survival is identical in the two populations and the difference you observed is due to random assignment of patients.
- It is rarely helpful to compute the mean time until death for two reasons. One reason is that you can't compute the mean until the last person has died. The other reason is that survival times are rarely Gaussian. As a result, the mean is not the best way to summarize the data.
- Computing the median survival is more useful than computing the mean survival, but that approach doesn't replace the need to view the survival curves.
- Most investigators analyze survival data from randomized trials using the intention-to-treat principle. In this approach, you analyze each person's data according to the treatment group to which they were assigned, even if they didn't actually get that treatment.

TERMS INTRODUCED IN THIS CHAPTER

- Five year survival (p. 258)
- Gehan–Breslow–Wilcoxon method (p. 257)
- Hazard (p. 254)
- Hazard ratio (p. 254)
- Intention to treat (ITT) (p. 259)
- Log-rank method (p. 256)
- Mantel–Cox method (p. 256)
- Median survival (p. 255)
- Per protocol approach (p. 258)
- Proportional hazard (p. 255)
- Will Rogers phenomenon (p. 254)

CHAPTER 30

Comparing Two Means:
Unpaired t Test

Researchers often want Bioinformaticians to be Biomagicians, people who can make significant results out of non-significant data, or Biomorticians, people who can bury data that disagree with the researcher's prior hypothesis.

DAN MASYS

This chapter explains the unpaired t test which compares the means of two groups, assuming the data were sampled from a Gaussian population. Chapter 31 will explain the paired t test, Chapter 39 will explain the nonparametric Mann–Whitney test (and computer-intensive bootstrapping methods), and Chapter 35 will show how an unpaired t test can be viewed as a comparison of the fit of two models.

INTERPRETING RESULTS FROM
AN UNPAIRED t TEST

Example: Maximum relaxation of bladder muscles

Frazier, Schneider, and Michel (2006) measured how well the neurotransmitter norepinephrine relaxes bladder muscles. Chapter 11 looked at the concentrations required to get half-maximal relaxation as an example of lognormal data. Here we will look at the maximal relaxation that can be achieved by large doses of norepinephrine and then compare old with young rats.

The raw data are listed in Table 30.1 and plotted in Figure 30.1. The two means are fairly far apart, but there is a lot of variation and the two sets of data overlap considerably. Do young and old rats differ, or is the difference observed here just a matter of chance? An *unpaired t test*—as shown in Table 30.2—can help you answer that question.

Confidence interval for difference between means

The mean maximum response in old rats is 23.5 lower than the mean maximum response in young rats. The 95% CI for the difference ranges from 9.3 to 37.8 (see Figure 30.2). Because the CI does not include zero, we are 95% sure that the mean response in old animals is less than the mean response in young ones.

261

OLD	YOUNG
20.8	45.5
2.8	55.0
50.0	60.7
33.3	61.5
29.4	61.1
38.9	65.5
29.4	42.9
52.6	37.5
14.3	

Table 30.1. Maximal relaxation of muscle strips of old and young rat bladders stimulated with high concentrations of norepinephrine (Frazier, Schneider, & Michel, 2006).

Larger values reflect more muscle relaxation. These values are graphed in Figure 30.1.

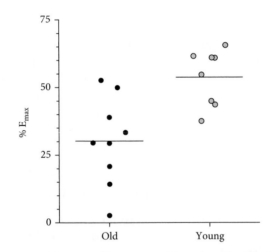

Figure 30.1. Maximal relaxation of muscle strips of old and young rat bladders stimulated with high concentrations of norepinephrine.

More relaxation is shown as larger numbers. Each symbol represents a measurement from one rat. The horizontal lines represent the means.

The width of the CI depends on three values:

- Variability. If the data are widely scattered (large SD), then the CI will be wider. If the data are very consistent (low SD), the CI will be narrower.
- Sample size. Everything else being equal, larger samples generate narrower CIs and smaller samples generate wider CIs.
- Degree of confidence. If you wish to have more confidence (i.e., 99% rather than 95%), the interval will be wider. If you are willing to accept less confidence (i.e., 90% confidence), the interval will be narrower.

UNPAIRED t TEST

P value	0.0030
P value summary	**
Are means significantly different? (P < 0.05)	Yes
One- or two-tailed P value?	Two-tailed
t, df	t = 3.531, df = 15

HOW BIG IS THE DIFFERENCE?

Mean ± SEM of Column A (old)	30.17 ± 5.365, n = 9
Mean ± SEM of Column B (young)	53.71 ± 3.664, n = 8
Difference between means	23.55 ± 6.667
95% CI	9.335 to 37.76
R^2	0.4540

F TEST TO COMPARE VARIANCES

F, DFn, Dfd	2.412, 8, 7
P value	0.2631
P value summary	ns
Are variances significantly different?	No

Table 30.2. Results of unpaired t test from GraphPad Prism.

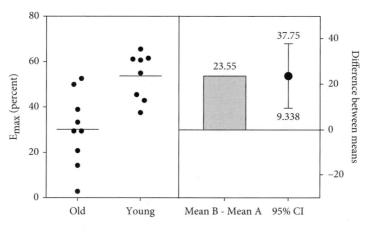

Figure 30.2. The CI for the difference between two means.
The first two columns show the actual data. The third column shows the difference between the two means, and the fourth column shows the 95% CI for that difference (plotted on the right axis).

P value

The null hypothesis is that both sets of data are randomly sampled from populations with identical means. The P value answers the question, If the null hypothesis were true, what is the chance of randomly observing a difference as large or larger than that observed in this experiment?

The two-tailed P value is 0.0030. Random sampling is unlikely to have created the difference, and the difference between the two means is statistically significant. The P value depends on the following:

- Difference between the means. Everything else being equal, the P value will be smaller when the means are farther apart.
- The SDs. Everything else being equal, the P value will be smaller when the data are more consistent (smaller SD).
- The sample size. Everything else being equal, the P value will be smaller when the sample sizes are larger.

R^2

Not all programs report R^2 with t test results, so you can ignore this value if it seems confusing. Chapter 35 will explain the concept in more detail. For these data, $R^2 = 0.45$. This means that a bit less than half (45%) of the variation among all the values is due to differences between the group means, and a bit more than half of the variation is due to differences within the groups.

t ratio

The P value is calculated from the *t ratio*, which is computed from the difference between the two sample means and the SD and sample size of each group. The t ratio has no units. In this example, t = 3.53. This number is part of the calculations of the t test, but its value is not very informative.

ASSUMPTIONS: UNPAIRED t TEST

When assessing any statistical results, it is essential to review the list of assumptions. The t test is based on a familiar list of assumptions (see Chapter 12).

- Random (or representative) samples
- Independent observations
- Accurate data
- Populations with Gaussian distributions (at least approximately)
- Equal standard deviation of the two populations, even if their means are distinct (more about this assumption below)

THE ASSUMPTION OF EQUAL VARIANCES

The unpaired t test depends on the assumption that the two data sets are sampled from populations that have identical standard deviations (and thus identical variances). Using statistical jargon, the assumption is of *homoscedasticity*.

Testing the assumption

In the sample of data collected in this experiment, you can see that there is more variability among the old animals than among the young. The standard deviations are 16.09 and 10.36. The old animals have an SD 1.55 times larger than that of the young animals. The square of that ratio (2.41) is called an *F ratio*.

The t test assumes that both groups are sampled from populations with equal variances. Are the data consistent with that assumption? To find out, let's compute another P value.

If that assumption were true, the distribution of the F ratio would be known, so a P value could be computed. This calculation depends on knowing the degrees of freedom for the numerator and denominator of the F ratio, abbreviated in Table 30.2 as DFn and DFd. Each df equals one of the sample sizes minus 1. The P value equals 0.26. If the assumption of equal variances were true, there would be a 26% chance that random sampling would result in as large a discrepancy between the two SD values as observed here (or larger still). The P value is large, so the data are consistent with the assumption.

Don't mix up this P value, which tests the null hypothesis that the two populations have the same SD, with the P value that tests the null hypothesis that the two populations have the same mean.

What if the assumption is violated?

What should you do if the P value (from the F test) is small, indicating that the assumption of equal variances is unlikely to be true? This is a difficult question, and there is no real consensus among statisticians. There are eight possible answers.

- Ignore the result. The t test is fairly robust to violations of the assumption of equal variances as long as the sample size isn't tiny and the two samples have an equal, or nearly equal, number of observations. This is probably the approach that most biologists use.
- Emphasize the result and conclude that the populations are different. If the standard deviations truly differ, then the populations are not the same. That conclusion is solid, whether the two means are close together or far apart.
- Transform the data (often to logarithms) in an attempt to equalize the variances and then run the t test on the transformed results.
- Instead of running the ordinary unpaired t test, run a modified t test that allows for unequal variance (the Welch modification to allow for different variances). If you always use this modified test, you don't have to think about the assumption of equal variances. The drawback is that this modified t test has less power to detect differences between means.
- Use the result of the test to compare variances to decide whether to run the usual unpaired t test or the modified t test that allows for unequal variance. Although this approach sounds sensible, it should not be used, because the results will be misleading (Moser & Stevens, 1992).
- Analyze the data using linear regression (as will be explained in Chapter 33) and weight by the reciprocal of each group's variability (as will be explained in Chapter 36).
- Use the result of the test that compares variances to decide whether to run a t test or the nonparametric Mann–Whitney test. This approach also sounds very reasonable but is not recommended. Chapter 41 will explain why.
- Use your experience from other studies of the same kind of data to decide how to best analyze this specific data set.

OVERLAPPING ERROR BARS AND THE t TEST

Figure 30.1 showed the individual data points of the t test example. More often (for no good reason), you'll see data plotted as a bar graph showing mean and error bars. These error bars can show the SD or the SEM (Chapter 14 explained the difference between the two).

It is tempting to look at whether two error bars overlap and try to reach a conclusion about whether the difference between means is statistically significant. Resist that temptation, because you really don't learn much by asking whether or not two error bars overlap (Lanzante, 2005). Table 30.3 summarizes the rules of thumb for interpreting these error bars.

SD error bars

Figure 30.3 graphs the mean and SD of the sample rat data. The graph on the left is typical of graphs you'll see in journals, with the error bar only sticking up. But of course, the scatter goes in both directions. The graph on the right shows this.

The two error bars overlap. The upper error bar for the data from the old rats is higher than the lower error bar for data from the young rats. What can you conclude from that overlap? Not much, because the t test also takes into account sample size (in this example, the number of rats studied). If the samples were larger (more rats), with the same means and same SDs, the P value would be much smaller. If the samples were smaller (fewer rats), with the same means and same SDs, the P value would be larger.

When the difference between two means is statistically significant ($P < 0.05$), the two SD error bars may or may not overlap. Likewise, when the difference between two means is not statistically significant ($P > 0.05$), the two SD error bars may or may not overlap. Knowing whether SD error bars overlap, therefore, does not let you conclude whether the difference between the means is statistically significant.

SEM error bars

Figure 30.4 is similar to Figure 30.3 but plots the mean and SEM of the rat data, instead of the SD.

TYPE OF ERROR BAR	CONCLUSION IF THEY OVERLAP	CONCLUSION IF THEY DON'T OVERLAP
SD	No conclusion	No conclusion
SEM	$P > 0.05$	No conclusion
95% CI	No conclusion	$P < 0.05$

Table 30.3. Rules of thumb for interpreting error bars that overlap or do not overlap.

These rules only apply when comparing two means with an unpaired t test, and when the two sample sizes are equal or nearly equal.

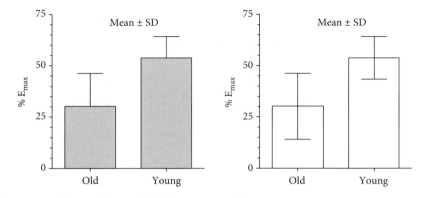

Figure 30.3. The same data plotted as mean and SD.

The left graph is similar to ones you'll often see in publications. The right graph shows the error bars going both up and down to emphasize that of course the scatter goes in both directions. To make the downward-pointing error bar obvious, the interior of the bar is kept white instead of shaded. The two error bars overlap. However, the fact that two SD error bars overlap tells you nothing about whether the difference between the two means is or is not statistically significant.

Figure 30.4. The same data plotted as mean and SEM.

Compare this figure with Figure 30.3. SEM error bars are always shorter than SD error bars. The left graph is similar to ones that you'll often see in publications. The right graph shows the error bars going both up and down to emphasize that of course the uncertainty goes in both directions. The two error bars do not overlap. However, the fact that two SEM error bars do not overlap tells you nothing about whether the difference between the two means is or is not statistically significant.

The two error bars do not overlap; the upper error bar for the data from the old rats is lower than the lower error bar for data from the young rats. What can you conclude from the lack of overlap?

The SEM error bars quantify how precisely you know the mean, taking into account both the SD and the sample size. Looking at whether the error bars overlap, therefore, lets you compare the difference between the mean with the precision of those means. This sounds promising. But in fact, you don't learn much by looking at whether SEM error bars overlap. By taking into account sample size and considering how far apart two error bars are, Cumming, Fidler, and Vaux (2007) came up with some rules for deciding when a difference is significant. However, these rules are hard to remember and apply.

Here is a rule of thumb that can be used when the two sample sizes are equal or nearly equal. If two SEM error bars overlap, the P value is (much) greater than 0.05, so the difference is not statistically significant. The opposite rule does not apply. If two SEM error bars do not overlap, the P value could be less than or greater than 0.05.

CI error bars

The error bars in Figure 30.5 show the 95% CI of each mean. Because the two 95% CI error bars do not overlap and the sample sizes are nearly equal, the P value must be a lot less than 0.05—probably less than 0.005 (Payton, Greenstone, & Schenker, 2003; Knol, Pestman, & Grobbee, 2011). But it would be a mistake to look at overlapping 95% CI error bars and conclude that the P value is greater than 0.05. That relationship, although commonly believed, is just not true. When two 95% CIs overlap, the P value might be greater than 0.05, but it also might be less than 0.05. What you can conclude is that the P value is probably greater than 0.005.

Figure 30.5. The same data plotted as means with 95% CIs.

Compare this figure with Figure 30.4. The 95% error bars are always longer than SEM error bars. The two error bars do not overlap, which means the P value must be much less than 0.05 and is probably less than 0.005.

HOW IT WORKS: UNPAIRED t TEST

Standard error of the difference between two means

Chapter 12 explained how the confidence interval of a mean is computed from the standard error of the mean (SEM). Similarly, the confidence interval of the difference between two means is computed from the standard error of the difference between the two means, which is computed by combining SEMs and both sample sizes. The standard error of the difference between two means will always be larger than either SEM but smaller than their sum. For the given example, the standard error of the difference between the two means (which is included in many programs' output) equals 6.67.

CI

The CI for the difference between the two population means is centered on the difference between the means of the two samples. The margin of error, the distance the CI extends in each direction, equals the standard error of the difference (see previous section) times a critical value from the t distribution (see Appendix D). For the sample data, the margin of error equals 6.67 (see previous section) times 2.1314 (the critical value from the t distribution for 95% CI and 15 df), or 14.22. The observed difference between means is 23.55, so the 95% CI extends from 23.55 − 14.22 to 23.55 + 14.22, or from 9.33 to 37.76.

t ratio

To determine the P value, first compute the t ratio, which equals the difference between the two sample means (23.5) divided by the standard error of that difference (6.67). In the rat example, t = 3.53. The numerator and denominator have the same units, so the t ratio has no units. Programs that compute the t test always report the t ratio, even though it is not particularly informative by itself.

Don't mix up the t ratio computed from the data with the critical value from the t distribution (see previous section) used to compute the confidence interval.

P value

The P value is computed from the t ratio and the number of df, which equals the total number of values (in both groups) minus two.

COMMON MISTAKES: UNPAIRED t TEST

Mistake: If the result is almost statistically significant, collect more data to increase the sample size and then recalculate the t test

You can only really interpret the results of any statistical calculation when the sample size was chosen in advance or you use special methods designed to handle sequential accumulation of data.

What's wrong with collecting more data when the P value is higher than 0.05? The problem is that it biases you toward getting small P values, even if the null hypothesis is true. If you stop when you get a P value you like but otherwise keep collecting more data, you can no longer interpret the P values. With this sequential approach, the chance of eventually reaching a conclusion that the result is statistically significant with $P < 0.05$ is much higher than 5%, even if the null hypothesis is true.

Mistake: If your experiment has three or more treatment groups, use the unpaired t test to compare two groups at a time

If you perform multiple t tests this way, it is too easy to reach misleading conclusions. Use one-way ANOVA, followed by multiple comparisons posttests, as explained in Chapters 39 and 40.

Mistake: If your experiment has three or more treatment groups, compare the largest mean with the smallest mean using a single unpaired t test

You won't know which group has the largest mean and which group has the smallest mean until after you have collected the data. So although you are formally just doing one t test to compare two groups, you are really comparing all the groups. Use one-way ANOVA.

Mistake: If the P value is larger than 0.05, try other tests to see whether they give a lower P value

If the P value from an unpaired t test is "too high," it is tempting to try a nonparametric test, or to run the t test after removing potential outliers or transforming the values to logarithms. If you try multiple tests and choose to only report the results you like the most, those results will be misleading. You can't interpret the P value at face value when it was chosen from a menu of P values computed using different methods. You should pick one test before collecting data and use it.

Q & A

Can a t test be computed from the mean, SD, and sample size of each group?

Yes. You don't need the raw data to compute an unpaired t test. It is enough to know the mean, SD or SEM, and sample size of each group.

Why is it called an unpaired test?

As its name suggests, this test is used when the values in the two groups are not matched. It should not be used when you compare two measurements in each subject (perhaps before and after an intervention) or if each measurement in one group is matched to a particular value in the other group. These kinds of data are analyzed using the paired t test, which will be explained in Chapter 31.

Is the unpaired t test sometimes called Student's t test because it is the first statistical test inflicted on students of statistics?

No. W. S. Gossett developed the t test when he was employed by a brewery that required him to publish anonymously in order to keep as a trade secret the fact that they were using statistics. Gossett's pen name was "Student." He is no longer anonymous, but the name Student is still attached to the test.

How is the z test different than the t test?

> The t test is computed using the t ratio, which is computed from the difference between the two sample means and the sample size and SD of the two groups. Some texts also discuss using the z test, based on the z ratio, which requires knowing the standard deviation of the population from which the data are sampled. This population's standard deviation cannot be computed from the data at hand but rather must be known precisely from other data. Since this situation (knowing the population SD precisely) occurs rarely (if ever), the z test is rarely useful.

Does it matter whether the two groups have different numbers of observations?

> No. The t test does not require equal n. However, the t test is more robust to violations of the Gaussian assumption when the sample sizes are equal or nearly equal.

Why is the t ratio sometimes positive and sometimes negative?

> The sign (positive or negative) of t depends on which group has the larger mean and in which order they are entered into the statistical program. Because the order in which you enter the groups into a program is arbitrary, the sign of the t ratio is irrelevant. In the example data, the t ratio might be reported as 3.53 or −3.53. It doesn't matter, because the CI and P value will be the same either way.

What would a one-tailed P value be in this example?

> As explained in Chapter 15, the P value can be one- or two-sided. Here, the P value is two-sided (two-tailed). If the null hypothesis were true, the probability of observing a difference as large as that observed here with the young rats having the larger mean is 0.0015. That's a one-tailed P value. The chance of observing a difference as large as observed here with the old rats having the larger mean is also 0.0015. That's the other one-tailed P value. The sum of those two one-tailed P values equals the two-tailed P value (0.0030).

How is the t ratio reported with the t test different than the critical value of the t distribution used to calculate the confidence interval?

> The t ratio is computed from the data and used to determine the P value. The critical value of the t distribution, abbreviated t* in this book, is computed from the sample size and the confidence level you want (usually 95%) and is used to compute the confidence interval. The value of t* does not depend on the data (except for sample size).

CHAPTER SUMMARY

- The unpaired t test compares the means of two unmatched groups.
- The results are expressed both as a confidence interval for the difference between means and as a P value testing the null hypothesis that the two population means are identical.
- In addition to the usual assumptions, the unpaired t test assumes that the two populations follow Gaussian distributions with the same SD in each population.
- Looking at whether error bars overlap or not is not very helpful. The one rule of thumb worth remembering is that if two SEM error bars overlap, the P value is greater than 0.05.

TERMS INTRODUCED IN THIS CHAPTER

- F ratio (p. 264)
- Homoscedasticity (p. 264)
- t ratio (p. 264)
- Unpaired t test (p. 261)

CHAPTER 31

❖❖❖

Comparing Two Paired Groups

To call in the statistician after the experiment is done may
be no more than asking him to perform a postmortem
examination: he may be able to say what the experiment
died of.

R. A. FISHER

This chapter explains the paired t test, which compares two matched or paired groups when the outcome is continuous, and the McNemar's test, which compares two matched or paired groups when the outcome is binomial. Chapter 30 explained the unpaired t test and Chapter 41 will explain the nonparametric Wilcoxon test (and computer-intensive bootstrapping methods).

WHEN TO USE SPECIAL TESTS FOR PAIRED DATA

Often, experiments are designed so that the same patients or experimental preparations are measured before and after an intervention. These data should not be analyzed with an unpaired t test or the nonparametric Mann–Whitney test. Unpaired tests do not distinguish variability among subjects that results from differences caused by treatment.

When subjects are matched or paired, you should use a special paired test instead. Paired analyses are appropriate in these types of protocols:

- A variable is measured in each subject before and after an intervention.
- Subjects are recruited as pairs and matched for variables such as age, postal code, or diagnosis. One of each pair receives one intervention, whereas the other receives an alternative treatment.
- Twins or siblings are recruited as pairs. Each receives a different treatment.
- Each run of a laboratory experiment has a control and treated preparation handled in parallel.
- A part of the body on one side is treated with a control treatment and the corresponding part of the body on the other side is treated with the experimental treatment (e.g., right and left eyes).

EXAMPLE OF PAIRED t TEST

Most of the examples in this book are from the fairly recent medical literature. For this example, however, we'll go back over a century to an example that was used by Ronald Fisher, one of the pioneers of statistics (Fisher, 1935). Fisher looked at an experiment by Charles Darwin that compared the growth of plants from seeds that were self-fertilized with the growth of plants from seeds that were cross-fertilized (Darwin, 1876).

Table 31.1 shows the data. Each row in the table represents one matched set, with the two kinds of seeds grown placed side by side. By designing the experiments this way, Darwin controlled for any changes in soil, temperature, or sunlight that would affect both kinds of seeds. Figure 31.1 graphs the data on a *before–after plot*. This kind of graph gets its name from its common use in experiments for which the same measurement is made for each subject before and after an experimental intervention. Here, the lines don't connect before and after measurements, but rather each of the 15 lines connects measurements of plants grown from matched seeds.

Figure 31.2 shows the same data as a bar graph, plotting the mean and SEM of each group. This kind of graph is commonly used, but it shows much less information than Figure 31.1. It shows you nothing about the pairing and only indirectly (via the SEM) gives you a sense of variation. When possible, avoid bar graphs and graph the actual data.

CROSS FERTILIZED	SELF-FERTILIZED	DIFFERENCE
23.500	17.375	6.125
12.000	20.375	−8.375
21.000	20.000	1.000
22.000	20.000	2.000
19.125	18.375	0.750
21.500	18.625	2.875
22.125	18.625	3.500
20.375	15.250	5.125
18.250	16.500	1.750
21.625	18.000	3.625
23.250	16.250	7.000
21.000	18.000	3.000
22.125	12.750	9.375
23.000	15.500	7.500
12.000	18.000	−6.000

Table 31.1. Sample data for paired t test.
These data were collected by Charles Darwin, who compared the growth of cross-fertilized and self-fertilized plants. The values are the height of the plants in inches (measured to the nearest eighth of an inch). Each row represents a matched pair of self- and cross-fertilized seeds, grown under the same conditions. The last column is the difference.

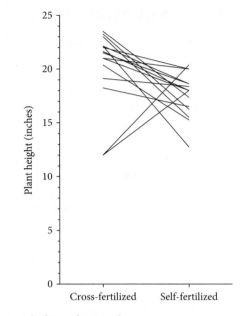

Figure 31.1. A before–after graph.

Usually, this kind of graph is used to plot values when the same measurement is made in each subject before and after an experimental intervention. Each of the 15 lines connects measurements from matched seeds.

Figure 31.2. An alternative, but inferior, approach.

This graph plots the same data as Figure 31.1 but shows only the mean and SEM of each group without accounting for pairing.

PAIRED t TEST

P value	0.0497
P value summary	*
Are means significantly different? (P < 0.05)	Yes
One- or two-tailed P value?	Two-tailed
t, df	t = 2.148, df = 14
Number of pairs	15

HOW BIG IS THE DIFFERENCE?

Mean of differences	2.617
95% CI	0.003639 to 5.230

HOW EFFECTIVE WAS THE PAIRING?

Correlation coefficient (r)	−0.3348
P value (one-tailed)	0.1113
P value summary	ns
Was the pairing significantly effective?	No

Table 31.2. Results of a paired t test as computed by GraphPad Prism.

INTERPRETING RESULTS FROM A PAIRED t TEST

Table 31.2 shows the results of a paired t test.

CI

A paired t test looks at the difference in measurements between two matched subjects (as in our example; see Figure 31.3) or a measurement made before and after an experimental intervention. For the sample data, the average difference in plant length between cross- and self-fertilized seeds is 2.62 inches. To put this in perspective, the self-fertilized plants have an average height of 17.6 inches, so this difference is approximately 15%.

The 95% CI for the difference ranges from 0.003639 inches to 5.230 inches. Because the CI does not include zero, we can be 95% sure that the cross-fertilized seeds grow more than the self-fertilized seeds. But the CI extends from a tiny difference (especially considering that plant height was only measured to the nearest eighth of an inch) to a fairly large difference.

The width of the CI depends on three values:

- Variability. If the observed differences are widely scattered, with some pairs having a large difference and some pairs a small difference (or a difference in the opposite direction), then the CI will be wider. If the data are very consistent, the CI will be narrower.
- Sample size. Everything else being equal, a sample with more pairs will generate narrower CIs, and a sample with fewer pairs will generate wider CIs.
- Degree of confidence. If you wish to have more confidence (i.e., 99% rather than 95%), the interval will be wider. If you are willing to accept less confidence (i.e., 90% confidence), the interval will be narrower.

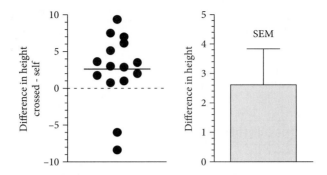

Figure 31.3. The difference between matched measurements.
Each of the circles on the left represents one paired experiment. The
length of the self-fertilized plant is subtracted from the length of the
cross-fertilized plant, and that difference is plotted. The cross-fertilized
seed grew more than the self-fertilized seed in 13 of the 15 paired
experiments, resulting in a positive difference. The other two
experiments resulted in more growth in the cross-fertilized plant, and
so the difference is negative. The graph on the right plots only the
mean and SEM. This graph takes up the same amount of space but
shows less information and isn't really easier to interpret.

P value

The null hypothesis is there is no difference in the height of the two kinds of plants.
In other words, the differences that we measure are sampled from a population for
which the average difference is zero. The P value answers the question, If the null
hypothesis were true, what is the chance of randomly observing a difference as
large as or larger than the difference observed in this experiment?

The P value is 0.0497. If the null hypothesis were true, 5% of random
samples of 15 pairs would have a difference this large or larger. Using the traditional definition, the difference is statistically significant, because the P value is
less than 0.05.

As explained in Chapter 15, the P value can be one- or two-sided. Here,
the P value is two-sided (two-tailed). Assuming the null hypothesis is true, the
probability of observing a difference as large as that observed here with the self-
fertilized plants growing more than the cross-fertilized plants is 0.0248. The
chance of observing a difference as large as that observed here with the cross-
fertilized plants growing more is also 0.0248. The sum of those two probabilities
equals the two-sided P value (0.0497).

The size of the P value depends on the following:

- Mean difference. Everything else being equal, the P value will be smaller
  when the average of the differences is far from zero.
- Variability. If the observed differences are widely scattered, with some
  pairs having a large difference and some pairs having a small difference,
  then the P value will be higher. If the data are very consistent, the P value
  will be smaller.

- Sample size. Everything else being equal, the P value will be smaller when the sample has more data pairs.

How effective was the pairing?

Darwin's experiment worked with pairs of seeds. Each pair of seeds was exposed to the same conditions. But different pairs were measured at different times over several years. The *paired t test* was created to account for that pairing. If the seeds were grown under good conditions, both seeds would grow tall. If grown under worse conditions, both seeds would grow less. In either case, the difference between the two kinds of seeds would be consistent.

Figure 31.4 and Table 31.2 show that, in fact, there are no positive correlations among the data. Because the whole point of a paired t test is to use the internal controls to get more consistent data, the paired t test actually has a larger P value than would an unpaired t test conducted with these data (the P value from an unpaired t test for these data is 0.02). This situation is rare. Usually, if you design an experiment with matched samples, you'll see a strong positive correlation.

With these data, is it fair to switch to an unpaired t test? It certainly is not fair to "shop around" for the P value you want. You'd need to establish the protocol for switching before collecting any data.

Assumptions

When assessing any statistical results, it is essential to review the list of assumptions. The paired t test is based on a familiar list of assumptions (see Chapter 12).

- The paired values are randomly sampled from (or at least are representative of) a population of paired samples.
- In that population, the differences between the matched values follow a Gaussian distribution.
- Each pair is selected independently of the others.

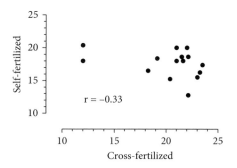

Figure 31.4. Testing the need for a paired t test.

The whole point of a paired t test is that you expect the paired values to be correlated—if one is higher, so is the other. This graph shows a negative correlation for sample data. This negative trend is far from convincing and is almost certainly a coincidence. But it is clear that there is no strong positive correlation. Chapter 32 explains correlation coefficients.

How a paired t test works

For each set of before–after measurements or each set of matched pairs, a paired t test first computes the differences between the two values for each pair. Only this set of differences is used to compute the CI and P value.

The CI for the mean difference is calculated exactly as explained in Chapter 12, except that the values used for the calculations are the set of differences.

The t ratio is computed by dividing the average difference (2.62 inches) by the SEM of those differences (1.22 inches). The numerator and denominator have the same units, so the t ratio is unitless. In the example, t = 2.15.

The P value is computed from the t ratio and the number of df, which equals the number of pairs minus one.

THE RATIO PAIRED t TEST

Example data: Does a treatment change enzyme activity?

Table 31.3 shows example data that test whether treating cultured cells with a drug increases the activity of an enzyme. Five different clones of the cell were tested. With each clone, control cells and treated cells were tested side by side. In all five clones, the treatment increased the activity of the enzyme. Figure 31.5 plots the data as a before–after graph and also as a graph of the differences.

CONTROL	TREATED
24	52
6	11
16	28
5	8
2	4

Table 31.3. Sample data for ratio t test.

These data (which are not real) test whether treating cultured cells with a drug increases the activity of an enzyme.

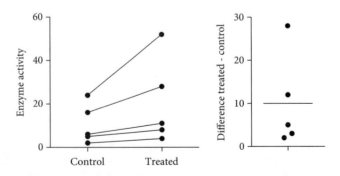

Figure 31.5. Enzyme activity in matched sets of control and treated cells.

The mean of the differences (treated minus control) is 10.0. The 95% CI for the average differences ranges from –3.4 to 23.4. Because that CI includes zero, you can't be sure (with 95% confidence) whether the treatment had any effect on the activity. The P value (two-tailed, from a paired t test) is 0.107, which is not low enough to be sure that the difference between control and treated is real; the difference could easily just be the result of random sampling. With only five pairs and inconsistent data, the data only lead to a very fuzzy conclusion.

Relative differences versus absolute differences

Nevertheless, look more carefully at the left panel of Figure 31.5. The difference between control and treated depends on where you start. The clones with higher enzyme activity in the control condition have a larger increment with treatment. The treatment multiplies the enzyme activity by a certain factor rather than adding a certain amount.

Figure 31.6 shows a simple way to align the multiplicative experimental model with the additive model used by the paired t test. This graph has a logarithmic Y-axis. Now, all five of the increases are about the same size. The treatment increases the logarithm of the outcome (enzyme activity) by a constant amount, because the treatment multiplies the activity by a constant factor rather than adding a constant factor. On a log scale, multiplication becomes addition.

If you are not comfortable working with logarithms, this process may seem a bit mysterious. But read on to see how logarithms are used practically to analyze these kinds of data.

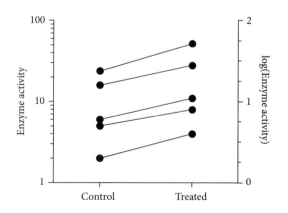

Figure 31.6. The same data as those shown in Figure 31.5 and Table 31.3, plotted on a logarithmic axis.

The right axis shows the logarithms. The left axis is labeled with the original values, logarithmically spaced. Both sets of data are plotted on both axes, which plot the same values but are expressed differently.

Results from a ratio paired t test

To analyze these data with a paired t test, first transform all values to their logarithms (base 10) and then perform the paired t test on those results (logarithms and antilogarithms are reviewed in Appendix E). The mean of the difference between logarithms is 0.26. Transform that back to its original ratio scale: $10^{0.26} = 1.82$. On average, the treatment multiplies the enzyme activity by a factor of 1.82. In other words, the treatment increases activity by 82%.

The 95% CI for the difference of logarithms is 0.21 to 0.33. Transform each of those values to their antilogarithms to find the 95% CI for the ratio (treated/control), which is 1.62 to 2.14. This interval doesn't include 1.0 (a ratio of 1.0 means no change) and doesn't even come close to including 1.0. Thus, the increase in activity with treatment is very unlikely to be a coincidence. The P value is 0.0003, low enough to convince you that the treatment effect is very unlikely to be caused by chance. Analyzed this way, the conclusion is very crisp.

I call this procedure (paired t test of data transformed to logarithms) a *ratio paired t test*, but that term is not widely used. However, the idea of transforming values to make the differences more consistent is entirely standard. Of course, you really should pick the analysis method before collecting data. It is not fair to analyze the data lots of different ways and then pick the P value you like the best.

McNEMAR'S TEST FOR A PAIRED CASE-CONTROL STUDY

McNemar's test analyzes matched data when the outcome is binomial (i.e., when there are two possible outcomes).

In a standard case-control study, the investigator compares a group of controls with a group of cases. As a group, the controls are chosen to be similar to the cases (except for the absence of disease and exposure to the putative risk factor). An alternative way to design a case-control study is to match each case with a specific control based on age, gender, occupation, location, or other relevant variables.

An example is shown in Table 31.4. The investigators studied 134 cases and 134 matched controls. Each entry in the table represents one pair (a case and a control). This is not a contingency table, and entering this table into a program designed to analyze contingency tables would lead to invalid results. Table 31.5 shows these data as a contingency table.

The 13 pairs for which both cases and controls were exposed to the risk factor provide no information about the association between risk factor and disease. Similarly, the 92 pairs for which both cases and controls were not exposed to the risk factor provide no useful information. The association between risk factor and disease depends on the other two values in the table, and their ratio is the odds ratio.

For the example, the odds ratio is the number of pairs for which the case was exposed to the risk factor but the control was not (25), divided by the number of pairs for which the control was exposed to the risk factor but the case was not (4), which equals 6.25.

| | CASES | | |
CONTROLS	RISK FACTOR +	RISK FACTOR −	TOTAL
Risk factor +	13	4	17
Risk factor −	25	92	117
Total	38	96	134

Table 31.4. A matched pair case-control study.

Each value in the table represents a matched case and control. The total number of the four values is 134, because the study used 134 pairs of subjects (so 268 subjects in all). This is not a contingency table, and analysis by programs designed for contingency tables would be invalid.

	RISK FACTOR +	RISK FACTOR −	TOTAL
Case	38	96	134
Control	17	117	134
Total	55	213	268

Table 31.5. Data from Table 31.4 expressed as a contingency table.

Each value in the table represents one subject. The study had 134 patient-control pairs, so 268 subjects in total. This is a contingency table. This arrangement of the data obscures the fact that the study was designed with a matched control for each case. Table 31.4 is more informative.

McNemar's test computes the 95% CI of the odds ratio and a P value from the two discordant values (25 and 4 for this example). Not all statistics programs calculate McNemar's test, but it can be computed using a free Web calculator in the QuickCalcs section of graphpad.com.

The 95% CI for the odds ratio ranges from 2.16 to 24.71. Assuming the disease is fairly rare, the odds ratio can be interpreted as a relative risk (see Chapter 28). These data show that exposure to the risk factor increases one's risk 6.25-fold and that there is 95% confidence that the range from 2.16 to 24.7 contains the true population odds ratio.

McNemar's test also computes a P value of 0.0002. To interpret a P value, the first step is to define the null hypothesis, which is that there really is no association between risk factor and disease, so the population odds ratio is 1.0. If that null hypothesis were true, the chance of randomly selecting 134 pairs of cases and controls with an odds ratio so far from 1.0 is only 0.02%.

COMMON MISTAKES: PAIRED t TEST

Mistake: If the result is almost statistically significant, collect more data to increase the sample size and then recalculate the test

You can only really interpret the results of any statistical calculation when the sample size is chosen in advance or you use special methods designed to handle sequential accumulation of data.

Mistake: If your experiment has three or more treatment groups, use the paired t test to compare two groups at a time

If you perform multiple t tests this way, it is too easy to reach misleading conclusions. Use repeated-measures one-way ANOVA, followed by multiple comparisons posttests, as explained in Chapters 39 and 40.

Mistake: If your experiment has three or more treatment groups, compare the largest mean with the smallest mean using a single paired t test

You won't know which group has the largest mean and which has the smallest mean until after you have collected the data. So although you are formally just doing one t test to compare two groups, you are really comparing all the groups. Use repeated-measures one-way ANOVA.

Mistake: Using the paired t test when the ratios, not the differences, are a more consistent measure of treatment effects

The ratio t test, explained above, is not commonly used, but it is often the best test for many situations.

Mistake: Analyzing the absolute value of the differences, rather than the actual differences

When computing the paired t test, the first step is to compute the difference for each pair. This must be done in a consistent way and must allow for negative values. If you are comparing measurements before and after an intervention, then you should compute the differences so that positive values mean the measurement went up and negative values mean the measurement went down. If you analyze absolute values (so that all differences are positive), the results will be meaningless.

Mistake: Deciding on the pairing only after seeing the data

Pairing or matching must be done as part of the experimental design and setup. It cannot be based on the values you are comparing with the t test.

Q & A

Can I decide how the subjects should be matched after collecting the data?

No. Pairing must be part of the experimental protocol that is decided before the data are collected. The decision about pairing is a question of experimental design and should be made long before the data are analyzed.

When computing the difference for each pair, in which order is the subtraction done?

It doesn't matter, so long as you are consistent. For each pair in the example, the length of the self-fertilized plant was subtracted from the length of the cross-fertilized plant. If the calculation were done the other way (if the length of the cross-fertilized plant was subtracted from the length of the self-fertilized plant), all the differences would have the opposite sign, as would the t ratio. The P value would be the same. It is very important, of course, that the subtraction be done the same way for every pair. It is also essential that the program doing the calculations doesn't lose track of which differences are positive and which are negative.

What if one of the values for a pair is missing?

The paired t test analyzes the differences between each set of pairs. If one value for a pair is missing, that pair can contribute no information at all, so it must be excluded from the analysis. A paired t test program will ignore pairs that lack one of the two values.

Can a paired t test be computed if you know only the mean and SD of the two treatments (or two time points) and the number of pairs?

No. These summary data tell you nothing about the pairing.

Can a paired t test be computed if you know only the mean and SD of the set of differences (as well as the number of pairs)? What about the mean and SEM?

Yes. All you need to compute a paired t test is the mean of the differences, the number of pairs, and the SD or SEM of the differences. You don't need the raw data.

Does it matter whether the two groups are sampled from populations that are not Gaussian?

Not necessarily. The paired t test only analyzes the set of paired differences, which it assumes is sampled from a Gaussian distribution. This doesn't mean that the two individual sets of values need to be Gaussian. If you are going to run a normality test on paired t test data, it only makes sense to test the list of differences (one value per pair). It does not make sense to test the two sets of data separately.

Will a paired t test always compute a lower P value than an unpaired test?

No. With the Darwin example, an unpaired test computes a lower P value than a paired test. When the pairing is effective (i.e., when the set of differences is more consistent than either set of values), the paired test will usually compute a lower P value.

CHAPTER SUMMARY

- The paired t test compares two matched or paired groups when the outcome is continuous.
- The paired t test only analyzes the set of differences. It assumes that this set of differences was sampled from a Gaussian population of differences.
- The paired t test is used when the *difference* between the paired measurements is a consistent measure of change. Use the ratio t test when the *ratio* of the paired measurements is a more consistent measure of the change or effect.
- McNemar's test compares two matched or paired groups when the outcome is binomial (categorical with two possible outcomes).

TERMS INTRODUCED IN THIS CHAPTER

- Before–after plot (p. 273)
- McNemar's test (p. 280)
- Paired t test (p. 275)
- Ratio paired t test (p. 280)

CHAPTER 32

❖❖❖

Correlation

The invalid assumption that correlation implies cause is
probably among the two or three most serious and common
errors of human reasoning.

STEPHEN JAY GOULD

The association between two continuous variables can be quantified
by the correlation coefficient, r. This chapter explains Pearson
correlation. Chapter 41 will explain nonparametric Spearman correlation.

INTRODUCING THE CORRELATION COEFFICIENT

An example: Lipids and insulin sensitivity

Borkman and colleagues (1993) wanted to understand why insulin sensitivity
varies so much among individuals. They hypothesized that the lipid composition of the cell membranes of skeletal muscle affects the sensitivity of the
muscle for insulin.

In their experiment, they determined the insulin sensitivity of 13 healthy men
by infusing insulin at a standard rate (adjusting for size differences) and quantifying how much glucose they needed to infuse to maintain a constant blood glucose
level. Insulin causes the muscles to take up glucose and thus causes the level of
glucose in the blood to fall, so people with high insulin sensitivity will require a
larger glucose infusion.

They also took a small muscle biopsy from each subject and measured its
fatty acid composition. We'll focus on the fraction of polyunsaturated fatty acids
that have between 20 and 22 carbon atoms (%C20–22).

Table 32.1 shows the data (interpolated from the author's published graph),
which are graphed in Figure 32.1. Note that both variables are scattered. The
mean of the insulin sensitivity index is 284, and the SD is 114 mg/m²/min. The CV
equals the SD divided by the mean, so 114/284, or 40.1%. This is quite high. The
authors knew that there would be a great deal of variability, and that is why they
explored the causes of the variability. The fatty acid content is less variable, with
a CV of 11.6%. If you don't look at the graph carefully, you could be misled. The
X-axis does not start at zero, so you might get the impression that the variability
is greater than it actually is.

INSULIN SENSITIVITY (MG/M²/MIN)	%C20–22 POLYUNSATURATED FATTY ACIDS
250	17.9
220	18.3
145	18.3
115	18.4
230	18.4
200	20.2
330	20.3
400	21.8
370	21.9
260	22.1
270	23.1
530	24.2
375	24.4

Table 32.1. Correlation between %C20–22 and insulin sensitivity.

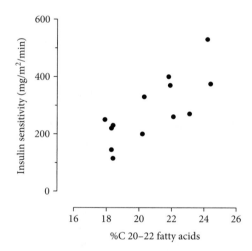

Figure 32.1. Correlation between %C20–22 and insulin sensitivity.

The graph shows a clear relationship between the two variables. Individuals whose muscles have more C20–22 polyunsaturated fatty acids tend to have greater sensitivity to insulin. The two variables vary together—statisticians say that there is a lot of *covariation* or *correlation*.

Correlation results

The direction and magnitude of the linear correlation can be quantified with a *correlation coefficient*, abbreviated r. Its value can range from −1 to 1. If the correlation coefficient is 0, then the two variables do not vary together at all. If the correlation coefficient is positive, the two variables tend to increase or decrease together. If the correlation coefficient is negative, the two variables are inversely

CORRELATION	
r	0.77.
95% confidence interval	0.3803 to 0.9275
r^2	0.5929
P (two-tailed)	0.0021
Number of XY pairs	13

Table 32.2. Correlation results.

related—that is, as one variable tends to decrease, the other one tends to increase. If the correlation coefficient is 1 or −1, the two variables vary together completely—that is, a graph of the data points forms a straight line.

The results of a computer program (GraphPad Prism 6) are shown in Table 32.2.

Correlation coefficient and its CI

In the example, the two variables increase together, so the correlation coefficient must be positive. But there is some scatter, so the correlation coefficient must be less than 1.0. In fact, the correlation coefficient equals 0.77.

It almost always makes sense to report any statistical result with a confidence interval. Here the 95% CI for the correlation coefficient ranges from 0.38 to 0.93. Assuming these data were randomly sampled from a larger population (and that other assumptions listed later in this chapter are true), there is a 95% chance that this range includes the population correlation coefficient.

Note that the CI is not symmetrical. In this example, it extends much further below the correlation coefficient than it extends above it. This makes sense. The correlation coefficient can never be larger than 1.0 or smaller than −1.0, and so the CI is usually asymmetrical. The asymmetry is especially noticeable when r is far from zero and when the sample size is small.

r^2

The square of the correlation coefficient is an easier value to interpret than r. For the example, r = 0.77, so r^2 (referred to as "r squared") = 0.59. Because r is always between −1 and 1, r^2 is always between 0 and 1. Note that when you square a positive fraction, the result is a smaller fraction.

If you can accept the assumptions listed in the next section, r^2 is the fraction of the variance shared between the two variables. In the example, 59% of the variability in insulin tolerance is associated with variability in lipid content. Knowing the lipid content of the membranes lets you explain 59% of the variance in the insulin sensitivity. That leaves 41% of the variance to be explained by other factors or by measurement error. X and Y are symmetrical in correlation analysis, so you can also say that 59% of the variability in lipid content is associated with variability in insulin tolerance.

P value

The P value is 0.0021. To interpret any P value, first define the null hypothesis. Here, the null hypothesis is that there is no correlation between insulin sensitivity and the lipid composition of membranes in the overall population. The two-tailed P value answers the question, If the null hypothesis were true, what is the chance that 13 randomly picked subjects would have an r greater than 0.77 or less than –0.77?

Because the P value is so low, the data provide strong evidence to reject the null hypothesis.

ASSUMPTIONS: CORRELATION

You can calculate the correlation coefficient for any set of data, and it may be a useful descriptor of the data. However, interpreting the CI and P value depends on the following assumptions.

Assumption: Random sample

Like all statistical analyses, you must assume that subjects are randomly selected from, or at least representative of, a larger population.

Assumption: Paired samples

Each subject (or each experimental unit) must have both X and Y values.

Assumption: Sampling from one population

Correlation assumes that all the points are randomly sampled from one population. If you sample some subjects from one population and some from another, the correlation coefficient and P value will be misleading.

Assumption: Independent observations

Correlation assumes that any random factor affects only one subject and not others. The relationship between all the subjects should be the same. In this example, the assumption of independence would be violated if some of the subjects were related (e.g., siblings). It would also be violated if the investigator purposely chose some people with diabetes and some without, or if the investigator measured each subject on two occasions and treated the values as two separate data points.

Assumption: X values are not used to compute Y values

The correlation calculations are not meaningful if the values of X and Y are not measured separately. For example, it would not be meaningful to calculate the correlation between a midterm exam score and the overall course score, because the midterm exam is one of the components of the course score.

Assumption: X values are not experimentally controlled

If you systematically control the X variable (e.g., concentration, dose, or time), you should calculate linear regression rather than correlation (see Chapter 33). You will get the same value for r^2 and the P value. However, the CI of r cannot be interpreted if the experimenter controlled the value of X.

Assumption: Both variables follow a Gaussian distribution

The X and Y values must each be sampled from populations that follow a Gaussian distribution, at least approximately. If this assumption isn't true, the P value cannot be interpreted at face value.

Assumption: All covariation is linear

The correlation coefficient would not be meaningful if, for example, Y increases as X increases up to a certain point but then Y decreases as X increases further. Curved relationships are common but are not quantified with a correlation coefficient.

Assumption: No outliers

Calculation of the correlation coefficient can be heavily influenced by one outlying point. Change or exclude that single point and the results may be quite different. Outliers can influence all statistical calculations, but especially when determining correlation. You should look at graphs of the data before reaching any conclusion from correlation coefficients. Don't instantly dismiss outliers as bad points that mess up the analysis. It is possible that the outliers are the most interesting observations in the study!

LINGO: CORRELATION

Correlation

When you encounter the word *correlation*, distinguish its strict statistical meaning from its more general usage. As used by statistics texts and programs, correlation quantifies the association between two continuous (interval or ratio) variables. However, the word *correlation* is often used much more generally to describe the association of any two variables, even if either (or both) of the variables are not continuous variables. It is not possible to compute a correlation coefficient to help you figure out whether survival times are correlated with choice of drug, or whether antibody levels are correlated with gender.

Coefficient of determination

Coefficient of determination is just a fancy term for r^2. Most scientists and statisticians just call it *r square* or *r squared*.

HOW IT WORKS: CALCULATING
THE CORRELATION COEFFICIENT

The calculation of the correlation coefficient is built into computer programs, so there is no reason to do it manually. The explanation below is provided to give you a feel for what the correlation coefficient means.

1. Calculate the average of all X values. Also calculate the average of all Y values. These two averages define a point at "the center of gravity" of the data.
2. Compare the position of each point with respect to that center. Subtract the average X value from each X value. The result will be positive for points to the right of the center and negative for points to the left. Similarly, subtract the average Y value from each Y value. The result will be positive for points above the center and negative for points below.
3. Standardize those X distances by dividing by the standard deviation of all the X values. Similarly, divide each Y distance by the standard

deviation of all Y values. Dividing a distance by the SD cancels out the units, so these ratios are fractions without units.

4. Multiply the two standardized distances for each data point. The product will be positive for points that are northeast (product of two positive numbers) or southwest (product of two negative numbers) of the center. The product will be negative for points that are to the northwest or southeast (product of a negative and a positive number).

5. Add up all the products computed in Step 4.

6. Account for sample size by dividing the sum by $(n - 1)$, where n is the number of XY pairs.

If X and Y are not correlated, then the positive products in Step 4 will approximately balance out the negative ones, and the correlation coefficient will be close to zero. If X and Y are correlated, the positive and negative products will not balance each other out, and the correlation coefficient will be far from zero.

Nonparametric Spearman correlation (see Chapter 41) adds one step to the beginning of this process. First, you would separately rank the X and Y values, with the smallest value getting a rank of one. Then you would calculate the correlation coefficient for the X ranks and the Y ranks, as explained above.

COMMON MISTAKES: CORRELATION

Mistake: Believing that correlation proves causation

Why do the two variables in this example correlate so well? There are five possible explanations:

- The lipid content of the membranes determines insulin sensitivity.
- The insulin sensitivity of the membranes somehow affects lipid content.
- Both insulin sensitivity and lipid content are under the control of some other factor, such as a hormone.
- Lipid content, insulin sensitivity, and other factors are all part of a complex molecular/biochemical/physiological network, perhaps with positive and/ or negative feedback components. In this case, the observed correlation would just be a peek at a much more complicated set of relationships.
- The two variables don't correlate in the population at all, and the observed correlation in our sample is a coincidence.

You can never rule out the last possibility, but the P value tells you how unlikely the coincidence would be. In this example, you would observe a correlation that strong (or stronger) in 0.21% (the P value) of experiments if there was no correlation in the overall population.

You cannot decide which of the first four possibilities is correct by analyzing only these data. The only way to figure out which is true is to perform additional experiments in which you manipulate the variables. Remember that this study simply measured both values in a set of subjects. Nothing was experimentally manipulated.

The authors, of course, want to believe the first possibility based on their knowledge of physiology. That does not mean they believe that the lipid composition is the *only* factor that determines insulin sensitivity, only that it is one factor.

Most people immediately think of the first two possibilities but ignore the rest. Correlation does not necessarily prove simple causality, however. Two variables can be correlated because both are influenced by the same third variable. Infant mortality in various countries is probably negatively correlated with the number of telephones per capita, but buying telephones will not make kids live longer. Instead, increased wealth (and thus increased purchases of telephones) is related to better plumbing, better nutrition, less crowded living conditions, more vaccinations, and so on.

Mistake: Focusing on the P value rather than the correlation coefficient

Arden and colleagues (2008) asked whether there is a correlation between intelligence and sperm count. Really!

They obtained data from a study of U.S. Vietnam War–era veterans that measured both intelligence and the number and motility of sperm in ejaculate. The variation in sperm count (measured as sperm per milliliter, or sperm per ejaculate) among individuals was very skewed, so the researchers transformed all the values to logarithms. This approach makes perfect sense, because the values approximated a lognormal distribution (see Chapter 11).

With data from 425 men, Arden and colleagues computed the correlation coefficient for a measure of intelligence and three assessments of sperm quality (logarithm of sperm count per volume, logarithm of sperm count per ejaculate, and the percentage of those sperm that were motile). The results are shown in Table 32.3, which demonstrates four points:

- The r values are all positive. This tells you that the trend is positive—that is, that increased intelligence is associated with increased sperm count and motility.
- No corrections were made for multiple comparisons (see Chapters 22 and 23). This is appropriate because the three variables that assess sperm are all somewhat related. These aren't three independent measurements. It makes perfect sense to report all three P values without correction for multiple comparisons (but it wouldn't be fair to report only the lowest P values and not mention the other two). The fact that these values all are tiny and that all correlations go in the same direction helps make the data convincing.

	r	r²	P VALUE
Log (sperm count per ml)	0.15	0.023	0.0019
Log (sperm count per ejaculate)	0.19	0.036	< 0.0001
Fraction of sperm that are motile (%)	0.14	0.020	0.0038

Table 32.3. Correlation between a measure of intelligence and three measures of sperm quality.
Data from Arden et al. (2008), with n = 425.

- The P values are all quite small. If there really was no relationship between sperm and intelligence, there would only be a tiny chance of obtaining a correlation this strong.
- The r values are all small. It is easier to interpret the r^2 values, which show that only 2 to 3% of the variation in intelligence in this study is associated with variation in sperm count and motility.

The last two points illustrate a common situation. With large samples, data can show small effects (in this case, r^2 values of 2–3%) and still have tiny P values. The fact that the P values are small tells you that the effect is unlikely to be a coincidence of random sampling. But the size of the P value does not tell you how large the effect is. The effect size is quantified by r and r^2. Here, the effect size is tiny.

Is an r^2 value of 2 to 3% large enough to be considered interesting and worth pursuing? That is a scientific question, not a statistical one. The investigators of this study seemed to think so. Other investigators might conclude that such a tiny effect is not worthy of much attention. I doubt that many newspaper reporters who wrote about the supposed link between sperm count and intelligence understood how weak the reported relationship really is.

Mistake: Interpreting the correlation coefficient without first looking at a graph

Don't try to interpret a correlation coefficient (or the corresponding P value) until you have looked at a graph of the data. Figure 32.2 shows why. This graph

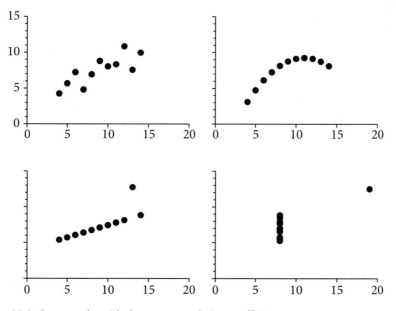

Figure 32.2. Four graphs with the same correlation coefficient.

In all four cases, the correlation coefficient (r) is 0.8164. Only the first graph seems to meet the assumptions of correlation.

shows four artificial data sets designed by Anscombe (1973). The correlation coefficients are identical (0.816), as are the P values (0.0022), but the data are very different.

Q & A

Does it matter which variable is called X and which is called Y?
No. X and Y are totally symmetrical in correlation calculations. But note this is not true for linear regression (see Chapter 33). When analyzing data with linear regression, you must carefully choose which variable is X and which is Y.

Do X and Y have to be measured in the same units to calculate a correlation coefficient? Can they be measured in the same units?
X and Y do not have to be measured in the same units, but they can be.

In what units is r expressed?
None. It is unitless.

Can r be negative?
Yes. It is negative when one variable tends to go down as the other goes up. When r is positive, one variable tends to go up as the other variable goes up.

Can correlation be calculated if all X values are the same? If all Y values are the same?
No. If all the X or Y values are the same, it makes no sense to calculate correlation.

Why are there no best-fit lines in Figure 32.1?
Correlation quantifies the relationship but does not fit a line to the data. Chapter 33 explains linear regression, which finds the best-fit line, and reviews the differences between correlation and regression.

If all the X or Y values are transformed to new units, will r change?
No. Multiplying by a factor to change units (e.g., inches to centimeters, milligrams per milliliter to millimolar) will not change the correlation coefficient.

If all the X or Y values are transformed to their logarithms, will r change?
Yes. A logarithmic transformation, or any transformation that changes the relative values of the points, will change the value of r. However, the nonparametric Spearman correlation coefficient (see Chapter 41), which depends only on the order of the values, will not change.

If you interchange X and Y, will r change?
No. X and Y are completely symmetrical in calculating and interpreting the correlation coefficient.

If you double the number of points but r doesn't change, what happens to the CI and P value?
With more data points, the CI will be narrower. With more data points and the same value of r, the P value will be smaller.

Can correlation be used to quantify how closely two alternative assay methods agree?
No. This is a common mistake. If you want to compare two different analysis methods, special methods are needed. Look up the Bland–Altman plot.

Is there a distinction between r^2 and R^2?
No. Upper- and lowercase mean the same thing in this context. However, the correlation coefficient, r, is always written in lower case.

CHAPTER SUMMARY

- The correlation coefficient, r, quantifies the strength of the linear relationship between two continuous variables.
- The value of r is always between –1.0 and 1.0; it has no units.
- A negative value of r means that the trend is negative—that is, Y tends to go down as X goes up.
- A positive value of r means that the trend is positive—that is, X and Y go up together.
- The square of r equals the fraction of the variance that is shared by X and Y. This value, r^2, is sometimes called the coefficient of determination.
- Correlation is used when you measure both X and Y variables, but it is not appropriate if X is a variable you manipulate.
- Note that correlation and linear regression are not the same. In particular, note that the correlation analysis does not fit or plot a line.
- You should always view a graph of the data when interpreting a correlation coefficient.
- Correlation does not imply causation. The fact that two variables are correlated does not mean that changes in one cause changes in the other.

TERMS INTRODUCED IN THIS CHAPTER

- Correlation (p. 288)
- Correlation coefficient (r) (p. 285)
- Coefficient of determination (r^2) (p. 288)
- Covariation (p. 285)

Fitting Models to Data

CHAPTER 33

✦✦✦

Simple Linear Regression

> In the space of 176 years, the Lower Mississippi has shortened
> itself 242 miles. This is an average of a trifle over one mile and
> a third per year. Therefore...any person can see that seven
> hundred and forty-two years from now, the lower Mississippi
> will be only a mile and three-quarter long.
>
> MARK TWAIN, *LIFE ON THE MISSISSIPPI*

One way to think about linear regression is that it fits the "best line" through a graph of data points. Another way to look at it is that linear regression fits a simple model to the data to determine the most likely values of the parameters that define that model (slope and intercept). Chapter 34 will introduce the more general concepts of fitting a model to data.

THE GOALS OF LINEAR REGRESSION

Recall the insulin example of Chapter 32. The investigators were curious to understand why insulin sensitivity varies so much among individuals. They measured insulin sensitivity and the lipid content of muscle obtained at biopsy in 13 men. You've already seen that the two variables (insulin sensitivity and the fraction of the fatty acids that are unsaturated with 20–22 carbon atoms [%C20–22]) correlate substantially.

In this example, the authors concluded that differences in the lipid composition affect insulin sensitivity and proposed a simple model—that insulin sensitivity is a linear function of %C20–22. As %C20–22 goes up, so does insulin sensitivity. This relationship can be expressed as an equation:

Insulin sensitivity = Intercept + Slope · %C20–22

Let's define the insulin sensitivity to be Y, %C20–22 to be X, the intercept to be b, and the slope to be m. Now the model takes this standard form:

$Y = b + m \cdot X$

The model is not complete, however, because it doesn't account for random variation. The investigators used the standard assumption that random variability from the predictions of the model will follow a Gaussian distribution.

BEST-FIT VALUES

Slope	37.21 ± 9.296
Y intercept when X = 0.0	−486.5 ± 193.7
X intercept when Y = 0.0	13.08
1/slope	0.02688

95% CONFIDENCE INTERVALS

Slope	16.75 to 57.67
Y intercept when X = 0.0	−912.9 to −60.17
X intercept when Y = 0.0	3.562 to 15.97

GOODNESS OF FIT

R^2	0.5929
Sy.x	75.90

IS SLOPE SIGNIFICANTLY NONZERO?

F	16.02
DFn, DFd	1.000, 11.00
P value	0.0021
Deviation from zero?	Significant

DATA

Number of X values	13
Maximum number of Y replicates	1
Total number of values	13
Number of missing values	0

Table 33.1. Linear regression results.
These results are from GraphPad Prism; other programs would format the results differently.

We can't possibly know the one true population value for the intercept or the one true population value for the slope. Given the sample of data, our goal is to find those values for the intercept and slope that are most likely to be correct and to quantify the imprecision with CIs. Table 33.1 shows the results of *linear regression*. The next section will review all of these results.

It helps to think of the model graphically (see Figure 33.1). A simple way to view linear regression is that it finds the straight line that comes closest to the points on a graph. That's a bit too simple, however. More precisely, linear regression finds the line that best predicts Y from X. To do so, it only considers the vertical distances of the points from the line and, rather than minimizing those distances, it minimizes the sum of the square of those distances. Chapter 34 explains why.

LINEAR REGRESSION RESULTS

The slope
The best-fit value of the slope is 37.2. This means that when %C20–22 increases by 1.0, the average insulin sensitivity is expected to increase by 37.2 mg/m²/min.

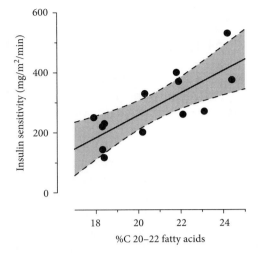

Figure 33.1. The best-fit linear regression line along with its 95% confidence band (shaded).

The CI is an essential (but often omitted) part of the analysis. The 95% CI of the slope ranges from 16.7 to 57.7 mg/m²/min. Although the CI is fairly wide, it does not include zero; in fact, it doesn't even come close to zero. This is strong evidence that the observed relationship between lipid content of the muscles and insulin sensitivity is very unlikely to be a coincidence of random sampling.

The range of the CI is reasonably wide. The CI would be narrower if the sample size were larger.

Some programs report the standard errors of the slope instead of (or in addition to) the CIs. The standard error of the slope is 9.30 mg/m²/min. CIs are easier to interpret than standard errors, but the two are related. If you read a paper that reports the standard error of the slope but not its CI, compute the CI using these steps:

1. Look in Appendix D to find the critical value of the t distribution. The number of df equals the number of data points minus two. For this example, there were 13 data points, so there are 11 df, and the critical t ratio for a 95% CI is 2.201.
2. Multiply the value from Step 1 by the SE of the slope reported by the linear regression program. For the example, multiply 2.201 times 9.30. The margin of error of the CI equals 20.47.
3. Add and subtract the value computed in Step 2 from the best-fit value of the slope to obtain the CI. For the example, the interval begins at $37.2 - 20.5 = 16.7$. It ends at $37.2 + 20.5 = 57.7$.

The intercept

A line is defined by both its *slope* and its Y-*intercept*, which is the value of the insulin sensitivity when the %C20–22 equals zero.

For this example, the Y-intercept is not a scientifically relevant value. The range of %C20–22 in this example extends from about 18 to 24%. Extrapolating back to zero is not helpful. The best-fit value of the intercept is –486.5, with a 95% CI ranging from –912.9 to –60.17. Negative values are not biologically possible, because insulin sensitivity is assessed as the amount of glucose needed to maintain a constant blood level and so must be positive. These results tell us that the linear model cannot be correct when extrapolated way beyond the range of the data.

Graphical results

Figure 33.1 shows the best-fit regression line defined by the values for the slope and the intercept, as determined by linear regression.

The shaded area in Figure 33.1 shows the 95% confidence bands of the regression line, which combine the CIs of the slope and the intercept. The best-fit line determined from this particular sample of subjects (solid black line) is unlikely to really be the best-fit line for the entire (infinite) population. If the assumptions of linear regression are true (which will be discussed later in this chapter), you can be 95% sure that the overall best-fit regression line lies somewhere within the shaded confidence bands.

The 95% confidence bands are curved but do not allow for the possibility of a curved (nonlinear) relationship between X and Y. The confidence bands are computed as part of linear regression, so they are based on the same assumptions as linear regression. The curvature simply is a way to enclose possible straight lines, of which Figure 33.2 shows two.

Figure 33.2. Although the confidence band is curved, it is for linear regression and only considers the fits of straight lines.

Two lines that fit within the confidence band. Given the assumptions of the analysis, you can be 95% sure that the true (population) line can be drawn within the shaded area.

The 95% confidence bands enclose a region that you can be 95% confident includes the true best-fit line (which you can't determine from a finite sample of data). But note that only six of the 13 data points in Figure 33.2 are included within the confidence bands. If the sample was much larger, the best-fit line would be determined more precisely, the confidence bands would be narrower, and a smaller fraction of data points would be included within the confidence bands. Note the similarity to the 95% CI for the mean, which does not include 95% of the values (see Chapter 12).

R^2

The R^2 value (0.5929) means that 59% of all the variance in insulin sensitivity can be accounted for by the linear regression model and the remaining 41% of the variance may be caused by other factors, measurement errors, biological variation, or a nonlinear relationship between insulin sensitivity and %C20–22. Chapter 35 will define R^2 more rigorously. The value of R^2 for linear regression ranges between 0.0 (no linear relationship between X and Y) and 1.0 (the graph of X vs. Y forms a perfect line).

P value

Linear regression programs report a P value, which in this case is 0.0021. To interpret any P value, the null hypothesis must be stated. With linear regression, the null hypothesis is that there really is no linear relationship between insulin sensitivity and %C20–22. If the null hypothesis were true, the best-fit line in the overall population would be horizontal with a slope of zero. In this example, the 95% CI for slope does not include zero (and does not come close), so the P value must be less than 0.05. In fact, it is 0.0021. The P value answers the question, If that null hypothesis were true, what is the chance that linear regression of data from a random sample of subjects would have a slope as far (or farther) from zero as that which is actually observed?

In this example, the P value is tiny, so we conclude that the null hypothesis is very unlikely to be true and that the observed relationship is unlikely to be caused by a coincidence of random sampling.

With this example, it makes sense to analyze the data both with correlation (see Chapter 32) and with linear regression (see Chapter 33). The two are related. The null hypothesis for correlation is that there is no correlation between X and Y. The null hypothesis for linear regression is that a horizontal line is correct. Those two null hypotheses are essentially equivalent, so the P values reported by correlation and linear regression are identical.

ASSUMPTIONS: LINEAR REGRESSION

Assumption: The model is correct

Not all relationships are linear (see Figure 33.3). In many experiments, the relationship between X and Y is curved, making simple linear regression inappropriate. Chapter 36 will explain nonlinear regression.

R^2 from linear regression = 0.01; P = 0.9

Figure 33.3. Linear regression only looks at linear relationships.
X and Y are definitely related here, just not in a linear fashion. The line
is drawn by linear regression and misses the relationship entirely.

The linear regression equation defines a line that extends infinitely in both directions. For any value of X, no matter how high or how low, the equation can predict a Y value. Of course, it rarely makes sense to believe that a model can extend infinitely. But the model can be salvaged by using the predictions of the model that only fall within a defined range of X values. Thus, we only need to assume that the relationship between X and Y is linear within that range. In the example, we know that the model cannot be accurate over a broad range of X values. At some values of X, the model even predicts that Y will be negative, a biological impossibility. But the linear regression model is useful within the range of X values actually observed in the experiment.

Assumption: The scatter of data around the line is Gaussian
Linear regression analysis assumes that the scatter of data around the model (the true best-fit line) is Gaussian. The CIs and P values cannot be interpreted if the distribution of scatter is far from Gaussian or if some of the values are outliers from another distribution.

Assumption: The variability is the same everywhere
Linear regression assumes that the scatter of points around the best-fit line has the same standard deviation all along the curve. This assumption is violated if the points with high or low X values tend to be farther from the best-fit line. The assumption that the standard deviation is the same everywhere is termed *homoscedasticity*. Linear regression can be calculated without this assumption by differentially weighting the data points, giving more weight to the points with small variability and less weight to the points with lots of variability.

Assumption: Data points (residuals) are independent
Whether one point is above or below the line is a matter of chance and does not influence whether another point is above or below the line.

Assumption: X and Y values are not intertwined

If the value of X is used to calculate Y (or the value of Y is used to calculate X), then linear regression calculations will be misleading. One example is a Scatchard plot, which is used by pharmacologists to summarize binding data. The Y value (drug bound to receptors divided by drug free in solution) is calculated from the X value (free drug), so linear regression is not appropriate. Another example would be a graph of midterm exam scores (X) versus total course grades (Y). Because the midterm exam score is a component of the total course grade, linear regression is not valid for these data.

Assumption: The X values are known precisely

Regression assumes that the true X values are known and that all the variation is in the Y direction. If X is something you measure (rather than control) and the measurement is not precise, the linear regression calculations might be misleading.

COMPARISON OF LINEAR REGRESSION AND CORRELATION

The example data have now been analyzed twice, using correlation (see Chapter 32) and linear regression. The two analyses are similar, yet distinct.

Correlation quantifies the degree to which two variables are related but does not fit a line to the data. The correlation coefficient tells you the extent to which one variable tends to change when the other one does, as well as the direction of that change. The CI of the correlation coefficient can only be interpreted when both X and Y are measured and when both are assumed to follow Gaussian distributions. In the example, the experimenters measured both insulin sensitivity and %C20–22. You cannot interpret the CI of the correlation coefficient if the experimenters manipulated (rather than measured) X.

With correlation, you don't have to think about cause and effect. You simply quantify how well two variables relate to each other. It doesn't matter which variable you call X and which you call Y. If you reversed the definition, all of the results would be identical. With regression, you do need to think about cause and effect. Regression finds the line that best predicts Y from X, and that line is not the same as the line that best predicts X from Y.

It made sense to interpret the linear regression line only because the investigators hypothesized that the lipid content of the membranes would influence insulin sensitivity and so defined %C20–22 to be X and insulin sensitivity to be Y. The results of linear regression (but not correlation) would be different if the definitions of X and Y were swapped (so that changes in insulin sensitivity would somehow change the lipid content of the membranes).

With most data sets, it makes sense to calculate either linear regression or correlation but not both. The example here (lipids and insulin sensitivity) is one for which both correlation and regression make sense. The R^2 is the same whether computed by a correlation or a linear regression program.

LINGO: LINEAR REGRESSION

Model

Regression refers to a method used to fit a model to data. A model is an equation that describes the relationship between variables. Chapter 34 will define the term more fully.

Parameters

The goal of linear regression is to find the values of the slope and intercept that make the line come as close as possible to the data. The slope and the intercept are called *parameters*.

Regression

Why the strange term *regression*? In the 19th century, Sir Francis Galton studied the relationship between parents and children. Children of tall parents tended to be shorter than their parents. Children of short parents tended to be taller than their parents. In each case, the height of the children reverted, or "regressed," toward the mean height of all children. Somehow, the term *regression* has taken on a much more general meaning.

Residuals

The vertical distances of the points from the regression line are called *residuals*. A residual is the discrepancy between the actual Y value and the Y value predicted by the regression model.

Least squares

Linear regression finds the value of the slope and intercept that minimizes the sum of the squares of the vertical distances of the points from the line. This goal gives linear regression the alternative name *linear least squares*.

Linear

The word *linear* has a special meaning to mathematical statisticians. It can be used to describe the mathematical relationship between model parameters and the outcome. Thus, it is possible for the relationship between X and Y to be curved but for the mathematical model to be considered linear.

Simple versus multiple linear regression

This chapter explained simple linear regression. The term *simple* is used because there is only one X variable. Chapter 37 will explain multiple linear regression. The term *multiple* is used because there are two or more X variables.

COMMON MISTAKES: LINEAR REGRESSION

Mistake: Concluding there is no relationship between X and Y when R^2 is low

A low R^2 value from linear regression means there is little linear relationship between X and Y. But not all relationships are linear. Linear regression in Figure 33.3 has an R^2 of 0.001, but X and Y are clearly related (just not linearly).

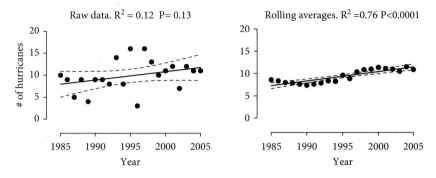

Figure 33.4. Don't smooth.

Smoothing, or computing a rolling average (which is a form of smoothing), appears to reduce the scatter of data and so gives misleading results.

Mistake: Fitting rolling averages or smoothed data

Figure 33.4 shows a synthetic example to demonstrate the problem of fitting smoothed data. The graph plots the number of hurricanes over time. The graph on the left shows actual "data," which jump around a lot. In fact, these are random values (from a Poisson distribution with a mean of 10), and each was chosen without regard to the others. There is no true underlying trend.

The graph on the right of Figure 33.4 shows a *rolling average* (adapted from Briggs, 2008b). This is also called *smoothing* the data. There are many ways to smooth data. For this graph, each value plotted is the average of the number of hurricanes for that year plus average of the number of hurricanes for the prior eight years (thus, a rolling average). The idea of smoothing is to reduce the noise so you can see underlying trends. Indeed, the smoothed graph shows a clear upward trend. The R^2 is much higher, and the P value is very low. But this trend is entirely an artifact of smoothing. Calculating the rolling average makes any random swing to a high or low value look more consistent than it really is, because neighboring values usually also become low or high.

One of the assumptions of linear regression is that each point contributes independent information. When the data are smoothed, the residuals are not independent. Accordingly, the regression line is misleading, and the P value and R^2 are meaningless.

Mistake: Fitting data when X and Y are intertwined

When interpreting the results of linear regression, make sure that the X- and Y-axes represent separate measurements. If the X and Y values are intertwined, the results will be misleading. Here is an example.

Figure 33.5 shows computer-generated data simulating an experiment in which blood pressure was measured before and after an experimental intervention. The graph on the top left shows the data. Each point represents an individual whose blood pressure was measured before or after an intervention. Each value was sampled from a Gaussian distribution with a mean of 120 and an SD of 10. The data are entirely random, and the two data sets (top-left graph) look about

Figure 33.5. Beware of regression to the mean.
(Top left) Actual data (which are simulated). The before and after values are indistinguishable. (Top right) Graph of one set of measurements versus the others. There is no trend at all. (Bottom) Graph of the starting (before) blood pressure versus the change (after – before). The apparent trend is an illusion. All the graph tells you is that if your blood pressure happens to be very high on one reading, it is likely to be lower on the next. And when it is low on one reading, it is likely to be higher on the next. Regressions in which X and Y are intertwined like this are not useful.

the same. Therefore, a best-fit regression line (top-right graph) is horizontal. Although blood pressure levels varied between measurements, there was no systematic effect of the treatment. The graph on the bottom shows the same data. But now the vertical axis shows the change in blood pressure (after–before) plotted against the starting (before) blood pressure. Note the striking linear relationship. Individuals who initially had low pressures tended to see an increase; individuals with high pressures tended to see a decrease. This is entirely an artifact of data analysis and tells you nothing about the effect of the treatment, only about the stability of the blood pressure levels between treatments. Chapter 1 gave other examples of regression to the mean.

Graphing a change in a variable versus the initial value of the variable is quite misleading. Attributing a significant correlation on such a graph to an

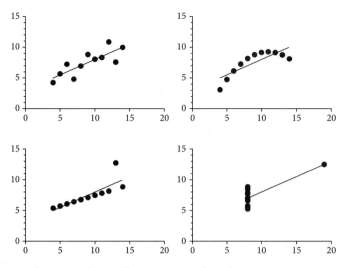

Figure 33.6. Look at a graph as well as at numerical results.

The linear regressions of these four data sets all have identical best-fit values for slope, best-fit values for the Y-intercept, and R^2s. Yet the data are far from identical.

experimental intervention is termed the *regression fallacy*. Such a plot should not be analyzed by linear regression, because these data (as presented) violate one of the assumptions of linear regression, that the X and Y values are not intertwined.

Mistake: Not thinking about which variable is X and which is Y

Unlike correlation, linear regression calculations are not symmetrical with respect to X and Y. Switching the labels X and Y will produce a different regression line (unless the data are perfect, with all points lying directly on the line). This makes sense, because the whole point is to find the line that best predicts Y from X, which is not the same as the line that best predicts X from Y.

Consider the extreme case, in which X and Y are not correlated at all. The linear regression line that best predicts Y from X is a horizontal line through the mean Y value. In contrast, the best line to predict X from Y is a vertical line through the mean of all X values, which is 90° different. With most data, the two lines are far closer than that but are still not the same.

Mistake: Looking at numerical results of regression without viewing a graph

Figure 33.6 shows four linear regressions designed by Anscombe (1973). The best-fit values of the slopes are identical for all four data sets. The best-fit values of the Y intercept are also identical. Even the R^2 values are identical. But a glance at the graph shows you that the data are very different!

Mistake: Using standard unweighted linear regression when scatter increases as Y increases

It is common for biological data to violate the assumption that variability is the same for all values of Y. Instead, it is common for variability to be proportional to Y. Few linear regression programs can handle this situation, but most nonlinear regression programs can. If you want to fit a linear model to data for which scatter increases with Y, use a nonlinear regression program to fit a straight-line model and choose to weight the data points to account for unequal variation (see Chapter 36).

Mistake: Extrapolating beyond the data

The best-fit values of the slope and intercept define a linear equation that defines Y for any value of X. So, for the example, it is easy enough to plug numbers into the equation and predict what Y (insulin sensitivity) would be if X (lipid composition as %C20–22) were 100%. But the data were only collected with X values between 18 and 24%, and there is no reason to think that the linear relationship would continue much beyond the range of the data. Predictions substantially beyond that range are likely to be very wrong.

Figure 33.7 shows why you must beware of predictions that go beyond the range of the data. The top graph shows data that are fit by a linear regression model very well. The data range from $X = 1$ to $X = 15$.

The bottom graphs predict what will happen at later times. The left graph on the bottom shows the prediction of linear regression. The other two graphs show predictions of two other models. Each of these models actually fit the data slightly better than does linear regression, so the R^2 values are slightly higher. The three models make very different predictions at late time points. Which prediction is most likely to be correct? It is impossible to answer that question without more data or at least some information and theory about the scientific context. The predictions of linear regression, when extended far beyond the data, may be very wrong.

Figure 33.8 is a cartoon that makes the same point.

Mistake: Overinterpreting a small P value

Figure 33.9 shows simulated data with 500 points fit by linear regression. The P value from linear regression answers the question of whether the regression line fits the data statistically significantly better than does a horizontal line (the null hypothesis). The P value is only 0.0302, so indeed, the deviation of the line from a horizontal line is statistically significant. But look at R^2, which is only 0.0094. That means that not even 1% of the overall variation can be explained by the linear regression model. The low P value should convince you that the true slope is unlikely to horizontal. But the discrepancy is tiny. There may possibly be some scientific field in which such a tiny effect is considered important or worthy of follow-up, but in most fields, this kind of effect is completely trivial, even though it is statistically significant. Focusing on the P value can lead you to misunderstand the findings.

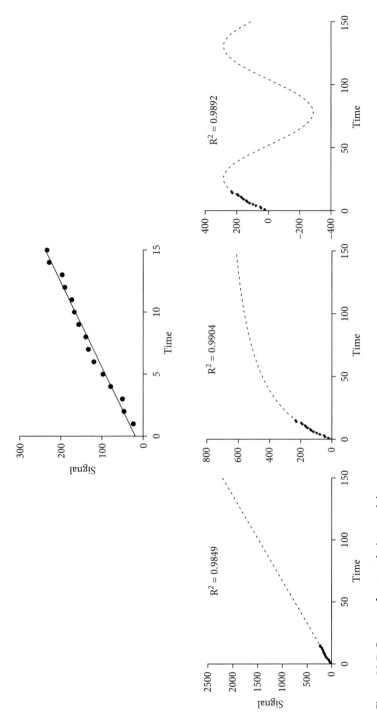

Figure 33.7. Beware of extrapolating models.

(Top) Data that fit a linear regression model very well. What happens at later time points? The three graphs on the bottom show three alternative predictions. (Left) The prediction of linear regression. (Middle and right) Predictions of two other models that actually fit the data slightly better than a straight line does. Predictions of linear regression that extend far beyond the data may be very wrong.

Figure 33.8. Beware of extrapolation.
Source: xkcd.com.

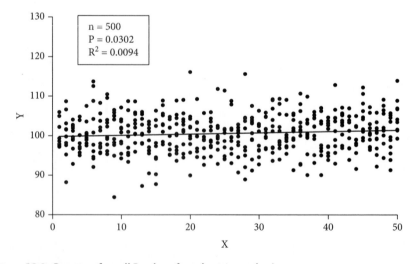

Figure 33.9. Beware of small P values from large sample sizes.
The line was fit by linear regression. It is statistically different from horizontal (P = 0.03).
But R^2 is only 0.0094, so the regression model explains less than 1% of the variance.
The P value is small, so these data would not be expected if the true slope were exactly
horizontal. However, the deviation from horizontal is trivial. You'd be misled if you focused
on the P value and the conclusion that the slope differs significantly from zero.

Mistake: Using linear regression when the data are complete with no sampling error

Figure 33.10 shows the number of high school students in the United States who took the advanced placement (AP) exam in statistics each year from 1998 to 2012. The graph shows the total number of students who took the exam. There is no sampling from a population and no random error. Fitting a linear regression line to the data would not be very helpful. The standard error and confidence interval of the slope would be meaningless. Interpolating between years would be meaningless as well, because the exam is only offered once each year. Computing a P value would be meaningless because there is no random sampling. The numbers really do speak for themselves, and the results of linear regression would not add any insights and might even be misleading.

What if you want to extrapolate to predict the number of students who will take the exam the next year (2013)? Using a linear regression line would not be unreasonable. But it might be just as reasonable to simply predict that the increase

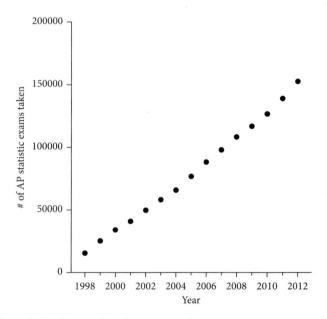

Figure 33.10. No need for linear regression.

This graph shows the number of high school students who took the advanced placement test in statistics each year from 1998 to 2012. These data are not sampled from a larger population. The graph shows the total number of American students who took the advanced placement statistics exam each year. Linear regression calculations would not be useful to describe the data.

Source: Data from Wikipedia (en.wikipedia.org/wiki/AP_Statistics).

from 2012 to 2013 (which you are trying to predict) will equal the increase from 2011 to 2012 (which you know). Any prediction requires assuming a model, and there really is no reason to assume that the increase will be linear indefinitely. You could also use a second- or third-order polynomial model (or some other model) to predict future years.

If I wanted to understand these data, I wouldn't bother with any analysis of the figure but would try to dig deeper. The number of students taking the exam depends on how many schools offer an AP statistics course, how many students (on average) take that course, and what fraction of the students who take the course end up taking the AP exam. If I wanted to understand the data, I'd try to find out how these three different factors have changed over the years.

Q & A

Do the X and Y values have to have the same units to perform linear regression?
No, but they can. In the example, X and Y are in different units.

Can linear regression work when all X values are the same? When all Y values are the same?
No. The whole point of linear regression is to predict Y based on X. If all X values are the same, they won't help predict Y. If all Y values are the same, there is nothing to predict.

Can linear regression be used when the X values are actually categories?
If you are comparing two groups, you can assign the groups to $X=1$ and $X=0$ and use linear regression. This is identical to performing an unpaired t test, as will be explained in Chapter 35. If there are more than two groups, using simple linear regression only makes sense when the groups are ordered and equally spaced and thus can be assigned sensible numbers. If you need to use a categorical variable with more than two possible values, read about indicator variables (also called dummy variables) in a text about multiple linear regression.

Is the standard error of the slope the same as the SEM?
No. The standard error is a way to express the precision of a computed value (parameter). The first standard error encountered in this book happened to be the standard error of the mean (see Chapter 14). The standard error of a slope is quite different. Standard errors can also be computed for almost any other parameter.

What does the variable $\hat{Y}$ (used in other statistics books) mean?
Mathematical books usually distinguish between the Y values of the data you collected and the Y values predicted by the model. The actual Y values are called Y_i, where i indicates the value to which you are referring. For example, Y_3 is the actual Y value of the third subject or data point. The Y values predicted by the model are called $\hat{Y}$, pronounced "Y hat." So $\hat{Y}_3$ is the value the model predicts for the third subject or data point. That value is predicted from the X value for the third data point but doesn't take into account the actual value of Y.

Will the regression line be the same if you exchange X and Y?
Linear regression fits a model that best predicts Y from X. If you swap the definitions of X and Y, the regression line will be different unless the data points line up perfectly so that every point is on the line. However, swapping X and Y will not change the value of R^2.

Can R^2 ever be zero? Negative?
R^2 will equal zero if there is no trend whatsoever between X and Y, so the best-fit line is exactly horizontal. R^2 cannot be negative with standard linear regression, but Chapter 36 explains that R^2 can be negative with nonlinear regression.

Do you need more than one Y value for each X value to calculate linear regression? Does it help?

Linear regression does not require more than one Y value for each X value. But there are three advantages to using replicate Y values for each value of X:

- With more data points, you'll determine the slope and intercept with more precision.
- Additional calculations (not detailed here) can test for nonlinearity. The idea is to compare the variation among replicates using the distances of the points from the regression line. If your points are "too far" from the line (given the consistency of the replicates), then you can conclude that a straight line does not really describe the relationship between X and Y.
- You can test the assumption that the variability in Y is the same at all values of X.

If you analyze the same data with linear regression and correlation (see Chapter 12), how do the results compare?

If you square the correlation coefficient (r), the value will equal R^2 from linear regression. The P value testing the null hypothesis that the population correlation coefficient is zero will match the P value testing the null hypothesis that the population slope is zero.

R^2 or r^2?

With linear regression, both forms are used and there is no distinction. With nonlinear and multiple regression, the convention is to always use R^2.

CHAPTER SUMMARY

- Linear regression fits a model to data to determine the value of the slope and intercept that makes a line best fit the data.
- Linear regression also computes the 95% confidence interval for the slope and intercept and can plot a confidence band for the regression line.
- Goodness of fit is quantified by R^2.
- Linear regression reports a P value, which tests the null hypothesis that the population slope is horizontal.

TERMS INTRODUCED IN THIS CHAPTER

- Intercept (p. 299)
- Linear regression (p. 297)
- Linear least squares (p. 304)
- Linear (p. 304)
- Parameter (p. 304)
- Regression (p. 304)
- Regression fallacy (p. 306)
- Residuals (p. 304)
- Rolling average (p. 305)
- Simple vs. multiple regression (p. 304)
- Slope (p. 298)
- Smoothed data (p. 305)

CHAPTER 34

❖❖❖

Introducing Models

> A mathematical model is neither a hypothesis nor a theory.
> Unlike scientific hypotheses, a model is not verifiable directly
> by an experiment. For all models are both true and false. . . .
> The validation of a model is not that it is "true" but that it
> generates good testable hypotheses relevant to important
> problems.
>
> R. LEVINS (1966)

Chapter 33 explained how linear regression fits a simple model to data. This chapter generalizes the concept of fitting a model to data. The goal in fitting a model is to find the best-fit values of the parameters that define the model. It is essential to understand these basic ideas of fitting models before reading about the various kinds of regression in Chapters 35 through 39.

LINGO: MODELS, PARAMETERS, AND VARIABLES

Model

In general, a *model* is a representation of something else. We study models because they are less expensive and more accessible than the real thing, as well as easier to manipulate (perhaps with fewer ethical issues).

A mathematical model is an equation or set of equations that describe, represent, or approximate a physical, chemical, or biological state or process. Using a model can help you think about chemical and physiological processes or mechanisms so you can design better experiments and comprehend the results.

Creating new and useful models from general principles can be difficult. In contrast, fitting models to data (a major theme of this book) and simulating data from models (a minor theme of this book) are much easier tasks.

When you fit a model to your data, you obtain best-fit values (parameter estimates) that you can interpret in the context of the model.

Random component

A mathematical model must specify both the ideal predictions and how the data will be randomly scattered around those predictions. The following two equivalent, crude equations point out that a full model must specify a random component (noise) as well as the ideal component (signal).

Data = Ideal + Random
Response = Signal + Noise

Parameters, statistics, and variables

A model is defined by an equation or a set of equations. The equation defines the outcome, called the *dependent variable*, as a function of one or more *independent variables*, one or more parameters, and often one or more true constants. The model in Chapter 33 defined insulin sensitivity (the dependent variable) as a function of %C20–22 (the independent variable) and the slope and intercepts (the parameters). It had no true constants.

When fitting a model, you will provide the computer program with a set of values for the dependent and independent varaiables, and then the program will find estimates of the parameters that make the model do the best possible job of predicting the dependent variable. In some cases models include true constants, such as 1, 2, 0.693, π, and so on, just come along for the ride and provide structure to the mathematical model.

Note that each data point has its own values for the independent and dependent variables. The population has one set of true parameters, which are unknown. Fitting the model to sample data generates an estimated value for each parameter in the model, along with a CI for each parameter.

The estimated values of each parameter are given several interchangeable names: parameter estimates, best-fit values of parameters, and sample statistics (which are distinct from population parameters).

Estimate

When regression is used to fit a model to data, it reports values for each parameter in the model. These can be called best-fit values or estimated values. Note that this use of the term *estimate* is very different from the word's conventional use to mean an informed guess. The estimates provided by regression are the result of calculations. The best-fit value is called a *point estimate*. The CI for each parameter is called an *interval estimate*.

Fitting a model to data

Regression fits a model to data. It does this by adjusting the values of the parameters in the model to make the predictions of the model come as close as possible to the actual data, taking into account the mathematical model for random error (a Gaussian distribution in the case of linear regression).

THE SIMPLEST MODEL

To compute an average, add up all the values and then divide by the number of values. That average, also called a mean, is a way to describe a stack of numbers with a single value. But what makes an average special?

When you sample values from a population that follows a Gaussian distribution, each value can be defined by this simple model:

$$Y = \mu + \varepsilon$$

where Y is the dependent variable, which is different for each value (data point); μ (the population mean) is a parameter with a single value, which you are trying to find out (it is traditional to use Greek letters to denote population parameters and regular letters to denote sample statistics); and ε (random error) is different for each data point, randomly drawn from a Gaussian distribution centered at zero. Thus, ε is equally likely to be positive or negative, so each Y is equally likely to be higher or lower than μ. The random variable ε is often referred to as *error*. As used in this statistical context, the term *error* refers to any random variability, whether caused by experimental imprecision or by biological variation.

This kind of model is often written like this:

$$Y_i = \mu + \varepsilon_i$$

The subscript i tells you that each Y and ε has a different value, but the population parameter μ has only a single value.

Note that the right side of this model equation has two parts. The first part computes the *expected* value of Y. In this case, it is always equal to the parameter μ, the population mean. More complicated models would have more complicated calculations involving more than one parameter. The second part of the model takes into account the random error. In this model, the random scatter follows a Gaussian distribution and is centered at zero. This is a pretty standard assumption but is not the only model of scatter.

Now you have a set of values, and you are willing to assume that this model is correct. You don't know the population value of μ, but you want to estimate its value from the sample of data. What value of the parameter μ is most likely to be correct? Mathematicians have proven that if the random scatter follows a Gaussian distribution, then the value of μ that is most likely to be correct is the sample mean or average. To use some mathematical lingo, the sample mean is the *maximum likelihood estimate* of μ.

What I've done so far is to take the simple idea of computing the average and turn it into a complicated process using statistical jargon and Greek letters. This doesn't help you understand the idea of a sample mean, but it warms you up for understanding more complicated models, for which the jargon really is helpful.

THE LINEAR REGRESSION MODEL

Recall the linear regression example of Chapter 33. The investigators wanted to understand why insulin sensitivity varies so much among individuals. They measured insulin sensitivity and the lipid content of muscle obtained at biopsy in

13 men. To find out how much insulin sensitivity increases for every percentage point increase in C20–22, linear regression was used to fit this model to the data:

Y = intercept + slope · X + scatter

This equation describes a straight line. It defines the dependent variable Y (insulin sensitivity) as a function of the independent variable X (%C20–22), two parameters (the slope and Y-intercept), and scatter (a random factor). It can be rewritten in a more standard mathematical form:

$$Y_i = \beta_0 + \beta_1 \cdot X_i + \varepsilon_i$$

The intercept (β_0) and slope (β_1) are parameters that each have a single true underlying population value. In contrast, the random component of the model takes on a different value for each data point. These random values are assumed to follow a Gaussian distribution with a mean of zero. Any point is just as likely to be above the line as below it but more likely to be close to the line than far from it. Remember that the term *error* refers to any random variability, whether caused by experimental imprecision or biological variation.

Note that this model is very simple and certainly not 100% correct. Although data were only collected with X values ranging from 18 to 24, the model predicts Y values for any value of X. Predictions beyond the range of the data are unlikely to be accurate or useful. This model even predicts Y values when X has values outside the range of zero to 100, although X quantifies the percentage of lipids that is C20–22, so values outside the range of 0 to 100% make no sense at all. Furthermore, the model assumes Gaussian random scatter. As discussed earlier in Chapter 10, this assumption is never 100% true.

It's OK that the model is simple and cannot possibly be 100% accurate. That is the nature of scientific models. If a model is too simple, it won't provide useful results. If a model is too complicated, it won't be possible to collect enough data to estimate all the parameters. Useful models are simple, but not too simple.

WHY LEAST SQUARES?

Linear regression finds the "best" values of the slope and intercept by minimizing the sum of the squares of the vertical distances of the points from the line. Why minimize the sum of the squared distances rather than the sum of the absolute values of the distances?

This question can't really be answered without delving into math, but the answer is related to the assumption that the random scatter follows a Gaussian distribution.

If the goal were to minimize the sum of the absolute distances, the regression would be indifferent to fitting a model for which the distances of two points from the line are each 5.0 versus a model for which the distances are 1.0 and 9.0. The sum of the distances equals 10.0 in both cases. But if the scatter is Gaussian, the first model is far more likely, so the model that places the line equidistant from both points should be preferred. Minimizing the sum of the squares accomplishes this. The sums of the squares from the two models are 50 and 82, so minimizing the sum of the squares makes the first model preferable.

A more rigorous answer is that the regression line determined by the *least-squares method* is identical to the line determined by maximum likelihood calculations. What does that mean? Given any hypothetical set of parameter values, it is possible to compute the chance of observing our particular data. Maximum likelihood methods find the set of parameter values for which the observed data are most probable. Given the assumption that scatter follows a Gaussian distribution (with a uniform standard deviation), it can be proven that the maximum likelihood approach and the least-squares approach generate identical results. More simply (perhaps a bit too simply), minimizing the sum of the squares finds values for the parameters that are most likely to be correct.

OTHER MODELS AND OTHER KINDS OF REGRESSION

Regression includes a large family of techniques beyond linear regression:
- Nonlinear regression (see Chapter 36). As in linear regression, Y is a measured variable and there is a single X variable. However, a graph of X versus Y is curved, and the model has a nonlinear relationship between the parameters and Y.
- Multiple linear regression (see Chapter 37). *Multiple* means there are two or more independent X variables.
- Multiple nonlinear regression.
- Logistic regression (see Chapter 37). Here, Y is a binary variable (or proportion) such as infected/not infected or cancer/no cancer. There may be only one X variable, but logistic regression is more commonly used with several X values.
- Proportional hazards regression. Here, the outcome is survival time. There may be only one X variable, but proportional hazards regression is more commonly used with several X variables.
- Poisson regression. The outcomes are counts that follow a Poisson distribution (see Chapter 6).

COMMON MISTAKES: MODELS

Mistake: Trying to find a perfect model

Your goal in using a model is not necessarily to describe your system perfectly. A perfect model may have too many parameters and variables to be useful. Rather, your goal is to find as simple a model as possible that comes close to describing your system for your intended purpose. You want a model to be simple enough so it is easy to collect enough data to which you can fit the model, yet complicated enough to do a good job of explaining the data. It is common for new fields of science to begin with simple models and then gradually use more and more complicated models as the systems become better understood.

Mistake: Saying that you will fit the data to the model

Regression does not fit data to a model, a phrase that implies fudging the data to make them comply with the predictions of the model. Rather, the model is fit to the data or the data are fit by the model.

Think of models as shoes in a shoe store and data as the pair of feet that you bring into the store. The shoe salesman fits a set of shoes of varying styles, lengths, and widths (parameter estimates) to your feet to find a good, comfortable fit. The salesman does not bring over one pair of universal shoes along with surgical tools to fit your feet to the shoes.

Mistake: Fitting a model without looking at a graph

It is too easy to be misled if you look only at best-fit parameter values and the R^2. Look at a graph of the data with the fit model superimposed.

Mistake: Being afraid to fit a nonlinear model

Fit the model that makes the most sense for your data. If linear regression does not make sense for your data, don't use it simply because it is easily available. Don't be afraid of nonlinear models (introduced in Chapter 36).

CHAPTER SUMMARY

- A mathematical model is an equation that describes a physical, chemical, or biological process.
- Models are simplified views of the world. As a scientific field progresses, it is normal for the models to get more complicated.
- Statistical techniques fit a model to the data. It is wrong to say that you fit data to the model.
- When you fit a model to data, you obtain the best-fit values of each parameter in the model along with a confidence interval showing you how precisely you have determined its value.
- If the scatter (error) is Gaussian, fitting the model entails finding the parameter values that minimize the sum of the square of the differences between the data and the predictions of the model. These are called least-squares methods.

TERMS INTRODUCED IN THIS CHAPTER

- Dependent variable (p. 315)
- Error (p. 316)
- Estimate (p. 315)
- Independent variable (p. 315)
- Interval estimate (p. 315)
- Least-squares method (p. 318)
- Maximum likelihood estimate (p. 316)
- Model (p. 314)
- Point estimate (p. 315)

CHAPTER 35

Comparing Models

Beware of geeks bearing formulas.
WARREN BUFFETT

A t first glance, comparing the fit of two models seems simple: just choose the model with predictions that come closest to the data. In fact, choosing a model is more complicated than that, and it requires accounting for both the number of parameters and the goodness of fit. The concept of comparing models provides an alternative perspective for understanding much of statistics.

COMPARING MODELS IS A MAJOR PART OF STATISTICS

So far, this book has explained the concepts of P values and hypothesis testing in the context of comparing two groups. Statistics is used in many other situations, and many of these can be thought of as comparing the fits of two models. Chapter 36 will briefly explain how to compare the fits of two different nonlinear models. Chapters 37 and 38 will explain multiple regression models (including logistic and proportional hazards regression models), and a major part of these analyses involves comparing alternative models. One-way ANOVA, explained in Chapter 39, can also be thought of as comparing models.

What's so hard about comparing models? It seems as though the answer is simple: pick the one that comes the closest to the data. In fact, that approach is not useful, because models with more parameters almost always fit the sample data better, even if they don't reflect what is actually going on in the population. A useful comparison must also take into account the number of parameters fit by each model.

A model with too few parameters won't fit the sample data well. A model with too many parameters will fit the sample data well, but the CIs of the parameters will be wide. If the goal of the model is to predict future values, a model with too many parameters won't make precise predictions. If the goal of the model is to obtain parameter values that can be interpreted scientifically, fitting a model with too many parameters will lead to estimated parameter values that are very uncertain (wide CIs).

Every P value can be viewed as the result of comparing the fit of two models. One model is the null hypothesis. The other model is a more general alternative hypothesis. The P value answers the question, If the null hypothesis were really true, what is the chance that the data would fit the alternative model so much better? Interpreting P values is easy if you can identify the two models being compared.

To set the stage for these later chapters, this chapter looks back at linear regression (Chapter 33) and the unpaired t test (Chapter 30) and presents them as a comparison of models.

LINEAR REGRESSION AS A COMPARISON OF MODELS

Chapter 33 explained how to informally interpret the P value and R^2 from linear regression. Here we use the same example (insulin sensitivity) but switch the focus to comparing two models.

The two models being compared

The linear regression model is shown (again) on the right side of Figure 35.1. The right side of Figure 35.2 shows the variability of points around the linear regression line. Each point in this graph shows the vertical distance of a point from the regression line. Points above the horizontal line at Y = 0 represent points that are above the regression line, whereas points below that line represent points that are below the regression line.

The alternative, null hypothesis model is a horizontal line. The left side of Figure 35.1 shows the horizontal line fit through the data. The left side of Figure 35.2 shows how well the null hypothesis model fits the data. Each point in the graph shows the difference between a Y value and the mean of all Y values. Points above the horizontal line at zero represent Y values that are greater than the mean, and points below that horizontal line represent Y values that are less than the mean.

Figure 35.1. Fitting the sample data with two alternative models.

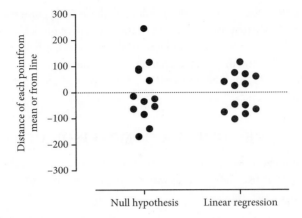

Figure 35.2. Residuals (distance from line) of each point from both models. The Y-axis has the same units as the Y-axis of Figure 35.1.

	HYPOTHESIS	SCATTER FROM	SUM OF SQUARES	PERCENTAGE OF VARIATION	
	Null	Horizontal line	155,642	100.0	
–	Alternative	Regression line	63,361	40.7	
=	Difference	Improvement	92,281	59.3	$R^2 = 0.593$

Table 35.1. Comparing the fit of a horizontal line versus the best-fit linear regression line. The points are closer to the regression line, so the sum of squares is lower.

The meaning of R^2

The linear regression model fits the data better. The variability of points around the regression line (right side of Figure 35.1) is less than the variability of points around the null hypothesis horizontal line (left side of Figure 35.1). How much less?

Table 35.1 compares the *sum of squares* (SS). The first row shows the sum of the squared distances of points from the horizontal line that is the null hypothesis. The second row shows the sum of the squared distances of points from the linear regression line.

The regression line fits the data better than a horizontal line, so the sum of squares is lower. The bottom row in Table 35.1 shows the difference between the fit of the two models. It shows how much better the linear regression model fits than the alternative null hypothesis model.

The fourth column shows the two sums of squares as a percentage of the total. Scatter around the regression line accounts for 40.7% of the variation. Therefore, the linear regression model itself accounts for $100\% - 40.7\% = 59.3\%$ of the variation. This is the definition of R^2, which equals 0.593.

P value

Table 35.2 is rearranged and relabeled to match other statistics books and programs. The focus is no longer on comparing the fit of two models, but on dividing

	SOURCE OF VARIATION	SUM OF SQUARES	DF	MS	F RATIO
	Regression	92,281	1	92,281.0	16.0
+	Random	63,361	11	5,760.1	
=	Total	155,642	12		

Table 35.2. Comparing the fit of a horizontal line versus the best-fit linear regression line. The points are closer to the regression line, so the F ratio is high. Table 35.2 uses the format of ANOVA, which will be explained in Chapter 39.

the total sum of squares into its components. The first row quantifies how well the linear regression model explains the data. The second row quantifies the scatter of data around the predictions of the model. The third row quantifies the scatter of the data from the predictions (horizontal line) of the null hypothesis.

The important points are that overall variation among data points is quantified with the sum of the squared distances between the point and a prediction of a model, and that the sum of squares can be divided into various sources of variation.

The third column of Table 35.2 shows the number of degrees of freedom. The bottom row shows the sum of squares of the distances from the fit of the null hypothesis model. There are 13 data points and only one parameter is fit (the mean), which leaves 12 df. The next row up shows the sum of squares from the linear regression line. Two parameters are fit (slope and intercept), so there are 11 degrees of freedom (13 data points minus two parameters). The top row shows the difference. The linear regression model has one more parameter than the null hypothesis model, so there is only one df in this row. The df, like the sums of squares, can be partitioned so the bottom row is the sum of the values in the rows above.

The fourth column of Table 35.2 divides the sums of squares by the number of df to compute the *mean square* (MS), which could also be called variances. Note that it is not possible to add the MS in the top two rows to obtain the MS in the bottom row.

Even if the null hypothesis were correct, you'd expect the sum of squares around the regression line to be a bit smaller than the sum of squares around the horizontal null hypothesis line. Dividing by the number of df accounts for this difference. If the null hypothesis were true, the two MS values would be expected to have similar values, so their ratio would be close to 1.0. In fact, for this example, the ratio equals 16.0.

This ratio of the two MS values is called the *F ratio*, named after the pioneering statistician Ronald Fisher. The distribution of F ratios is known when the null hypothesis is true. So for any value of F, and for particular values of the two df values, a P value can be computed. When using a program to find the P value from F, be sure to distinguish the df of the numerator (1 in this example) and the df for the denominator (11 in this example). If you mix up those two df values, you'll get the wrong P value.

The P value, which you already saw in Chapter 33, is 0.0021. From the point of view of probability distributions, this P value answers the question, If the null hypothesis were true, and given an experimental design with one and 11 df, what is the chance that random sampling would result in data with such a strong linear trend that the F ratio would be 16.0 or higher?

UNPAIRED t TEST RECAST AS COMPARING THE FIT OF TWO MODELS

Chapter 30 explained how to compare the means of two unpaired groups with an unpaired t test. Here we present that same example (bladder relaxation in young and old rats) with a different mindset—comparing the fit of the data to two models.

Unpaired t test as linear regression

To view the unpaired t test as a comparison of the fits of two models, consider it a special case of linear regression.

The example compares two groups, old and young. Let's call the variable that defines age X and assign X=0 to the old group and X=1 to the young group (those values are arbitrary). Figure 35.3 shows the data analyzed by linear regression.

The slope of the regression line is the increase in Y when the X value increases by one unit. The X values denoting old and young groups are one unit

Figure 35.3. Comparing two groups by linear regression.

The data are maximal bladder relaxation in percentages, as explained in Chapter 30. The two groups were assigned arbitrary X values, and the data were fit by linear regression. The best-fit slope of the regression line equals the difference between the two group means, and the CI of the slope equals the CI for the difference between two means. Linear regression reports a two-tailed P value, testing the null hypothesis that the true slope is zero.

apart, so the slope of the best-fit regression line equals the difference between means. The best-fit value of the slope is 23.5%, with a 95% CI ranging from 9.338% to 37.75%. These values match the results reported by the unpaired t test.

The P value from linear regression tests the null hypothesis that the slope is horizontal. This is another way to state the null hypothesis of the unpaired t test (that the two populations share the same mean). Because the two null hypotheses are equivalent, the P value determined by linear regression (0.0030) is identical to the P value reported by the t test.

Goodness of fit and R^2

The top row of Table 35.3 and the left side of Figure 35.4 quantify how well the null hypothesis model (a horizontal line) fits the data. Goodness of fit is quantified by the sum of squares of the difference between each value and the grand mean

	HYPOTHESIS	SCATTER FROM	SUM OF SQUARES	PERCENTAGE OF VARIATION	
	Null	Grand mean	5,172	100.0	
−	Alternative	Group means	2,824	54.6	
=	Difference	Improvement	2,348	45.4	$R^2 = 0.454$

Table 35.3. The t test example of Chapter 30 recast as a comparison of models.

Values are closer to their group means than the grand mean, so the sum of squares of the alternative hypothesis is lower.

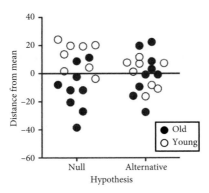

Figure 35.4. How well the t test example data are fit by the two models.

(Left) The fit of the model defined by the null hypothesis that both groups share the same population mean, showing the difference between each value and the grand mean. (Right) How well the data fit the alternative model that the groups have different population means. This part of the graph shows the difference between each value and the mean of its group.

SOURCE OF VARIATION	SUM OF SQUARES	DF	MS	F RATIO	P VALUE
Between groups	2,348	1	2,348.0	12.47	0.0030
+ Within groups	2,824	15	188.3		
= Total	5,172	16			

Table 35.4. The t test example of Chapter 30 recast as a comparison of models.
Values are closer to their group means than the grand mean, so the F ratio is high and the P value is low.

(calculated by ignoring any distinctions between the two groups). Figure 35.4 plots the distance of each value from the grand mean, and Table 35.3 shows the sum of squares.

The second row of Table 35.3 and the right side of Figure 35.4 show how well the alternative hypothesis fits the data. Figure 35.4 plots the distance of each value from its own group mean, and Table 35.3 reports the sum of squares of the difference between each value and the mean of the group from which that value was drawn.

The third row shows the difference in variation. Of all the variation (sum of squares from the null hypothesis grand mean), 54.6% is caused by scatter within the groups, and 45.5% of the total variation is caused by a difference between the two group means. Therefore, $R^2 = 0.454$.

P value

Determining the P value requires more than partitioning the variance into its components. It also requires accounting for the number of values and the number of parameters fit by each model.

The third column of Table 35.4 shows the number of df. The bottom row shows the fit of the null hypothesis model. There are 17 data points and only one parameter is fit (the grand mean), which leaves 16 df. The next row up quantifies the fit of the alternative model. Two parameters are fit (the mean of each group), so there are 15 df (17 data points minus 2 parameters). The top row shows the difference. The alternative model (two distinct means) has one more parameter than the null hypothesis model (one mean for both groups), so there is only 1 df in this row. The df, like the sum of squares, can be partitioned so the bottom row is the sum of the values in the two rows above. The fourth column divides the sum of squares by the number of df to compute the MS, which could also be called variance. Note that it is not possible to add the MS in the top two rows to obtain the MS in the bottom row.

If the null hypothesis were correct, you'd expect the sum of squares around the individual means to be a bit smaller than the sum of squares around the grand mean. But after dividing by df, the MS values would be expected to be about the same if the null hypothesis were in fact true. Therefore, if the null hypothesis were true, the ratio of the two MS values would be expected to be close to 1.0. If fact, for this example, the ratio (called the F ratio) equals 12.47.

The F distribution under the null hypothesis is known, and so the P value can be computed. The P value is 0.0030. It is the answer to the question, If the simpler (null hypothesis) model were correct, what is the chance that randomly chosen values would have group means far enough apart to yield an F ratio of 12.47 or higher?

Summary

The t test data can be viewed as comparing how well the data are fit by two models. One model is the null hypothesis—that the two groups of data are sampled from two populations with the same mean. Viewed in terms of linear regression, this model is a horizontal line with a slope equal to zero. The alternative model is that the population means differ. Viewed as linear regression, the slope is not zero. The goodness of fit of the two models is compared to see whether there is substantial evidence to reject the simpler (null hypothesis) model and accept the more complicated alternative model.

COMMON MISTAKES: COMPARING MODELS

Mistake: Comparing the fits of models that don't make scientific sense

Only use a statistical approach to compare models that are scientifically sensible. It rarely makes sense to blindly test a huge number of models. If a model isn't scientifically sensible, it probably won't be useful—no matter how high its R^2 may be for a particular data set.

Mistake: Using the F test method to compare unrelated models

The approach described in this chapter can only compare two related models. One model must be a simpler case of the other. The linear regression example does compare two related models, because the null hypothesis model is the same as the other model, with the slope fixed to equal zero. Another way to say this is that the two models are *nested*. Nonnested models can be compared, but they require methods that are beyond the scope of this book. To learn about alternative methods, start by reading Burnham and Anderson (2003).

Mistake: Comparing the fits of models to different data sets

When comparing models, both models must be fit to the same data sets. In some cases, two data sets are combined and both models fit both data sets. That's fine. But it is critical that both models "see" all the data. You can't use the methods described here to compare the fit of a model to a data set with the fit of a model to the same data set minus some outliers. You can't compare the fit of a model to a data set with the fit of a model to the same data after being transformed (say to logarithms). You can't compare the fit of a model to data using one weighting scheme with the fit of the model using a different weighting scheme. When comparing the fit of two models, only the model can differ, not the data.

Mistake: Comparing the fits of models whose predictions are indistinguishable in the range of the data

Figure 33.7 shows the predictions of three alternative models (straight line, hyperbola, and sine wave). In the range of the data ($X < 15$), the three models are indistinguishable. If you had scientific reason to compare the fits of these models, you'd be stuck. Spending your time struggling with statistical software wouldn't help a bit. The data simply cannot distinguish between those three models.

At larger X values ($X > 15$), the predictions of the models diverge considerably. If you had reason to compare those models, it would be essential to collect data at X values (times) for which the predictions of the model differ.

Mistake: Fitting lots of models, and accepting the one that fits best

This is a form of multiple comparisons as explained in Chapters 22 and 23. It can be a good start, but you'll want to test the model with new data.

CHAPTER SUMMARY

- Much of statistics can be thought of as comparing the fits of two models.
- This chapter looked at examples from Chapter 33 (linear regression) and Chapter 30 (unpaired t test) and recast the examples as comparisons of models.
- Viewing basic statistical tests (t test, regression, ANOVA) as a comparison of two models is a bit unusual, but it makes statistics more sensible for many people.
- Every P value can be viewed as the result of comparing the fit of the null hypothesis with the fit of a more general alternative hypothesis. The P value answers the question, If the null hypothesis were true, what is the chance that the data would fit the alternative model so much better? Interpreting P values is easy once you identify the two models being compared.
- It is only fair to compare the fits of two models to data when the data are exactly the same in each fit. If one of the fits has fewer data points or transformed data, the comparison will be meaningless.

TERMS INTRODUCED IN THIS CHAPTER

- F ratio (p. 323)
- Mean square (MS) (p. 323)
- Nested models (p. 327)
- Sum of squares (SS) (p. 322)

CHAPTER 36

⁘⁙⁘

Nonlinear Regression

Models should be as simple as possible, but not more so.

A. EINSTEIN

Nonlinear regression is more general than linear regression. It can fit any model that defines Y as a function of X and one or more parameters. Don't skip this chapter because it seems advanced and complicated. In fact, analyzing data with nonlinear regression is not much harder than analyzing data with linear regression. Be sure to read Chapters 33 (on linear regression) and 35 (on comparing models) before reading this one.

INTRODUCING NONLINEAR REGRESSION

From a scientist's point of view, nonlinear regression is a simple extension of linear regression. While linear regression can only fit one model (a straight line), nonlinear regression can fit any model that defines Y as a function of X. As the name suggests, the relationship between Y and X can be curved.

From a statistician's point of view, nonlinear regression is quite different from linear regression. The methodology is much more complicated, and you really can't understand how it works without using matrix algebra and calculus. Even the term *nonlinear* is more complicated than it seems, as the mathematical definition of *nonlinear* is not simply that the graph of X versus Y is curved. Perhaps for these reasons, nonlinear regression is a topic excluded from almost all introductory statistics books.

Nonlinear regression is commonly used in many fields of science such as pharmacology and physiology. Using nonlinear regression and understanding the results is not much more difficult than using linear regression.

AN EXAMPLE OF NONLINEAR REGRESSION

The data

Chapters 11 and 30 have already discussed data from Frazier, Schneider, and Michel (2006), who measured the degree to which the neurotransmitter norepinephrine relaxes bladder muscle in old and young rats. Strips of bladder muscle were exposed to various concentrations of norepinephrine, and muscle relaxation

LOG[NOREPINEPHRINE, M]	% RELAXATION
−8.0	2.6
−7.5	10.5
−7.0	15.8
−6.5	21.1
−6.0	36.8
−5.5	57.9
−5.0	73.7
−4.5	89.5
−4.0	94.7
−3.5	100.0
−3.0	100.0

Table 36.1. Bladder muscle relaxation data for one young rat.

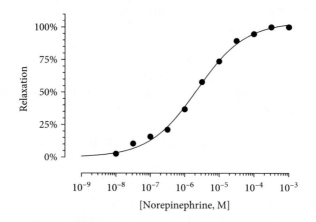

Figure 36.1. Bladder muscle relaxation data for one young rat.
The circles show the data from Table 36.1. The curve was fit by
nonlinear regression.

was measured. The data from each rat were analyzed to determine the maximum
relaxation and the concentration of norepinephrine that relaxes the muscle half
that much (EC50).

Table 36.1 and Figure 36.1 show the data from one young rat. Note that the
X-axis of Figure 36.1 is logarithmic. Going from left to right, each major tick on the
axis represents a concentration of norepinephrine 10-fold higher than the previous
major tick.

The model

The first step in fitting a model is choosing a model. In many cases, like this one, a
standard model will work fine. Pharmacologists commonly model dose-response
(or concentration-effect) relationships using the equation below.

$$Y = \text{Bottom} + \frac{\text{Top} - \text{Bottom}}{1 + 10^{(\text{LogEC50} - X) \cdot \text{HillSlope}}}$$

To enter this equation into a computer program, use the syntax below. Note that the asterisk (*) denotes multiplication and the carrot (^) denotes exponentiation.

Y = Bottom + (Top − Bottom)/(1 + 10^((LogEC50 − X)*HillSlope))

In this equation, X represents the logarithm of the concentration of norepinephrine and Y is the response the investigators measured, the amount of muscle relaxation. Y is defined as a function of X and four parameters:

- Bottom. This is the value of Y when X is very low, the baseline value of Y when there is no drug added.
- Top. This is the value of Y when X is very high. As the concentration gets higher and higher, the response plateaus at a value called Top.
- logEC50. This is the value of X (the logarithm of concentration) that provokes a response halfway between Bottom and Top.
- HillSlope. This quantifies how steep the curve is. The two words ("Hill" and "Slope") are smashed together to form a single parameter in the equation.

After choosing a model, the next step is to decide which of the parameters in the model should be fit to the data and which should be fixed to constant values. In this case, muscle relaxation must be zero with no added norepinephrine. So we'll tell the program not to find a best-fit value of Bottom, but rather to set it to a constant value of zero. We don't want to fix the top to 100, because the one goal of the study is to see whether that top plateau differs in old and young rats.

NONLINEAR REGRESSION RESULTS

Interpreting nonlinear regression results is quite straightforward. The whole point of nonlinear regression is to fit a curve to the data, so first look at the curve (Figure 36.1) and then look at the numerical results (Table 36.2).

Best-fit values of parameters
The best-fit values of the parameters should be interpreted in a scientific context.
    For the bladder relaxation example, let's focus only on the logEC50. The program fits the logEC50 and also transforms it back to the EC50, which is reported as 2.3e−006. This notation is commonly used by many programs and means 2.3 times 10 to the power of −6. Because X values are the logarithms of concentration in molar, the EC50 is in molar units (M). The best-fit EC50 is 0.0000023 M, which can also be written as 2.3 μM. This concentration of norepinephrine relaxes the bladder muscle half as much as the maximum possible relaxation.

CI of the parameters
The program also reports the standard error and the 95% CI of each parameter. The CIs are much easier to interpret and are an essential part of the nonlinear regression results. Given all the assumptions of the analysis (listed later in the chapter), you can be 95% confident that the interval contains the true parameter value.

BEST-FIT VALUES

Bottom	0.0
Top	104
LogEC50	−5.64
HillSlope	0.622
EC50	2.30e−006

STANDARD ERRORS

Top	2.06
LogEC50	0.0515
HillSlope	0.0358

95% CONFIDENCE INTERVALS

Top	99.3 to 109
LogEC50	−5.76 to −5.52
HillSlope	0.540 to 0.705
EC50	1.75e−006 to 3.02e−006

GOODNESS OF FIT

df	8
R^2	0.997
Absolute sum of squares	43.0

Table 36.2. Results of nonlinear regression as reported by GraphPad Prism.

The 95% CI for the logEC50 ranges from −5.76 to −5.52. Pharmacologists are accustomed to thinking in log units of concentration. Most people prefer to see the values in concentration units. The 95% CI of the EC50 ranges from 1.75 to 3.02 µM. The CI is fairly narrow and shows us that the EC50 has been determined to within a factor of two. That is more than satisfactory for this kind of experiment.

If the CI had been very wide, then you would not have determined the parameter very precisely and would not be able to interpret its value. How wide is too wide? It depends on the context and goals of the experiment.

Some programs don't report the confidence intervals but instead report the standard error of each parameter. Sometimes this is simply labeled the error, and sometimes it is labeled the standard deviation of the parameter. The confidence interval of each parameter is computed from its standard error. When you have plenty of data, the 95% CI extends about two standard errors in each direction.

R^2

R^2 is the fraction of the total variance of Y that is explained by the model. In this example, the curve comes very close to all the data points, so the R^2 is very high, at 0.997.

When $R^2 = 0.0$, the best-fit curve fits the data no better than a horizontal line at the mean of all Y values. When $R^2 = 1.0$, the best-fit curve fits the data perfectly, going through every point. If you happen to fit a really bad model (maybe you made a mistake when choosing a model in a nonlinear regression program or set an invalid constraint), R^2 can be negative. That would tell you that the fit of the selected model is even worse than the fit of a horizontal line at the mean of all Y values.

Upper- or lowercase? With linear regression, you'll see both r^2 and R^2. This is just a matter of style. With nonlinear and multiple regression, it is always written R^2.

How high should R^2 be? There are no general guidelines. If you are performing a routine set of experiments, you will learn what range of R^2 values to expect and then can troubleshoot if the value is too low. R^2 can be low for many reasons—the presence of outliers, fitting the wrong model, or the presence of lots of experimental or biological variation.

HOW NONLINEAR REGRESSION WORKS

Both linear and nonlinear regression find the values of the parameters that minimize the sum of the squares of the difference between the actual Y value and the Y value predicted by the regression model. Linear regression fits one simple model. Nonlinear regression can fit any model you choose. Linear regression can be thought of as a special case of nonlinear regression.

Although linear and nonlinear regression have the same goal, they work differently. The linear regression method can be completely explained with simple algebra. In contrast, nonlinear regression uses a computationally intensive approach that can only be explained using calculus and matrix algebra.

Nonlinear regression works via an *iterative* or *stepwise approach*. In this method, each parameter is initially provided with an estimated value. Nonlinear regression programs may provide these values automatically, or you may need to enter them yourself. The idea is to generate an initial curve that goes somewhere in the vicinity of the data points. The nonlinear regression method then changes the values of the parameters to move the curve closer to the points. Then it changes the values again. It repeats, or iterates, these steps many times, which is why the method is called *iterative*. When any possible changes to the parameters would make the curve fit more poorly (or fit the same), the method finishes and reports the results.

Because the foundations of nonlinear regression cannot be understood without using matrix algebra and calculus, nonlinear regression has a reputation of being complicated and advanced. But this reputation only applies to the mathematical foundation of the method. Using a nonlinear regression program and interpreting the results is just a tiny bit harder than using linear regression.

ASSUMPTIONS: NONLINEAR REGRESSION

Assumption: The model is correct
When fitting one model, nonlinear regression assumes the model is correct. The next section explains how to compare the fits of two (or more) models.

Assumption: The independent variable has no variability, or at least much less variability than the dependent variable
This assumption is quite reasonable for most experimental work. For example, there will be (or should be) very little error in pipetting different amounts of

drugs into wells of a 96-well plate or recording the time at which a blood sample is taken.

Assumption: The scatter of data around the curve is Gaussian

Nonlinear regression analysis assumes that the scatter of data around the curve is Gaussian. The CIs and P values cannot be interpreted if the distribution of the residuals (the difference between the actual and predicted Y value) is far from Gaussian, or if some of the values are outliers from another distribution.

Assumption: The variability is the same everywhere

Nonlinear regression assumes that scatter of points around the best-fit line has the same standard deviation all along the curve. This assumption is referred to as *homoscedasticity*. This assumption is violated if, for example, the points with high Y values tend to be farther from the best-fit line than points with small Y values.

Assumption: Data points are independent

Regression assumes that each point is randomly situated above or below the true curve and that no factor influences a batch of points collectively.

Assumption: The value of X is not used to compute Y

If the value of X is used to calculate Y (or the value of Y is used to calculate X), then linear and nonlinear regression calculations will be misleading. One example is a linearized Scatchard plot, used by pharmacologists to summarize binding data. The Y value (drug bound to receptors divided by drug free in solution) is calculated from the X value (free drug), so linear regression is not appropriate. If a Scatchard plot is curved, it is a mistake to fit it with nonlinear regression. Instead, fit an appropriate model to the nonlinearized raw data.

COMPARING TWO MODELS

Chapter 35 explained the general idea of comparing the fit of two models. Applying this idea to nonlinear regression is straightforward.

Figure 36.2 compares the fit of two dose-response models. The solid curve shows the fit explained above, for which the program determined the best-fit value of the Hill slope (0.622). The dotted curve represents a fit by a simpler model, in which the Hill slope was fixed to a standard value of 1.0. It is obvious that this curve does not fit the data as well.

Figure 36.3 shows the residuals of the data points from each curve. You can see that the residuals from the fixed-slope model are, on average, much larger. You can also see that the residuals from that fit (on the right) are not random. In contrast, the residuals from the variable slope model are smaller (the curve comes closer to the points) and random (no obvious pattern).

Figure 36.4 makes the comparison of the size of the residuals easier to see. Tables 36.3 and 36.4 compare the two fits using a format similar to that used in examples in Chapter 35. The null hypothesis is that the simpler model (fixed

Figure 36.2. Comparing the fit of two models.
The solid curve is the same as that in Figure 36.1. The Hill slope was fit by nonlinear regression and equals 0.622. The data were then fit again, but with the constraint that the Hill slope must equal 1.0. The dotted curve shows this fit, which does not do a good job of fitting the data.

Figure 36.3. Residuals from the two models.
Each circle represents the residual (distance from the curve) of one data point. (Left) The residuals from the fit shown in Figure 36.1, with the Hill slope fit by nonlinear regression (variable slope). (Right) The residuals from the dotted curve of Figure 36.2, in which the Hill slope is fixed to a standard value of 1.0.

slope, one fewer parameter to fit) is correct. In fact, the alternative model fits much better (lower sum of squares) but has one fewer df. The calculations balance the difference in df with the difference in sum of squares.

If the fixed-slope model were correct, it is possible that random scatter of points made the curve fit so badly, but since the P value is less than 0.0001, this would happen less than 0.01% of the time. The low P value is strong evidence that the simple model is inadequate, and that more complicated model (in which the program fits the Hill slope) is preferable.

Figure 36.4. Residuals from the two models.

Each circle represents the residual (distance from the curve) of one data point. (Left) The residuals from the solid curve of Figure 36.2, with the Hill slope fit by nonlinear regression. (Right) The residuals from the dotted curve of Figure 36.2, in which the Hill slope is fixed to a standard value of 1.0. This curve does not come as close to the points, so the residuals are larger. Note that the points are displaced horizontally to avoid overlap. Because the residuals from the variable slope model are closer to each other, there is more potential for overlap, so those points have more horizontal displacement.

	HYPOTHESIS	SCATTER FROM	SUM OF SQUARES	DF
	Null	Fixed slope	358.1	9
−	Alternative	Variable slope	43.0	8
=	Difference	Improvement	315.1	1

Table 36.3. The variable slope model fits the data much better, so the sum of squares is much smaller.

SOURCE OF VARIATION	SUM OF SQUARES	DF	MS	F RATIO	P VALUE
Difference	315.1	1	315.1	58.6	<0.0001
Variable slope model	43.0	8	5.4		
Fixed slope model	358.1	9			

Table 36.4. Computing an F ratio and P value from the fits of the two models.
The F ratio is high, so the P value is tiny.

TIPS FOR UNDERSTANDING MODELS

The first step in nonlinear regression is to choose a model (equation). These tips will help you understand what an equation means. As an example, let's use the Michaelis–Menten equation that describes enzyme activity (Y) as a function of substrate concentration (X):

$$Y = \frac{V_{max} \cdot X}{K_m + X}$$

Tip: Make sure you know the meaning and units of X and Y

Here Y is enzyme activity, which can be expressed in various units, depending on the enzyme. X is the substrate concentration expressed in units of concentration.

Tip: Figure out the units of the parameters

In the example equation, the parameter K_m is added to X. It only makes sense to add things that are expressed in the same units, so K_m must be expressed in the same concentration units as X. This means that the units cancel out in the term $X/(K_m + X)$, so V_{max} is expressed in the same units of enzyme activity as Y.

Tip: Figure out the value of Y at extreme values of X

Because X is a concentration, it cannot be negative. It can, however, be zero. Substitute $X = 0$ into the equation, and you will see that Y is also zero.

What happens when X is very large? As X gets large compared with K_m, the denominator $(X + K_m)$ approaches a value very similar to X. So the ratio $X/(X + K_m)$ approaches 1.0 and Y approaches V_{max}. Therefore, the graph of the model must level off at $Y = V_{max}$ as X gets very large. V_{max} is the maximum enzyme velocity.

Tip: Figure out the value of Y at special values of X

Because K_m is expressed in the same units as X, what is Y when X equals K_m? The ratio $X/(K_m + X)$ equals 0.5, so Y equals half of V_{max}. This means K_m is the concentration of substrate that leads to a velocity equal to half the maximum velocity, V_{max}.

LEARN MORE ABOUT NONLINEAR REGRESSION

Few statistics texts even mention nonlinear regression, and the advanced texts on nonlinear regression tend to be very mathematical. Start with Glantz and Slinker (2000), which includes a great chapter on nonlinear regression. If you like the style of this book, you'll also want to read my book on curve fitting (Motulsky & Christopoulos, 2004) or the GraphPad Curve Fitting Guide (Motulsky, 2012).

COMMON MISTAKES: NONLINEAR REGRESSION

Mistake: Choosing a polynomial model because it seems simpler

Some computer programs offer only one kind of curve fit, *polynomial regression*, which fits the following model to data:

$$Y = \alpha + \beta_1 \cdot X + \beta_2 \cdot X^2 + \beta_3 \cdot X^3 + \beta_4 \cdot X^4 \ldots$$

If the equation ends with the β_2 term, it is called a *second-order* or *quadratic equation*. If you stop after the β_3 term, it is called a *third-order* or *cubic equation*.

Mathematicians and programmers prefer polynomial regression, because the method used to fit polynomial models is much more straightforward than the method used to fit nonlinear models. In fact, although a graph of the polynomial equation is curved, mathematicians consider it a linear model (because Y is linear with respect to each of the parameters).

Few biological or chemical processes follow models described by polynomial equations, so the best-fit parameters can rarely be interpreted in terms of chemistry or biology. Avoid using polynomial regression just because it seems simpler. Instead, use nonlinear regression to fit the model that makes the most scientific sense. This will rarely be a polynomial model.

Mistake: Transforming curved data to a linear form and then using linear regression

Before nonlinear regression became readily available in the early 1980s, scientists often transformed their data to make the graph linear, plotted the transformed linearized data by hand on a sheet of graph paper, drew a straight line by hand with a ruler, calculated the slope and Y-intercept by hand, and then back-calculated the parameter estimates of the original nonlinear model. Later, scientists substituted computerized linear regression for the manual calculations.

Examples of linearizing transforms include Scatchard plots used by pharmacologists, double reciprocal Lineweaver–Burk plots used in enzymology, and the log transform used in kinetic studies. These methods are outdated and should not be used to analyze data. The problem is that the assumptions of linear regression are often badly violated after the data are transformed. Nonlinear regression gives more precise results and is not much harder to perform.

Mistake: Letting a computer program choose the model

Choosing a model should be a scientific decision based on chemistry (or physiology, or genetics, etc.). The choice should not be based solely on the shape of the graph.

Some people don't like thinking about equations, so they want a computer program to automatically fit thousands of equations and choose the model that fits the data best. Because a computer program cannot understand the scientific context of your experiment, a model chosen in this way is unlikely to be scientifically meaningful. The best-fit values of the parameters will probably have no scientific interpretation, so the fit is unlikely to be useful. Don't use a computer program as a way to avoid making scientific decisions.

Mistake: Fitting rolling averages, or smoothed data

As discussed in Chapter 33, smoothing the data (or computing rolling averages) violates the assumption that the residuals are independent and misleads the regression program about the amount of scatter. If you fit smoothed data, R^2 will be falsely high and the CIs will be falsely narrow. The results will be misleading.

Mistake: Using unweighted nonlinear regression when scatter increases as Y increases

It is common for biological data to violate the assumption that variability is the same for all values of Y. Instead, variability is often proportional to Y. Nonlinear regression programs offer weighting, and it is important to choose an appropriate weighting scheme.

Mistake: Using a standard model without thinking about which parameters should be fit and which should be held constant

A crucial step in using nonlinear regression is to decide which parameters should be estimated by nonlinear regression and which should be fixed to constant values based on control data. For example, in the experiment discussed earlier in this chapter, it made sense to constrain the Bottom parameter to a constant value of zero. If you skip this step and fit all the parameters into the model, you may end up with unhelpful results.

Mistake: Using R^2 to assess whether the fit is useful

A high value of R^2 tells you that the curve comes close to the data points. If your goal is to create a standard curve for interpolations, then this fit is useful. But if your goal is to determine best-fit values of parameters, a fit can have a high value of R^2 but still be quite useless. You need to look at whether the best-fit values are scientifically meaningful and have reasonably narrow CIs.

Mistake: Accepting the best-fit values of the parameters, without first looking at the corresponding CIs.

Nonlinear regression reports a CI with each parameter. Look at it to see if the parameter has been determined with reasonable precision. If the CI extends, say, over twelve orders of magnitude, you won't really care about the best-fit value of the parameter.

Mistake: Comparing two R^2 values to compare two fits

Comparing fits is tricky, and just choosing the fit with the larger R^2 is rarely appropriate. A model with more parameters will almost always fit with a larger R^2. You should not use R^2 to compare the fit of two models. Instead, use the method outlined in Chapter 35, which takes into account the number of parameters being fit to the model.

Q & A

Is nonlinear regression mean the same as *curve fitting*?
> *Curve fitting* is a more general phrase that can include anything from using a flexible ruler, to using cubic spline to nonlinear regression. In contrast, *nonlinear regression* means fitting a specified model to your data.

Is polynomial regression the same as nonlinear regression?
> No. Polynomial models looked curved when you plot X vs. Y, but they are actually considered to be linear by mathematicians. Polynomial models can be fit by nonlinear regression programs, but most nonlinear models cannot be fit by polynomial regression programs.

How can I use nonlinear regression is I don't know what model to fit?

> You can't. The idea of nonlinear regression usually is to fit a model that makes scientific sense with the goal of obtaining best-fit values for the parameters which are scientifically meaningful. If you goal is simply to draw an attractive curve, you could try fitting a polynomial model, or just try a bunch of standard models.

Can R^2 be negative? Really?

> R^2 is not actually calculated as the square of anything, but rather as the difference between two values, so yes it can be negative. This can only happen when the model you chose fits worse than a horizontal line. This could happen if you chose the wrong model by mistake, or set an incorrect constraint.

CHAPTER SUMMARY

- Like linear regression, nonlinear regression fits a model to your data to determine the best-fit values of parameters.
- As the name suggests, nonlinear regression fits nonlinear models. For this reason, the details of how the method works are far more complicated than those of linear regression.
- Using a nonlinear regression program is not much harder than using a linear regression program.
- The most important result from nonlinear regression is a graph of the data with the best-fit curve superimposed.
- The most important numerical results are the best-fit values of the parameters and the confidence intervals of each parameter.

TERMS INTRODUCED IN THIS CHAPTER

- Iterative/stepwise approach (p. 333)
- Nonlinear regression (p. 329)
- Polynomial regression (p. 337)
- Second-order/quadratic equation (p. 337)
- Third-order/cubic equation (p. 337)

❖❖❖

CHAPTER 37

Multiple Regression

An approximate answer to the right problem is worth a good
deal more than an exact answer to an approximate problem.

JOHN TUKEY

In laboratory experiments, you can generally control all the variables. You change one variable, measure another, and then analyze the data with one of the standard statistical tests. But in some kinds of experiments, and in many observational studies, you must analyze how one variable is influenced by several variables. Before reading this chapter, which introduces the idea of multiple regression, first read Chapters 33 (simple linear regression) and 34 (fitting models to data). This chapter explains multiple regression, in which the outcome is a continuous variable. The next chapter explains logistic regression (in which the outcome is binary) and proportional hazards regression (in which the outcome is survival time).

GOALS OF MULTIVARIABLE REGRESSION

Multiple regression extends simple linear regression to allow for multiple independent (X) variables. This is useful in several contexts:

- To assess the impact of one variable after accounting for others. Does a drug work after accounting for age differences between the patients who received the drug and those who received a placebo? Does an environmental exposure increase the risk of a disease after taking into account other differences between people who were and were not exposed to that risk factor?
- To create an equation for making useful predictions. Given the data we know now, what is the chance that this particular man with chest pain is having a myocardial infarction (heart attack)? Given several variables that can be measured easily, what is the predicted cardiac output of this patient?
- To understand scientifically how much changes in each of several variables contribute to explaining an outcome of interest. For example, how do the concentrations of high-density lipoproteins (HDL, good cholesterol), low-density lipoproteins (LDL, bad cholesterol), triglycerides, C-reactive protein, and homocysteine predict the risk of heart disease? One goal might be

to generate an equation that can predict the risk for individual patients (as explained in the previous point). But another goal might be to understand the contributions of each risk factor in order to aid public health efforts and help prioritize future research projects.

Different regression methods are available for different kinds of data. This chapter explains *multiple linear regression*, in which the outcome variable is continuous. The next chapter explains logistic regression (binary outcome) and proportional hazards regression (survival times).

LINGO

Variables

A regression model predicts an outcome, called the *dependent variable* (Y), from one or more predictors, called the *independent variables* (X). Each independent variable can be:

- Continuous (e.g., age, blood pressure, weight).
- Binary. An independent variable might, for example, code for gender by defining zero as male and one as female. These codes, of course, are arbitrary. When there are only two possible values for a variable, it is called a *dummy variable*.
- Categorical, with two or more categories (e.g., four medical school classes or three different countries). Consult more advanced books if you need to do this, because it is not straightforward. Several dummy variables are needed.

Parameters

The multiple regression model defines the dependent variable as a function of the independent variables and a set of parameters, or *regression coefficients*. Regression methods find the values of each parameter that make the model predictions come as close as possible to the data. This approach is analogous to linear regression, which determines the values of the slope and intercept (the two parameters or regression coefficients of the model) to make the model predict Y from X as closely as possible.

Simple regression versus multiple regression

Simple regression refers to models with a single X variable, as explained in Chapter 33. *Multiple regression*, also called *multivariable regression*, refers to models with two or more X variables.

Univariate versus multivariate regression

Although they are beyond the scope of this book, methods do exist that can simultaneously analyze several outcomes (Y variables) at once. These are called *multivariate methods*, and they include factor analysis, cluster analysis, principal components analysis, and multiple ANOVA (MANOVA). These methods contrast with *univariate methods*, which deal with only a single Y variable.

Note that the terms *multivariate* and *univariate* are sometimes used inconsistently. Sometimes *multivariate* is used to refer to multivariable methods for which there is one outcome and several independent variables (i.e., multiple and logistic regression). And sometimes *univariate* is used to refer to simple regression with only one independent variable.

AN EXAMPLE OF MULTIPLE LINEAR REGRESSION

As you learned in Chapter 33, simple linear regression determines the best linear equation to predict Y from a single variable X. Multiple linear regression extends this approach to find the linear equation that best predicts Y from multiple independent (X) variables.

Study design and questions

Staessen and colleagues (1992) investigated the relationship between lead exposure and kidney function. Heavy exposure to lead can damage kidneys. Kidney function decreases with age, and most people accumulate small amounts of lead as they get older. These investigators wanted to know whether accumulation of lead could explain some of the decrease in kidney function with aging.

The researchers studied 965 men and measured the concentration of lead in blood, as well as creatinine clearance to quantify kidney function. The people with more lead tended to have lower creatinine clearance, but this is not a useful finding. Lead concentration increases with age, and creatinine clearance decreases with age. Differences in age are said to *confound* any investigation between creatinine clearance and lead concentration. To adjust for the confounding effect of age, the investigators used multiple regression and included age as an independent variable. They also included three other independent variables: body mass, logarithm of γ-glutamyl transpeptidase (a measure of liver function), and previous exposure to diuretics (coded as 0 or 1).

The X variable about which they cared most was the logarithm of lead concentration. They used the logarithm of concentration rather than the lead concentration itself because they expected the effect of lead to be multiplicative rather than additive—that is, they expected that a doubling of lead concentration (from any starting value) would have an equal effect on creatinine clearance. So why logarithms? The regression model is intrinsically additive. Note that the sum of the two logarithms is the same as the logarithm of the product: $\log(A) + \log(B) = \mathrm{Log}(A \cdot B)$. Therefore, transforming a variable to its logarithm converts a multiplicative effect to an additive one (Appendix E reviews logarithms).

Mathematical model

This book is mostly nonmathematical. I avoid explaining the math of how methods work and only provide such explanation when it is necessary to understand the questions the statistical methods answer. But you really can't understand multiple regression at all without understanding the model that is being fit to the data.

	β_0
+	$\beta_1 \times$ log(serum lead)
+	$\beta_2 \times$ Age
+	$\beta_3 \times$ Body mass index
+	$\beta_4 \times$ log(GGT)
+	$\beta_5 \times$ Diuretics?
+	ε (Gaussian random)
=	Creatinine clearance

Table 37.1. The multiple regression model for the example.
β_0 through β_5 are the six parameters fit by the multiple regression program. Each of these parameters has different units (listed in Table 37.2). The product of each parameter times the corresponding variable is expressed in the units of the Y variable (creatinine clearance), which are milliliters per minute (ml/min). The goal of multiple regression is to find the values for the six parameters of the model that make the predicted creatinine clearance values come as close as possible to the actual values.

The multiple regression model is shown in Table 37.1. The dependent (Y) variable is creatinine clearance. The model predicts its value from a baseline value plus the effects of five independent (X) variables, each multiplied by a regression coefficient, also called a regression parameter.

The X variables were the logarithm of serum lead, age, body mass, logarithm of γ-glutamyl transpeptidase (a measure of liver function), and previous exposure to diuretics (coded as 0 or 1). This last variable is designated as the dummy variable (or indicator variable) because those two particular values were chosen arbitrarily to designate two groups (people who have not taken diuretics and those who have).

The final component of the model, ε, represents random variability (error). Like ordinary linear regression, multiple regression assumes that the random scatter (individual variation unrelated to the independent variables) follows a Gaussian distribution.

The model of Table 37.1 can be written as an equation:

$$Y_i = \beta_0 + \beta_1 \cdot X_{i,1} + \beta_2 \cdot X_{i,2} + \beta_3 \cdot X_{i,3} + \beta_4 \cdot X_{i,4} + \beta_5 \cdot X_{i,5} + \varepsilon_i \qquad (37.1)$$

Notes on this equation:

- Y_i. The subscript i refers to the particular patient. So Y_3 is the predicted creatinine clearance of the third patient.
- $X_{i,j}$. The first subscript (i) refers to the particular subject/patient, and the second subscript enumerates a particular independent variable. For example, $X_{3,5}$ encodes the use of diuretics (the fifth variable in Table 37.2) for the third patient. These values are data that you collect and enter into the program.
- The regression coefficients (β_1 to β_5). These will be fit by the regression program. One of the goals of regression is to find the best-fit value of each regression coefficient, along with its confidence interval. Each regression

VARIABLE	MEANING	UNITS
X_1	log(serum lead)	Logarithms are unitless. Untransformed serum lead concentration was in micrograms per liter.
X_2	Age	Years
X_3	Body mass index	Kilograms per square meter
X_4	log(GGT)	Logarithms are unitless. Untransformed serum GGT level was in units per liter.
X_5	Diuretics?	Unitless. 0 = never took diuretics. 1 = took diuretics.
Y	Creatinine clearance	Milliliters per minute

Table 37.2. Units of the variables used in the multiple regression examples.

coefficient represents the average change in Y when you change the corresponding X value by 1.0. For example, β_5 is the average difference in the logarithm of creatinine between those who have diabetes ($X_{i,5} = 1$) and those who don't ($X_{i,5} = 0$).

• The intercept, β_0. This is the predicted average value of Y when all the X values are zero. In this example, the intercept has only a mathematical meaning, but no practical meaning, because setting all the X values to zero means that you are looking at someone who is zero years old with zero weight! β_0 is fit by the regression program.
• ε. This is a random variable that is assumed to follow a Gaussian distribution. Predicting Y from all the X variables is not a perfect prediction because there also is a random component, designated by ε.

Goals of regression

Multiple regression fits the model to the data to find the values for the parameters that will make the predictions of the model come as close as possible to the actual data. Like simple linear regression, it does so by finding the values of the parameters (regression coefficients) in the model that minimize the sum of the squares of the discrepancies between the actual and predicted Y values. Like simple linear regression, multiple regression is a least-squares method.

Best-fit values of the parameters

The investigator's goal was to answer the question, After adjusting for effects of the other variables, is there a substantial linear relationship between the logarithm of lead concentration and creatinine clearance?

The best-fit value of β_1 (the parameter for the logarithm of lead concentration) was –9.5 ml/min. This means that on average, and after accounting for differences in the other variables, an increase in log(lead) of one unit (so that the lead concentration increases 10-fold, since they used common base 10 logarithms) is associated with a decrease in creatinine clearance of 9.5 ml/min. The 95% CI ranges from –18.1 to –0.9 ml/min.

Understanding these values requires some context. The study participants' average creatinine clearance was 99 ml/min. So a 10-fold increase in lead concentrations is associated with a reduction of creatinine clearance (reduced renal

Figure 37.1. The prediction of multiple regression.
One of the variables entered into the multiple regression model was the logarithm of lead concentration. The best-fit value for its coefficient was −9.5. For every one-unit increase in the logarithm of lead concentration, the creatinine clearance is predicted to go down by 9.5 ml/min. When the logarithm of lead concentration increases by one unit, the lead concentration itself increases 10-fold. The solid line shows the best-fit value of that slope. The two dashed lines show the range of the 95% CI.

function) of about 10%, with a 95% CI ranging from about 1% to 20%. Figure 37.1 illustrates this model.

The authors also report the values for all the other parameters in the model. For example, the β_5 coefficient for the X variable "previous diuretic therapy" was −8.8 ml/min. That variable is coded as zero if the patient had never taken diuretics and one if the patient had taken diuretics. It is easy to interpret the best-fit value. On average, after taking into account differences in the other variables, subjects who had taken diuretics previously had a mean creatinine clearance that was 8.8 ml/min lower than that of subjects who had not taken diuretics.

Statistical significance

Multiple regression programs can compute a P value for each parameter in the model testing the null hypothesis that the true value of that parameter is zero. Why zero? When a regression coefficient (parameter) equals zero, then the corresponding independent variable has no effect in the model (because the product of the independent variable times the coefficient always equals zero).

The authors of this paper did not include P values and reported only the best-fit values with their 95% confidence intervals. We can figure out which ones are less than 0.05 by looking at the confidence intervals.

The CI of β_1 runs from a negative number to another negative number and does not include zero. Therefore, you can be 95% confident that increasing lead concentration is associated with a drop in creatinine clearance (poorer kidney function). Since the 95% confidence interval does not include the value defining the null hypothesis (zero), the P value must be less than 0.05. The authors don't quantify it more accurately than that, but most regression programs would report the exact P value.

R^2: How well does the model fit the data?

R^2 equals 0.27. This means that only 27% of the variability in creatinine clearance is explained by the model. The remaining 73% of the variability is explained by variables not included in this study, variables included in this study but not in the forms entered in the mode, and random variation.

With simple linear regression, you can see the best-fit line superimposed on the data and visualize goodness of fit. This is not possible with multiple regression. With two independent variables, you could visualize the fit on a three-dimensional graph, but most multiple regression models have more than two independent variables.

Figure 37.2 shows a way to visualize how well a multiple regression model fits the data and to understand the meaning of R^2. Each point represents one subject.

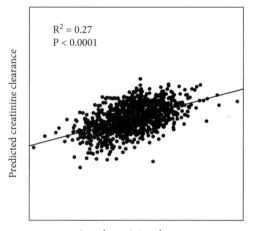

$R^2 = 0.27$
$P < 0.0001$

Predicted creatinine clearance

Actual creatinine clearance

Figure 37.2. The meaning of R^2 in multiple regression.

The authors did not post the raw data, so this graph does not accurately represent the data in the example. Instead, I simulated some data that look very much like what the actual data would have looked like. Each of the 965 points represents one man in the study. The horizontal axis shows the actual creatinine clearance for each subject. The vertical axis shows the creatinine clearance computed by the multiple regression model from that subject's lead level, age, body mass, log γ-glutamyl transpeptidase, and previous exposure to diuretics. The prediction is somewhat useful, because generally people who actually have higher creatinine clearance levels are predicted to have higher levels. However, there is a huge amount of scatter. If the model were perfect, each predicted value would be the same as the actual value, all the points would line up on a 45-degree line, and R^2 would equal 1.00. Here the predictions are less accurate, and R^2 is only 0.27.

The horizontal axis plots each subject's measured creatinine clearance (the Y variable in multiple regression). The vertical axis plots the creatinine clearance value predicted by the model. This prediction is computed from the other variables for that subject and the best-fit parameter values computed by multiple regression, but this calculation does not use the measured value of creatinine clearance. So the graph shows how well the model predicts the actual creatinine clearance. If the prediction were perfect, all the points would align on a 45-degree line with the predicted creatinine clearance matching the actual creatinine clearance. You can see that the predictions for the example data are far from perfect. The predicted and actual values are correlated, with R^2 equal to 0.27. By definition, this is identical to the overall R^2 computed by multiple regression.

R^2 versus adjusted R^2

R^2 is commonly used as a measure of goodness of fit in multiple linear regression, but it can be misleading. Even if the independent variables are completely unable to predict the dependent variable, R^2 will be greater than zero. The expected value of R^2 increases as more independent variables are added to the model. This limits the usefulness of R^2 as a way to quantify goodness of fit, especially with small sample sizes.

In addition to reporting R^2, which quantifies how well the model fits the data being analyzed, most programs also report an *adjusted R^2,* which estimates how well the model is expected to fit new data. This measure accounts for the number of independent variables and is always smaller than R^2. How much smaller depends on the relative numbers of subjects and variables. This study has far more subjects (965) than independent variables (5), so the adjusted R^2 is only a tiny bit smaller than the unadjusted R^2, and the two are equal to two decimal places (0.27).

ASSUMPTIONS

Assumption: Sampling from a population

This is a familiar assumption of all statistical analyses. The goal in all forms of multiple regression is to analyze a sample of data in order to make more general conclusions (or predictions) about the population from which the data were sampled.

Assumption: Linear effects only

The multiple regression model assumes that increasing an X variable by one unit increases (or decreases) the value of Y (multiple regression) by the same amount, regardless of the values of the X variables.

In the multiple regression (lead exposure) example, the model predicts that (on average) creatinine clearance will decrease by a certain amount when the logarithm of lead concentration increases by 1.0 (which means the lead concentration increases 10-fold, since the common or base 10 logarithm of 10 is 1). The assumption of linear effects means that increasing the logarithm of lead concentration by 2.0 will have twice the impact on creatining clearlance as increasing it by 1.0, and that the decrease in creatinine clearance by a certain concentration of lead does not depend on the values of the other variables.

Assumption: No interaction beyond what is specified in the model

Two of the independent variables in the multiple linear regression model are the logarithm of lead concentration and age. What if lead has a bigger impact on kidney function in older people than it does in younger people? If this were the case, the relationship between the effects of lead concentration and the effects of age would be called an *interaction*. Standard multiple regression models assume there is no interaction. It is possible to include interaction terms in the model, and this process is explained later in the chapter.

Assumption: Independent observations

The assumption that data for each subject provide independent information about the relationships among the variables should be familiar by now. This assumption could be violated, for example, if some of the subjects were identical twins (or even siblings).

Assumption: The random component of the model is Gaussian

For any set of X values, multiple linear regression assumes that the random component of the model follows a Gaussian distribution, at least approximately. Furthermore, it assumes that the standard deviation of that scatter is always the same and is unrelated to any of the variables.

AUTOMATIC VARIABLE SELECTION

How variable selection works

The authors of our example stated that they collected data for more variables for each subject and that the fit of the model was not improved when the model also accounted for smoking habits, mean blood pressure, serum ferritin level (a measure of iron storage), residence in urban versus rural areas, or urinary cadmium levels. Consequently, they omitted these variables from the model whose fit they reported. In other words, they computed a P value for each independent variable in the model, removed variables for which P values were greater than 0.05, and then reran the model without those variables.

The appeal of automatic variable selection

Many multiple regression programs can choose variables automatically. One approach (called *all-subsets regression*) is to fit the data to every possible model (recall that each model includes some X variables and may exclude others) and then find the one that is the best. With many variables and large data sets, the computer time required for this approach is prohibitive. To conserve computer time when working with huge data sets, other algorithms use a stepwise approach. One approach (called *forward-stepwise selection* or a *step-up procedure*) is to start with a very simple model and add new X variables one at a time, always adding the X variable that most improves the model's ability to predict Y. Another approach (*backward-stepwise selection* or a *step-down procedure*) is to start with the full model (including all X variables) and then sequentially eliminate those X variables that contribute the least to the model.

The problems of automatic variable selection

The appeal of automatic variable selection is clear. You just put all the data into the program, and it makes all the decisions for you. The problem is multiple comparisons. How many models does a multiple regression program compare when given data with k independent variables and instructed to use the all-subsets method to compare the fit of every possible model? Each variable can be included or excluded from the final model, so the program will compare 2^k models. For example, if the investigator starts with 20 variables, then automatic variable selection compares 2^{20} models (more than a million), even before considering interactions.

When you read a paper presenting results of multiple regression, you may not even know the number of variables with which the investigator started. Peter Flom (2007) explains why this ignorance makes it impossible to interpret the results of multiple regression with stepwise variable selection:

> If you toss a coin ten times and get ten heads, then you are pretty sure that something weird is going on. You can quantify exactly how unlikely such an event is, given that the probability of heads on any one toss is 0.5. If you have 10 people each toss a coin ten times, and one of them gets 10 heads, you are less suspicious, but you can still quantify the likelihood. But if you have a bunch of friends (you don't count them) toss coins some number of times (they don't tell you how many) and someone gets 10 heads in a row, you don't even know how suspicious to be. That's stepwise.

The consequences of automatic variable selection are pervasive and serious (Harrell 2001; Flom 2007):

- The final model fits too well. R^2 is too high.
- The best-fit parameter values are too far from zero. This makes sense. Since variables with low absolute values have been eliminated, the remaining variables tend to have absolute values that are higher than they should be.
- The confidence intervals are too narrow, so you think you know the parameter values with more precision than is warranted.
- When you test whether the parameters are statistically significant, the P values are too low.

Simulated example of variable selection

Chapter 23 already referred to a simulated study by Freedman (1983) that demonstrated the problem with this approach (his paper was also reprinted as an appendix in a text by Good & Hardin [2006]). He simulated a study with 100 subjects and recorded data from 50 independent variables for each. All values were simulated, so it is clear that the outcome is not associated with any of the simulated independent variables. As expected, the overall P value from multiple regression was high, as were most of the individual P values (one for each independent variable).

He then chose the 15 independent variables that had the lowest P values (less than 0.25) and reran the multiple regression program using only those variables. The resulting overall P value from multiple regression was tiny (0.0005). The contributions of six of the 15 independent variables were statistically significant (P<0.05).

If you didn't know these were all simulated data with no associations, the results might seem impressive. The tiny P values beg you to reject the null hypotheses and conclude that the independent variables can predict the dependent variable.

The problem is essentially one of multiple comparisons, an issue already discussed in Chapter 23. With lots of variables, it is way too easy to be fooled. You can be impressed with high R^2 values and low P values, even though there are no real underlying relationships in the population.

Should you mistrust all analyses with variable selection?

It may make sense to use statistical methods to decide whether to include or exclude one or a few carefully selected independent variables. But it does not make sense to let a statistics program test dozens or hundreds (or thousands) of possible models in hope that the computer can work magic.

When reading the results of research that depends on multiple regression, read the paper carefully to see how many variables were collected but not used in the final model. If a lot of variables were collected but not ultimately used, be skeptical of the results. In some cases, there is no problem, as the extra variables were collected for other purposes and the variables used in the model were selected based on a plan created according to the goal of the study. But watch out if the investigators first included many variables in the model but report results with variables that were selected because they worked the best.

In some cases, the goal of the study is exploration. The investigators are not testing a hypothesis but rather are looking for a hypothesis to test. Variable selection can be part of the exploration. But any model that emerges from an exploratory study must be considered a hypothesis to be tested with new data. Deciding how to construct models is a difficult problem. Says Gelman (2012), "This is a big open problem in statistics: how to make one's model general enough to let the data speak, but structured enough to hear what the data have to say."

SAMPLE SIZE FOR MULTIPLE REGRESSION

How many subjects are needed to perform a useful multiple regression analysis? It depends on the goals of the study and your assumptions about the distributions of the variables. Chapter 26 explained the general principles of sample size determination, but applying these principles to a multiple regression analysis is not straightforward.

One approach to determining sample size is to follow a rule of thumb. The only firm rule is that you need more cases than variables, a lot more. Beyond that, rules of thumb specify that the number of subjects (n) should be somewhere between 10 and 20 (or even 40) times the number of variables. You must count the number of variables you have at the start of the analysis. Even if you plan to use automatic variable selection to reduce the number of variables in the final model, you can't use that lower number as the basis to calculate a smaller sample size.

The rules of thumb are inconsistent because they don't require you to state the goal of the study. More sophisticated approaches to calculating sample size for multiple regression require you to specify your goal (Kelley & Maxwell, 2008). Is your goal to test the null hypothesis that the true R^2 is zero? Is it to test the null hypothesis that one particular regression coefficient equals zero? Is it to determine particular regression coefficients with a margin of error smaller than a specified quantity? Is it to predict future points within a specified margin of error? Once you have articulated your goal, along with corresponding alpha and beta as explained in Chapter 26, you can then use specialized software to compute the required sample size.

MORE ADVANCED ISSUES WITH MULTIPLE REGRESSION

Multicollinearity

When two X variables are highly correlated, they both convey essentially the same information. For example, one of the variables included in the multiple regression example of Chapter 37 is BMI, which is computed from an individual's weight and height. If the investigators had entered weight and height separately into the model, they probably would have encountered *collinearity*, because people who are taller also tend to be heavier.

The problem is that neither variable adds much to the fit of the model after the other one is included. If you removed either height or weight from the model, the fit probably would not change much. But if you removed both height and weight from the model, the fit would be much worse.

The multiple regression calculations assess the *additional* contribution of each variable after accounting for all the other independent variables. When variables are collinear, each variable makes little individual contribution. Therefore the CIs for the corresponding regression coefficients are wider and the P values for those parameters are larger.

One way to reduce collinearity (or *multicollinearity*, when three or more variables are entangled) is to avoid entering related or correlated independent variables into your model. An alternative is the approach used in the example study. In this case, the researchers combined height and weight into a single variable (BMI).

Interactions among independent variables

Two of the independent variables in the multiple linear regression model are the logarithm of lead concentration and age. What if the effects of lead concentration matter more with older people? This kind of relationship, as mentioned earlier, is called an *interaction*.

To include interaction between age and the logarithm of serum lead concentration, add a new term to the model equation with a new parameter multiplied by the product of age (X_2) times the logarithm of lead (X_1):

NAME	EXAMPLE
Longitudinal or repeated measures	Multiple observations of the same subject at different times
Crossover	Each subject first gets one treatment, then another.
Bilateral	Studying both knees in arthritis.
Cluster	Study pools results from three cities. Patients from the same city are more similar to each other than they are to patients from another city.
Hierarchical	A clinical study of a surgical procedure uses patients from three different medical centers. Within each center, several different surgeons may do the procedure. For each patient, results might be collected at several time points.

Table 37.3. Study designs that violate the assumption of independence.

$$Y = \beta_0 + \beta_1 \cdot X_1 + \beta_2 \cdot X_2 + \beta_3 \cdot X_3 + \beta_4 \cdot X_4 + \beta_5 \cdot X_5 + \beta_{1,2} \cdot X_1 \cdot X_2 + \varepsilon$$

If the CI for the new parameter ($\beta_{1,2}$) does not include zero, then you will conclude that there is a significant interaction between age and log(lead). This means that the effects of lead change with age. Equivalently, the effect of age depends on the lead concentrations.

Correlated observations

One of the assumptions of multiple regression is that each observation is independent. In other words, the deviation from the prediction of the model is entirely random. Table 37.3 (adapted from Katz, 2006) is a partial list of study designs that violate this assumption and so require specialized analysis methods.

COMMON MISTAKES: MULTIPLE REGRESSION

Mistake: Comparing too many model variations

In addition to choosing which variables to include or exclude from the model (explained above), multiple regression analyses also let investigators choose among many variations of the same model:

- With and without interaction terms
- With and without transforming some independent variables
- With and without transforming the dependent variable
- Including and excluding outliers or influential points
- Pooling all the data or analyzing some subsets separately
- Rerunning the model after defining a different variable to be the dependent (outcome) variable

When you fit the data many different ways and then report only the model that fits the data best, you are likely to come up with conclusions that are not valid. This is essentially the same problem as choosing which variables to include in the model, discussed above.

Mistake: Preselecting variables to include in the model

To decide which variables to enter into a multiple regression program, some investigators first look at the correlation between each possible independent variable and the outcome variable, and then only enter the variables that are strongly correlated with the outcome. Selecting variables this way is not so different than letting a program select variables. The results will be affected the same way, with an R^2 that is too high, confidence intervals that are too narrow, and P values that are too low.

Mistake: Too many independent variables

The goal of regression, as in all of statistics, is to analyze data from a sample and make valid inferences about the overall population. That goal cannot always be met using multiple regression techniques. It is too easy to reach conclusions that apply to the fit of the sample data but are not really true in the population. When the study is repeated, the conclusions will not be reproducible.

This problem is called *overfitting* (Babyak, 2004). It happens when you ask more questions than the data can answer—for example, when you have too many independent variables in the model compared to the number of subjects.

How many independent variables is too many? For multiple regression, a rule of thumb is to have at least 10–20 subjects per independent variable. Fitting a model with five independent variables thus requires at least 50 to 100 subjects.

Mistake: Dichotomizing without reason

A common mistake is to convert a continuous variable to a binary one without good reason. For example, if you want to include blood pressure in the model, blood pressure itself could be entered into the model or the data could be dichotomized to a binary variable that encodes whether the patient has hypertension (high blood pressure) or not. One problem with the latter approach is that it requires deciding on a somewhat arbitrary definition of whether or not someone has hypertension. Another problem is that it treats patients with mild and severe hypertension as the same, and so information is lost.

Mistake: Assuming that regression proves causation

In the example from the beginning of this chapter, the investigators chose to place creatinine clearance (a measure of kidney function) on the left side of the model (Y) and the concentration of lead on the right (X) and therefore concluded that lead affects kidney function. It is conceivable that the relationship goes the opposite direction, and that damaged kidneys somehow accumulate more lead. This is a fundamental problem of observational studies. The best way to overcome any doubts about cause and effect is to do an experiment. While it wouldn't be ethical to expose people to lead to see what would happen to their renal function, such experiments can be done with animals.

Also beware of confounding variables. An absurd example mentioned by Katz (2006) makes this concept clear. A multiple regression model to find risk factors associated with lung cancer might identify carrying matches as a significant risk factor. This would prove that people who carry matches in their pocket are more likely to get lung cancer than people who don't. But, of course, carrying matches doesn't cause cancer. Rather, people who carry matches are also likely to smoke, and that does cause cancer.

Mistake: Not thinking about the model

It is too easy to run a multiple regression program without thinking about the model. Does the assumption of linearity make sense? Does it make sense to assume that the variables don't interact? You probably know a lot about the scientific context of the work. Don't forget all that when you pick a standard multiple regression model.

Q & A

Do you always have to decide which variable is the outcome (dependent variable) and which variables are the predictors (independent variables) at the time of data collection?

No. In some cases, the independent and dependent variables may not be distinct at the time of data collection. The decision is sometimes made only at the time of data analysis. But beware of these analyses. The more ways you analyze the data, the more likely you are to be fooled by overfitting and multiple comparisons.

Does it make sense to compare the value of one best-fit parameter with another?

No. The units of each parameter are different, so they can't be directly compared. If you want to compare, read about standardized parameters in a more advanced book. Standardizing rescales the parameters so they become unitless and can then be compared. A variable with a larger standardized parameter has a more important impact on the dependent variable.

Do all the independent variables have to be expressed in the same units?

No. Usually they are in different units.

In what units are the parameters expressed?

Each parameter is expressed in the Y units divided by the X units of the variable associated with that parameter.

How is regression related to other statistical methods?

Chapter 35 pointed out that an unpaired t test can be recast as linear regression. Similarly, ANOVA can be recast as multiple regression. It is also possible to use logistic or proportional hazards regression to compare two groups, replacing the methods explained in Chapters 27 through 29 (the results won't be exactly the same, but they should be close). Essentially all of statistics can be recast as a form of fitting some kind of model using an appropriate kind of regression.

How does ANCOVA fit in?

This book does not discuss ANCOVA. It is a model that is equivalent to multiple linear regression when at least one independent variable is categorical and at least one is continuous.

How are the CIs calculated?

Some programs report the standard error of each parameter instead of its CI. Computing the CI of the best-fit value of a model parameter works just like computing the CI of the mean if you know the standard error of the mean (see Chapter 12). Compute the margin of error by multiplying the reported standard error by a value obtained from the t distribution (see Appendix D). This value depends only on the level of confidence desired

(95% is standard) and the number of df (equal to the number of subjects in the study minus the number of parameters fit by the model). For 95% confidence and with plenty of df (common in multiple regression), the multiplier approximates 2.0. Add and subtract this margin of error from the best-fit value to obtain the CI.

Where can I learn more about multiple regression and related analyses?
The texts by Katz (2006) and Campbell (2006) are concise, clear, practical, and nonmathematical. The books by Glantz and Slinker (2000) and Vittinghoff and colleagues (2007) have more depth and more math, but they remain clear and practical.

CHAPTER SUMMARY

- Multiple variable regression is used when the outcome you measure is affected by several other variables.
- This approach is used when you want to assess the impact of one variable after correcting for the influences of others in order to predict outcomes from several variables, or to try to tease apart complicated relationships among variables.
- Multiple linear regression is used when the outcome (Y) variable is continuous. Chapter 38 explains methods used when the outcome is binary.
- Beware of the term *multivariate*, which is used inconsistently.
- Automatic variable selection is appealing but the results can be misleading. It is a form of multiple comparisons.

TERMS INTRODUCED IN THIS CHAPTER

- Adjusted R^2 (p. 348)
- All-subsets regression (p. 349)
- Automatic variable selection (p. 349)
- Backwards-stepwise selection (or step-down procedure) (p. 349)
- Collinearity (p. 352)
- Confounding variable (p. 343)
- Dummy variable (p. 342)
- Forward-stepwise selection (or step-up procedure) (p. 349)
- Interaction (p. 349)
- Multicollinearity (p. 352)
- Multiple linear regression (p. 341)
- Multivariable/multiple regression (p. 342)
- Multivariate methods (p. 342)
- Overfitting (p. 354)
- Regression coefficients (p. 342)
- Simple linear regression (p. 342)
- Univariate methods (p. 342)

CHAPTER 38

❖

Logistic and Proportional Hazards Regression

The plural of anecdote is not data.

ROGER BRINNER

This chapter briefly explains regression methods used when the dependent variable is not continuous. Logistic regression is used when the dependent variable is binary (i.e., has two possible values). Proportional hazards regression is used when the dependent variable is survival time. Read Chapter 37 (multiple regression) before reading this one.

LOGISTIC REGRESSION

When is logistic regression used?

As has been discussed in Chapter 37, methods that fit regression models with two or more independent variables are called *multiple regression methods*. Multiple regression is really a family of methods, with the specific type of regression used depending on what kind of outcome is being measured (see Table 38.1). *Logistic regression* is used when the outcome (dependent variable) has two possible values. Because it is almost always used with two or more independent variables, it should be called *multiple logistic regression*. However, the word *multiple* is often omitted but assumed.

Study design of example study

Bakhshi and colleagues (2008) studied factors that influence the prevalence of obesity in Iranian women. They collected data from 14,176 women. For each woman, they recorded height and weight and computed BMI. When the BMI exceeded a threshold, that person was defined to be obese. In this way, every woman was defined to be obese or not obese. Since there are two categories, these data are appropriately analyzed by logistic regression.

The researchers also collected data on the variables listed in Table 38.2: urban versus rural, age, years of education, married or not, and economic class. The only thing confusing about Table 38.2 is the encoding of economic class. They defined

TYPE OF REGRESSION	WHERE?	TYPE OF DEPENDENT (Y) VARIABLE	EXAMPLE OF Y VARIABLES
Linear	Chapters 33 and 37	Continuous (interval or ratio)	• Enzyme activity • Renal function (creatinine clearance) • Weight
Logistic	Chapter 38	Binary or dichotomous	• Death during surgery • Graduation • Recurrence of cancer
Proportional hazards	Chapter 38	Elapsed time to a one-time event	• Months until death • Days until patient is weaned from ventilator • Quarters in school before graduation

Table 38.1. Different kinds of regression for different kinds of outcomes.

VARIABLE	MEANING	UNITS
X_1	Place of residence	Rural = 0; urban = 1
X_2	Age	Years
X_3	Years of education	Years
X_4	Smoking	No = 0; yes = 1
X_5	Married	No = 0; yes = 1
X_6	Lower-middle economic index	No = 0; yes = 1
X_7	Upper-middle economic index	No = 0; yes = 1
X_8	Higher economic index	No = 0; yes = 1

Table 38.2. Variables used in the logistic regression example.

four economic classes (lower, lower middle, upper middle, higher) and encoded them using three binary (also called *dummy*) variables. The first variable encodes whether or not a person belongs to lower middle class (1=yes; 0=no). The second variable encodes membership in the upper middle class, and the third variable encodes membership in the higher economic class. Lower-class status is encoded by setting all three of those variables to zero. This is one of several ways to encode categorical variables with more than two possible outcomes.

The goal of logistic regression
The goal of logistic regression for this study is to create an equation that can be used to estimate the probability that someone will be obese as a function of the independent variables.

Review: Odds versus probabilities
Logistic regression actually works with odds rather than probability. Chapter 2 explained how the two are related. The key point is that odds and probability are two alternative ways to express precisely the same concept. The odds of an event occuring equals the probability that that event will occur divided by the

probability that it will not occur. Every probability can be expressed as odds. Every odds can be expressed as a probability.

The model: Multiplying odds ratios

The logistic regression model computes the odds that someone is obese (in this example) from baseline odds and from odds ratios computed for each independent variable. Conceptually, the model looks like this:

$$\text{Odds} = (\text{Baseline Odds}) \cdot OR_1 \cdot OR_2 \cdot OR_3 \cdot OR_4 \cdot OR_5 \cdot OR_6 \cdot OR_7 \cdot OR_8$$

$$(38.1)$$

The baseline odds answers this question, If every single independent X variable equalled zero, what are the odds of obesity? For this example, this baseline odds value is not really useful on its own, because one of the variables is age. So the baseline odds, interpreted literally, would be the odds of obesity for someone who is zero years old. Sometimes investigators define the X variable differently so as to make the baseline odds more meaningful. For this example, they could have encoded that variable as Age $-$ 20, so X $=$ 0 would encode people who are 20 years old, and X $=$ 1 would encode people who are 21 years old, and so on.

Each of the other odds ratios correspond to one independent variable. The first independent variable is for place of residence, coded as zero for rural and one for urban. The corresponding odds ratio answers the question, how much does knowing whether a particular person is urban or rural change the odds of obesity? If $X_1 = 0$, the person in question lives in a rural area and the baseline odds do not change. If $X_1 = 1$, the person in question lives in an urban area and the odds ratio estimated by logistic regression will tell us if the urban residents have the same odds of being obese as the rural residents (OR $=$ 1), if the odds are greater (OR $>$ 1), or if the odds are smaller (OR $<$ 1). The goal of logistic regression is to estimate a value for that odds ratio, and all the others.

The next independent variable (X_2) is age, which is a continuous variable. The corresponding odds ratio (OR_2) answers this question: for each additional year of age, by how much do the odds of obesity increase or decrease? If the OR associated with age equals 1.0, then age is not related to obesity, since multiplying the baseline odds by 1.0 does not change its value. If the OR $>$ 1, then the odds of obesity increase by a set percentage for each additional year of age. For instance, an OR of 1.03 would mean that the odds of obesity increase by 3% as a person grows older by one year. Conversely, an OR $<$ 1 indicates a decline in the odds of obesity associated with each additional year of age.

How logistic regression works

The goal of logistic regression is to find values for all the coefficients in an equation that estimates the odds that someone will have the outcome being tracked as a function of all the independent variables. However logistic regression equations do not use Equation 38.1. Instead, the equation is transformed so it is almost identical to Equation 37.1, used for multiple regression.

$$Y_i = \beta_0 + \beta_1 \cdot X_{i,1} + \beta_2 \cdot X_{i,2} + \beta_3 \cdot X_{i,3} + \beta_4 \cdot X_{i,4} + \beta_5 \cdot X_{i,5}$$
$$+ \beta_6 \cdot X_{i,6} + \beta_7 \cdot X_{i,7} + \beta_8 \cdot X_{i,8} \qquad\qquad (38.2)$$

The equation is easy to understand:

- Y_i is the natural logarithm of the odds for a particular subject. Logistic regression is based on logarithms, because the logarithm of the odds can take on any value, either negative or positive, whereas odds must be zero or positive, and the probability itself must be between zero and one.
- $X_{i,j}$. The first subscript (i) refers to the particular subject/patient, and the second subscript enumerates a particular independent variable. For example, $X_{3,5}$ encodes whether or not the third patient is married (the fifth variable in Table 38.2). These values are data that you collect and enter into the program.
- β_0 is the natural logarithm of the baseline odds.
- β_1 is the natural logarithm of the odds ratio for the first independent variable.
- β_2 is the natural logarithm of the odds ratio for the second independent variable.
- β_3 to β_8 are the natural logarithms of the odds ratios for the corresponding independent variables.

Equation 37.1 for multiple linear regression ended with a term ε that added Gaussian scatter to the result. The Y value in Equation 38.2 is the natural logarithm of odds, which can be transformed to a probability. Since it implicitly embodies uncertainty, there is no need to explicitly add a random term to the model.

Because there is no Gaussian distribution, the method of least squares, is not used to determine the values of the odds ratios. Instead, logistic regression finds the values of the odds ratios using what is called a *maximum likelihood method* (briefly explained in Chapter 34).

Interpreting the results: Odds ratios and confidence intervals

The key results of logistic regression are shown in Table 38.3. For each independent variable, logistic regression reports an odds ratio, along with a 95% CI.

SYMBOL	VARIABLE	ODDS RATIO	95% CI
OR_1	Urban?	2.13	1.915 to 2.369
OR_2	Age	1.02	1.017 to 1.026
OR_3	Years of education	0.98	0.968 to 0.993
OR_4	Smoking?	0.65	0.468 to 0.916
OR_5	Married?	1.48	1.312 to 1.668
OR_6	Lower middle class?	1.37	1.206 to 1.554
OR_7	Upper middle class?	1.29	1.136 to 1.468
OR_8	Higher class?	1.25	1.094 to 1.425

Table 38.3. Results of logistic regression.

The first independent variable has two possible values (rural = 0; urban = 1) and so is a binary, or dummy, variable. The corresponding odds ratio is 1.0 for rural women and 2.13 for urban women. This means that a woman who lives in a city has a bit more than twice the odds of being obese as someone who lives in a rural area but shares other attributes (age, education, married or not, social class). The 95% CI ranges from 1.9 to 2.4.

The next independent variable is age. The odds ratio for this variable is 1.02. Every year, the odds of obesity goes up about 2%. Don't forget that odds ratios multiply. The odds ratio for a 40 year old compared to a 25 year old is 1.02^{15}, or 1.35. The exponent is 15 because $40 - 25 = 15$. That means a 40 year old person has about 35% greater odds of being obese than a 25 year old.

The next independent variable is education. The odds ratio is 0.98. Since that is less than 1.0, having more education (in this population) makes one less slightly likely to be obese. For each additional year of formal education, the odds of being obese decline by 2% (since we multiply by the OR of 0.98).

Interpreting the results: P values for each independent variable

The investigators in this study did not report any P values, because the confidence intervals were sufficient to clearly present their findings. However, it is easy to learn something about the P values by looking at the confidence intervals of the odds ratios. Remember that if a null hypothesis were true, the corresponding odds ratio would equal 1.0. None of the eight confidence intervals in Table 38.3 include 1.0, so all of the corresponding P values must be less than 0.05. Each P value would answer the question, Assuming the model is correct and the population odds ratio (for this particular X variable) equals 1.0, what is the chance that random sampling would lead to an odds ratio as far (or farther) from 1.0 as that observed in this study?

Interpreting the results: Relationship between confidence intervals and P values

The relationship between confidence intervals and P values should be familiar by now. If the 95% confidence interval of the odds ratio does not contain 1.0 (the value that denotes no effect), then the P value will be less than 0.05. If the 95% confidence interval of the odds ratio does contain 1.0, then the P value must be greater than 0.05.

PROPORTIONAL HAZARDS REGRESSION

Proportional hazards regression is similar to logistic regression, but is used when the outcome is elapsed time to an event and is often used for analyses of survival times. To understand the brief discussion below, first read the discussions about analyzing survival data in Chapters 4 and 29.

An example

Rosman and colleagues (1993) investigated whether diazepam (also known as Valium) would prevent febrile seizures in children. They recruited about

400 children who had had at least one febrile seizure. Their parents were instructed to give medication to the children whenever they had a fever. Half were given diazepam and half were given a placebo (because there was no standard therapy).

The researchers asked whether treatment with diazepam would delay the time of first seizure. Their first analysis was simply to compare the two survival curves with a log-rank test (see Chapter 29). The P value was 0.064, so the difference between the two survival curves was not considered to be statistically significant. They did not report the hazard ratio for this analysis.

Next, the researchers did a more sophisticated analysis that adjusted for differences in age, number of previous febrile seizures, the interval between the last seizure and entry into the study, and developmental problems. The proportional hazards regression, which accounts for all such differences among kids in the study, found that the relative risk was 0.61, with a 95% CI extending from 0.39 to 0.94. This means that at any time during the two-year study period, kids treated with diazepam had only 61% of the risk of having a febrile seizure compared with those treated with a placebo. This reduction was statistically significant, with a P value of 0.027. If diazepam were truly ineffective, there would have been only a 2.7% chance of seeing such a low relative risk in a study of this size. This example shows that the results of proportional hazards regression are easy to interpret, although the details of the analysis are complicated.

The proportional hazards model

Note some confusing lingo. The method used to analyze these data is given the name *survival analysis*, although the event being tracked is time to a febrile seizure, not death. Most descriptions of proportional hazards regression assume that the event is death, but the same principles apply to studying time to any event—in this example, time to first seizure.

The slope of the survival curve, called the *hazard*, is the rate of death (or whatever event is being tracked) in a short time interval. For example, if 20% of patients with a certain kind of cancer are expected to die in the first year after diagnosis, then the average hazard during that time is 20% per year. When comparing two groups, investigators often assume that the *ratio* of hazard functions is constant over time. For example, the hazard among treated patients might be one-half the hazard among control patients. The hazards (death rates) may change over the course of the study, but at any particular time, the treated patients' risk of dying is one-half the risk of the control patients. Another way to say this is that the two hazard functions are proportional to one another—hence the name *proportional hazards regression*. If the assumption is true, then the difference between two survival curves can be quantified by a single number, a relative risk. If the ratio is 0.5, then the relative risk of dying in one group is half the risk of dying in the other group.

Proportional hazards regression, which is also called *Cox regression* after the person who developed the method, uses regression methods to fit the relative risk associated with each independent variable, along with a confidence interval and a P value testing the null hypothesis that the population relative risk is 1.0.

ASSUMPTIONS: LOGISTIC REGRESSION

Assumption: Sampling from a population
This is a familiar assumption of all statistical analyses. The goal in all forms of multiple regression is to analyze a sample of data in order to make more general conclusions (or predictions) about the population from which the data were sampled.

Assumption: Linear effects only
The multiple regression model assumes that increasing an X variable by one unit increases (or decreases) the value of Y (the logarithm of odds) by the same amount, regardless of the values of the X variables.

Assumption: No interaction beyond what is specified in the model
It is possible to include interaction terms in the model, and this process was briefly explained in Chapter 37.

Assumption: Independent observations
The assumption that data for each subject provide independent information about the relationships among the variables should be familiar by now. This assumption could be violated, for example, if some of the subjects were identical twins (or even siblings).

Assumption: The random component of the model is binomial
For any set of X values, multiple linear regression assumes that the random component of the model follows a binomial distribution, at least approximately. All subjects with the same set of X values share the same probabilities of having each of the two possible outcomes.

COMMON MISTAKES: LOGISTIC REGRESSION

Most common mistakes: Same as listed in Chapter 37
The list of common mistakes for logistic and proportional hazards regression are the same as those for multiple linear regression (see Chapter 37):

- Comparing too many model variations
- Preselecting variables to include in the model
- Too many independent variables
- Dichotomizing independent variables without reason
- Assuming that regression proves causation
- Not thinking about the model

Mistake: Converting from logarithms incorrectly
By tradition, the logistic regression model is written using natural logarithms. If you want to convert from log(odds ratio) to odds ratio, you therefore need to

use the exponential function. In Excel, this is the exp() function. Most calculators are labeled the same way. Don't take 10 to the power of the log(odds ratios). That would work if the logarithms were common (base 10) logs, but logistic regression always uses natural logs.

Mistake: Mixing up odds ratios for a particular subject vs. odds ratios for an independent variable

The Y variable in the logistic regression equation (Equation 38.2) is the natural logarithm of the odds that a subject with a defined set of independent (X) variables will have the outcome being tabulated. Each parameter on the right side of the equaton is the odds ratio for that parameter. If you increase that X value by 1.0, it is the amount by which Y (log odds for that subject) increases. Don't confuse the odds ratio for a subject with the odds ratio for a particular independent variable.

Q & A

As its name suggests, proportional hazards regression assume proportional hazards. Is this assumption always reasonable?
> No. For example, you would not expect the hazard functions of medical and surgical therapy for cancer to be proportional. You might expect surgical therapy to have the higher hazard at early times (because of deaths during the operation or soon thereafter) and medical therapy to have the higher hazard at later times. In such situations, proportional hazards regression should be avoided or used only over restricted time intervals for which the assumption is reasonable.

In what units are the parameters expressed?
> Each parameter is expressed as an odds ratio.

Do all the independent variables have to be expressed in the same units?
> No. Usually they are in different units.

How does one decide how many subjects one needs for a study to be analyzed by logistic regression?
> Chapter 37 explained the principles of figuring out the necessary sample size for data to be analyzed by multiple regression. The principles are the same for data to be analyzed by logistic or proportional hazards regression, with one big difference: the calculated value is not the number of subjects, but the number of subjects with the rarer of the two possible outcomes being monitored.

> For example, imagine that you have five independent variables and decide to use the rule of thumb for multiple regression that you need 15 times more subjects per treatment group than independent variables. For multiple regression, you'd need 75 subjects per group. But for logistic regression, you would need to go a step further. Say you are planning a study for which the outcome is whether or not the patient had a stroke (in a defined period of time). If you expect about 10% of the subjects to have strokes, you will need 750 subjects per group so you can expect 75 strokes per group. This would also be the case if you expected 90% of the patients to have strokes, as the calculation must be based on the rarer of the two outcomes.

Does R^2 quantify goodness-of-fit in logistic and proportional hazards regression?
> No. There are several other ways to quantify goodness of fit. Read a more advanced book for details.

Does the discussion on variable selection in multiple regression in Chapter 37 also apply to logistic and proportional hazards regression?

> Yes. The problems of automatic variable selection apply to all kinds of regression with multiple independent variables.

How about the discussions in Chapter 37 about collinearity, interactions and correlated observations?

> Those discussions apply to all regressions with multiple independent variables.

CHAPTER SUMMARY

- Logistic regression is used when the outcome is binary (has two possible values).
- Proportional hazards regression is used when the outcome is survival time.
- These methods are used for the same purposes for which multiple regression is used: to assess the impact of one variable after correcting for influences of others, to predict outcomes from several variables, or to try to tease apart complicated relationships among variables.
- The assumptions and traps of logistic and proportional hazards regression are about the same as those for multiple regression.

TERMS INTRODUCED IN THIS CHAPTER

- Logistic regression (or multiple logistic regression) (p. 357)
- Proportional hazards regression (or Cox regression) (p. 361)

The Rest of Statistics

CHAPTER 39

❖❖❖

Analysis of Variance

In science, consensus is irrelevant. What is relevant is
reproducible results. The greatest scientists in history are great
precisely because they broke with the consensus.

MICHAEL CRICHTON (2005)

One-way ANOVA compares the means of three or more groups,
assuming that all values are sampled from Gaussian populations.
This chapter explains the idea of ANOVA and the specifics of one-way
ANOVA. Chapter 40 will explain the follow-up multiple comparisons
tests.

COMPARING THE MEANS OF THREE OR MORE GROUPS

Example data

Hetland and coworkers (1993) were interested in hormonal changes in women
runners. Among their investigations, they measured the level of luteinizing
hormone (LH) in nonrunners, recreational runners, and elite runners. Because
the distribution of hormone levels appeared to be lognormal (see Chapter 11),
the investigators transformed all the values to the logarithm of concentration
and performed all analyses on the transformed data. This was a smart decision,
because it made the values actually analyzed (the logarithms) come close to
approximating a Gaussian distribution.

They didn't publish the raw data, but the summarized data are shown in
Table 39.1 and Figure 39.1. The left side of Figure 39.1 shows the means and
SDs. Assuming the data (as shown, already converted to logarithms) are sampled
from a Gaussian distribution, the SD error bars should include approximately
two-thirds of the data points. You can see that there is a lot of variation within
each group and a huge amount of overlap between groups. Certainly, you could
not use the LH level to predict whether someone was a runner.

The right side of Figure 39.1 shows the means with SEM error bars.
Because the sample sizes are fairly large, you can view these SEM error bars as
approximately 68% CIs (see Chapter 14). Looked at this way, it seems as though
the mean of the nonrunners might be distinct from the mean of the two groups of
runners.

369

GROUP	LOG(LH)	SD	SEM	N
Nonrunners	0.52	0.25	0.027	88
Recreational runners	0.38	0.32	0.034	89
Elite runners	0.40	0.26	0.049	28

Table 39.1. LH levels in three groups of women.

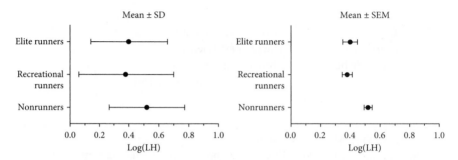

Figure 39.1. Data from Table 39.1 shown as a graph of the mean with SD (left) or SEM (right).

What's wrong with computing several t tests?

Your first thought when analyzing these data might be to use unpaired t tests (see Chapter 30). You'd need one t test to compare nonrunners with recreational runners, another to compare nonrunners with elite runners, and yet another to compare recreational runners with elite runners. The main problem with this approach is multiple comparisons (see Chapters 22 and 23). As you include more groups in the study, you increase the chance of observing one or more significant P values by chance. If the null hypothesis were true (all three populations have the same mean), there would be a 5% chance that each particular t test would yield a significant P value. But with three comparisons, the chance that any one (or more) of them will be significant would be far higher than 5%.

Interpreting a P value from one-way ANOVA

The data were not analyzed with t tests, but rather with *one-way ANOVA*, which compares all the groups at once. The one-way ANOVA reports that the P value is 0.0039.

To interpret any P value, the first step is to articulate the null hypothesis. Here, the null hypothesis is that the mean concentration of LH is the same in all three populations. The P value answers this question: Assuming the null hypothesis is true, what is the chance of randomly picking samples with mean values as different (or more different) than those observed in this study?

If the three sets of data were really sampled from identical populations, there is only a 0.4% chance that the three means would be as far apart as actually observed.

ASSUMPTIONS: ONE-WAY ANOVA

One-way ANOVA is based on the same assumptions as the unpaired t test:

- The samples are randomly selected from, or at least representative of, the larger populations.
- The observations within each sample are obtained independently. The relationships between all the observations in a group should be the same. You don't have independent measurements if some of the LH measurements are from the same person measured on several occasions, or if some of the subjects are twins (or even sisters).
- The data are sampled from populations that approximate a Gaussian distribution. In this case, the assumption applies to the logarithms of LH, so the assumption is that LH values are sampled from a lognormal distribution.
- The SDs of all populations are identical. This assumption is less important with large samples and when the sample sizes are equal. In this example, the data were all transformed to logarithms before the analysis was done, so the assumption refers to the log-transformed values.

HOW IT WORKS: ONE-WAY ANOVA

You can interpret ANOVA without knowing how it works, but you are less likely to use ANOVA inappropriately if you have some idea of how it works. The descriptions below give you a general idea of the process. The first two sections explain alternative ways to look at the method. These approaches are equivalent and produce exactly the same results. I find the first approach easier to understand, but the second is more traditional.

Determine sum of squares by fitting alternative models

Comparing three or more means with one-way ANOVA can be viewed as comparing the fit of the data to two different models using the ideas presented in Chapter 35. The two models are as follows:

- Null hypothesis: all populations share the same mean, and nothing but random sampling causes any differences between sample means.
- Alternative hypothesis: all populations do not share the same means. At least one population has a mean different than the rest.

The top row of Table 39.2 quantifies how well the null hypothesis fits the data. Goodness of fit is quantified by the sum of squares of the difference between each value and the grand mean (which ignores any distinctions among the three groups).

The second row of Table 39.2 shows how well the alternative hypothesis fits the data. Table 39.2 reports the sum of squares of the difference between each value and the mean of the sample from which that value came.

The third row of Table 39.2 shows the difference. Of all the variation (sum of squares from the null hypothesis grand mean), 94.7% is the result of scatter within

HYPOTHESIS	SCATTER FROM	SUM OF SQUARES	PERCENTAGE OF VARIATION	
Null	Grand mean	17.38	100.0	
− Alternative	Group means	16.45	94.7	
= Difference	Improvement	0.93	5.3	$R^2 = 0.053$

Table 39.2. One-way ANOVA as a comparison of models.

SOURCE OF VARIATION	SUM OF SQUARES	df	MS	F RATIO	P VALUE
Between groups	0.93	2	0.46	5.69	0.0039
+ Within groups (error, residual)	16.45	202	0.081		
= Total	17.38	204			

Table 39.3. ANOVA table showing the F ratio.

the groups, leaving 5.3% of the total variation as the result of differences between the group means.

To determine the P value requires more than dividing the variance into its components. It also requires accounting for the number of values and the number of groups. This is done in Table 39.3, which will be discussed later, after an explanation of an alternative approach to thinking about the sum of square values.

Alternative approach: Partitioning the sum of squares

A more common way to think about ANOVA is that it portions the variability into different components.

The example data come from three groups. The first step in ANOVA is to compute the total variability among all the values (ignoring which value came from which group). This is done by summing the squares of the difference of each value from the grand mean. This is shown in the bottom row of Table 39.3. The total sum of squares is 17.38.

Some of the variation comes from differences *among* the group means because the group means are not all the same. Sum the squares of each group mean from the grand mean and weight by sample size to get the sum of squares resulting from treatment. This is shown in the top row of Table 39.3. The sum of squares between group means is 0.93.

The rest of the variation comes from variably *within* each group, quantified by summing the squares of the differences of each value from its group mean. These numbers are shown in the second row of Table 39.3. The sum of squares within groups is 16.45. This is also called the *residual sum of squares* or the *error sum of squares*.

It isn't obvious (although it can be proven with simple algebra), but the sum of squares resulting from treatment and the sum of squares within the groups always add up to the total sum of squares.

Determining P from F

The third column of Table 39.3 shows the number of df. The bottom row, labeled "Total," is for the null hypothesis model. There are 205 values and only one parameter was estimated (the grand mean), which leaves 204 df. The next row up shows the sum of squares from the group means. Three parameters were fit (the mean of each group), so there are 202 df (205 data points minus three parameters). The top row shows the difference. The alternative model (three distinct means) has two more parameters than the null hypothesis model (one grand mean), so there are two df in this row. The df, like the sums of squares, can be partitioned so that the bottom row is the sum of values in the rows above.

The fourth column divides the sum of squares by the number of df to compute the MS, which could also be called the variance. Note that it is not possible to add the MSs of the top two rows to obtain a meaningful MS for the bottom row. Because the MS for the null hypothesis is not used in further calculations, it is left blank.

To compute a P value, you must take into account the number of values and the number of groups. This is done in the last column of Table 39.3.

If the null hypothesis were correct, each MS value would estimate the variance among values, so the two MS values would be similar. The ratio of those two MS values is called the *F ratio*, after Ronald Fisher, the pioneering statistician who invented ANOVA and much of statistics.

If the null hypothesis were true, F would be likely to have a value close to one. If the null hypothesis were not true, F would probably have a value greater than one. The probability distribution of F under the null hypothesis is known for various df and can be used to calculate a P value that answers the question, If the null hypothesis were true, what is the chance that randomly selected data (given the total sample size and number of groups) would lead to an F ratio this large or larger? For this example, F = 5.690, with 2 df in the numerator, 202 df in the denominator, and P = 0.0039.

R^2

Look at the first column of Table 39.2, which partitions the total sum of squares into its two component parts. Divide the sum of squares resulting from differences between groups (0.93) by the total sum of squares (17.38) to determine the fraction of the sum of squares resulting from differences between groups, 0.053. This is eta squared, $\square^2$, which is interpreted in the same way as R^2. Only 5.3% of the total variability in this example is the result of differences between group means. The remaining 94.7% of the variability is the result of differences within the groups.

The low P value means that the differences among group means are very unlikely to be caused by chance. The low R^2 means that the differences among group means are only a tiny fraction of the overall variability. You could also reach this conclusion by looking at the left side of Figure 39.1, where the SD error bars overlap so much.

REPEATED-MEASURES ANOVA

The difference between ordinary and *repeated-measures ANOVA* is the same as the difference between the unpaired and paired t tests. Use repeated-measures ANOVA to analyze data collected in three kinds of experiments:

- Measurements are made repeatedly for each subject, perhaps before, during, and after an intervention, and these repeated measurements make up the groups being compared by ANOVA.
- Subjects are recruited as matched sets (often called blocks), which are matched for variables such as age, postal code, or diagnosis. Each subject in the set receives a different intervention (or placebo).
- A laboratory experiment is run several times, each time with several treatments (or a control and several treatments) handled in parallel. More generally, you should use a repeated-measures test whenever the value of one subject in the first group is expected to be closer to a particular value in the other group than to a random subject in the other group.

When the experimental design incorporates matching, use the repeated-measures test, because it is usually more powerful than ordinary ANOVA. Of course, the matching must be done as part of the protocol before the results are collected. The decision about matching is a question of experimental design and should be made long before the data are analyzed.

Although the calculations for the repeated-measures ANOVA are different from those for ordinary ANOVA, the interpretation of the P value is the same. Similar kinds of multiple comparisons tests (see Chapter 39) are performed.

TWO-WAY ANOVA AND BEYOND

Although this chapter only explains one-way ANOVA, many other forms of ANOVA are used in statistics. This section will briefly describe two-way ANOVA, without a full example.

The term *one-way* means that the subjects are categorized in one way. In the example above, they were categorized by how much exercise they did. If they were also divided into age groups, the data would be analyzed using *two-way ANOVA*, also called *two-factor ANOVA*. If the subjects were simultaneously divided into a group that is pregnant and a group that is not pregnant, you'd need three-way ANOVA.

Two-way ANOVA simultaneously tests three null hypotheses and so computes three P values:

1. Interaction. The null hypothesis is that there is no interaction between the two factors. In our example, the null hypothesis is that the difference in log(LH) caused by exercise is the same for all age groups. Equivalently, the null hypothesis is that the difference in log(LH) among age groups is consistent for all levels of exercise. If the interaction P value is small, then there usually is no point in trying to interpret the other two P values. If you know the effect of exercise differs for different age groups, then trying to quantify one P value for the effects of age and one for the effects of exercise is usually pointless.
2. First factor. The null hypothesis is that the population means are identical for each category of the first factor. In our example, the null hypothesis is

that the mean log(LH) is the same for each level of exercise in the overall population and that all observed differences are caused by chance.

3. Second factor. The null hypothesis is that the population means are identical for each category of the second factor. In our example, the null hypothesis is that the mean log(LH) is the same for all age groups in the overall population and that all observed differences are caused by chance.

As with one-way ANOVA, special methods have been developed to deal with repeated measures in one or both factors.

Q & A

Can one-way ANOVA be done with two groups?

One-way ANOVA is usually only done for three or more groups, but it could be done for only two groups. Many programs won't allow this, but it certainly is mathematically possible. Although the approach seems very different, one-way ANOVA for two groups is mathematically equivalent to an unpaired t test and will compute an identical P value.

Are the results valid if sample size differs between groups?

Yes. ANOVA does not require that all samples have the same number of values. Two of the assumptions of ANOVA—that the data come from Gaussian populations and that these populations have equal standard deviations—matter much more when sample size varies a lot between groups. If you have very different sample sizes, a small P value from ANOVA may be caused by non-Gaussian data (or unequal variances) rather than differences among means.

Is the F ratio always positive?

Yes. Because the ANOVA calculations deal with sums of squares, the F ratio is always positive. It can, however, be a fraction less than 1.0.

Does ANOVA really compare variances?

Don't be misled by the term *variance*. *Variance* refers to the statistical method being used, not the hypothesis being tested. ANOVA tests whether group means differ significantly from each other. ANOVA, as usually performed, does not test whether the variances of the groups are different.

If the P value is small, can we be sure that all the means are distinct?

No. A P value can be small when all groups are distinct or if one is distinct and the rest are indistinguishable. Chapter 40 will show how to compare individual pairs of means.

Does ANOVA account for the order of groups?

No. The order of the groups is entirely irrelevant when computing the ANOVA table and P value. In the example, the three groups have a natural order that the ANOVA calculations do not take into account. ANOVA treats the groups as categories, and the calculations don't take into account how the groups are related to each other. Chapter 40 briefly explains a posttest for trend that does consider the group order.

Is the P value from one-way ANOVA one- or two-tailed?

Neither. With ANOVA, the concept of one- and two-tailed P values does not really apply. Because the means of the groups can be in many different orders, the P value has many tails.

If I need to use a program to compute a P value from F, does the order in which the two df values are entered matter?

>Yes. The calculation of a P value from F requires knowing the number of df for the numerator of the F ratio, and also the number of df for the denominator. If you mistakenly swap the two df values, the P value will be incorrect.

Do the one-way ANOVA calculations require raw data?

>No. One-way ANOVA (but not repeated-measures ANOVA) can be computed without raw data, so long as you know the mean, sample size, and SD (or SEM) of each group.

Can all kinds of ANOVA also be done using regression techniques?

>Yes. ANOVA compares the fit of several models to the data, and this can be done with regression techniques as well. The answers will be fundamentally identical, but they will look very different.

What is the difference between fixed-effects and random-effects ANOVA?

>Most programs perform Type I ANOVA, also known as fixed-effects ANOVA. This method tests for differences among the means of the particular groups from which you have collected data. Type II ANOVA, also called random-effects ANOVA, is less commonly used. It assumes that you have randomly selected groups from an infinite (or at least large) number of possible groups, and that you want to reach conclusions about differences among *all* the groups, not just the groups from which you collected data.

CHAPTER SUMMARY

- One-way ANOVA compares the means of three or more groups.
- The P value tests the null hypothesis that all the population means are identical.
- ANOVA assumes that all data are randomly sampled from populations that follow a Gaussian distribution and have equal standard deviations.
- The difference between ordinary and repeated-measures ANOVA is the same as the difference between the unpaired and paired t tests. Repeated-measures ANOVA is used when measurements are made repeatedly for each subject or when subjects are recruited as matched sets.
- Two-way ANOVA, also called two-factor ANOVA, determines how a response is affected by two factors. For example, you might measure a response to three different drugs in both men and women.

TERMS INTRODUCED IN THIS CHAPTER

- F ratio (p. 373)
- One-way analysis of variance (ANOVA) (p. 369)
- Repeated-measures ANOVA (p. 373)
- Residual sum of squares (or error sum of squares) (p. 372)
- Two-way ANOVA (or two-factor ANOVA) (p. 374)

CHAPTER 40

❖

Multiple Comparison Tests
after ANOVA

When did "skeptic" become a dirty word in science?
MICHAEL CRICHTON (2003)

One-way ANOVA computes a single P value that tests the null hypothesis that all groups were sampled from populations with identical means. This chapter explains how multiple comparison tests let you dig deeper to see which pairs of groups are statistically distinguishable. Before reading this chapter, read Chapters 22 and 23 on multiple comparisons and Chapter 39 on one-way ANOVA.

MULTIPLE COMPARISON TESTS FOR
THE EXAMPLE DATA

Goal

Chapter 39 analyzed a sample data set comparing LH levels (actually, their logarithms) in three groups of women. One-way ANOVA reported a very small P value. If the null hypothesis that all groups were sampled from populations with equal means were true, it would be quite rare for random sampling to result in so much variability among the sample means.

Multiple comparison tests dig deeper to find out which group means differ from which other group means, taking into account multiple comparisons (see Chapter 22). To avoid getting fooled by bogus statistically significant conclusions, the significance level is redefined to apply to an entire family of comparisons, rather than to each individual comparison (see Table 40.1). This is one of the most confusing parts of statistics. Using this new definition of statistical significance will reduce the chance of obtaining false reports of statistical significance (i.e., fewer Type I errors) but at the cost of reducing the power to detect real differences (i.e., more Type II errors).

For this example, the goal is to compare every mean with every other mean. One appropriate test is called *Tukey's test* (or, more generally, the *Tukey–Kramer test*, which allows for unequal sample size). The results include both CIs and conclusions about statistical significance.

377

SITUATION	MEANING OF A 5% SIGNIFICANCE LEVEL	ERROR RATE
One comparison	If the null hypotheses were true, there is a 5% chance that random sampling would lead to the incorrect conclusion that there is a real difference.	Per comparison
A family of comparisons	If all null hypotheses were true, there is a 5% chance that random sampling would lead to one or more incorrect conclusions that there is a real difference between the mean of two groups.	Per experiment or familywise

Table 40.1. Statistical significance is redefined for multiple comparisons.

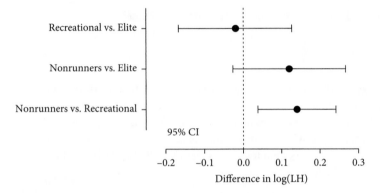

Figure 40.1. 95% CIs for Tukey's multiple comparisons test.
The 95% confidence level applies to the entire family of comparisons rather than to an individual comparison.

TUKEY'S MULTIPLE COMPARISONS TEST	DIFFERENCE BETWEEN MEANS	95% CI OF DIFFERENCE
Recreational runners vs. elite runners	−0.0200	−0.17 to 0.13
Nonrunners vs. elite runners	0.1200	−0.027 to 0.27
Nonrunners vs. recreational runners	0.1400	0.038 to 0.24

Table 40.2. Multiple comparisons confidence intervals.
The 95% confidence level applies to the entire family of intervals, rather than to each interval individually. The original data were expressed as the logarithm of LH concentration, so the mean differences and both confidence limits shown here are in those same units (logarithm of concentration).

Multiple comparison confidence intervals

Figure 40.1 plots the 95% CI for the difference between each mean and every other mean. These are tabulated in Table 40.2.

These are multiple comparisons CIs, so the 95% confidence level applies to the entire family of comparisons, rather than to each individual interval. Given the assumptions of the analysis (listed in Chapter 39), there is a 95% chance that all three of these CIs include the true population value, leaving only a 5% chance that any one or more of the intervals does not include the population value. Because the 95% confidence level applies to the entire set of intervals, it is impossible to correctly interpret any individual interval without seeing the entire set.

CIs as ratios

For this example, the data were entered into the ANOVA program as the logarithm of the concentration of LH (shown in Chapter 39). Therefore, Figure 40.1 and Table 40.2 show differences between two logarithms. Many find it easier to think about these results without logarithms, and it is easy to convert the data to a more intuitive format.

The trick is to note that the difference between the logarithms of two values is mathematically identical to the logarithm of the ratio of those two values (logarithms and antilogarithms are reviewed in Appendix E). Transform each of the differences (and each confidence limit) to its antilogarithm, and the resulting values can be interpreted as the ratio of two LH levels. Written as equations:

$$\log(A) - \log(B) = \log\left(\frac{A}{B}\right)$$

$$10^{(\log(A) - \log(B))} = \frac{A}{B}$$

Each row in Table 40.3 shows the ratio of the mean LH level in one group divided by the mean in another group, along with the 95% CI of that ratio. These are plotted in Figure 40.2.

Statistical significance

If a 95% CI for the difference between two means includes zero (the value specified in the null hypothesis), then the difference is not statistically significant ($P > 0.05$). Two of the three 95% CIs shown in Table 40.2 and Figure 40.1 include zero, and so these comparisons are not statistically significant at the 5% significance level. The CI for the remaining comparison (nonrunners vs. recreational runners) does not include zero, so that difference is statistically significant.

TUKEY'S MULTIPLE COMPARISONS TEST	RATIO	95% CI OF RATIO
Recreational runners vs. elite runners	0.96	0.68 to 1.35
Nonrunners vs. elite runners	1.32	0.94 to 1.86
Nonrunners vs. recreational runners	1.38	1.09 to 1.74

Table 40.3. Multiple comparisons confidence intervals, expressed as ratios rather than differences.

The values in Table 40.2 are the differences between two logarithms, which are mathematically identical to the logarithms of the ratios. Transforming to antilogarithms creates the table of ratios and CIs of ratios, shown here.

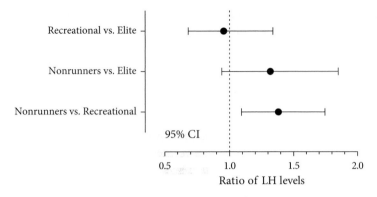

Figure 40.2. 95% CIs for Tukey's multiple comparisons test, expressed as ratios of LH levels rather than as differences between mean log(LH) concentrations.
The values correspond to Table 40.3.

The same conclusion can be reached using the ratios in Table 40.3 and Figure 40.2. The null hypothesis of identical populations corresponds to a ratio of 1.0. Only one comparison (nonrunners vs. recreational runners) does not include 1.0, so that comparison is statistically significant.

The 5% significance level is a famiylwise significance level, meaning that it applies to the entire set, or family, of comparisons (previously defined in Chapter 22). If the overall null hypothesis (values from all groups were sampled from populations with identical means) is true, there is a 5% chance that one or more of the comparisons will be statistically significant and a 95% chance that none of the comparisons will be statistically significant.

Table 40.4 shows the conclusions about statistical significance at three different significance levels.

THE LOGIC OF MULTIPLE COMPARISONS TESTS

Multiple comparisons tests account for multiple comparisons

If all null hypotheses are true and you make several comparisons without any special corrections, then about 5% of all those comparisons will generate a statistically significant result. The chance of making one or more Type I errors (declaring statistical significance when the null hypotheses really are all true) will be greater than 5%. This problem of multiple comparisons was discussed in Chapter 22.

The confidence level applies to the entire family of CIs. Given certain assumptions, there is a 95% chance that *all* of the CIs include the true population value, leaving only a 5% chance that one or more of the intervals do not include the population value.

Similarly, the 5% significance level applies to the entire family of comparisons. If the null hypothesis is true (all data were sampled from populations with identical means), there is a 95% chance that none of the comparisons will be declared

	DIFFERENCE	STATISTICALLY SIGNIFICANT?		
		a = 0.05	a = 0.01	a = 0.001
Recreational – elite	−0.02000	No	No	No
Nonrunners – elite	0.1200	No	No	No
Nonrunners – recreational	0.1400	Yes	Yes	No

Table 40.4. Statistical significance of multiple comparisons.

The significance levels (α) apply to the entire family of three comparisons, but the yes/no conclusions apply to each comparison individually.

NO. OF GROUPS	NO. OF PAIRWISE COMPARISONS
3	3
4	6
5	10
6	15
7	21
8	28
9	36
10	45
11	55
12	66
13	78
14	91
15	105
16	120
17	136
18	153
19	171
20	190
k	k(k − 1)/2

Table 40.5. Number of possible pairwise comparisons between group means as a function of the number of groups.

The number of possible pairwise comparisons is very large when you have many groups.

statistically significant, leaving a 5% chance that one or more comparisons will (erroneously) be declared statistically significant.

It is hard to think about significance levels that apply to a family of comparisons. Don't think about whether one particular comparison is statistically significant. Instead, think about dividing the family of comparisons into two piles: statistically significant and not statistically significant.

With more groups, CIs between means become wider

The example includes three groups and three possible pairwise comparisons of means. If there were five groups, then there would be 10 possible comparisons between sample means (AB, AC, AD, AE, BC, BD, BE, CD, CE, DE). Table 40.5 and Figure 40.3 show how the number of possible comparisons increases with the number of groups.

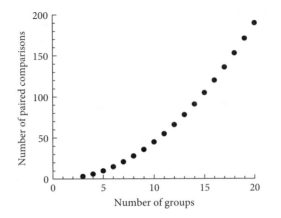

Figure 40.3. Number of possible pairwise comparisons between group means as a function of the number of groups.

If you include more groups in your experiment, the confidence intervals will be wider and a difference must be larger before it will be deemed statistically significant. When you include more treatment groups in your experimental design, you lose statistical power to detect differences between selected group means.

Multiple comparisons tests use data from all groups, even when only comparing two groups

The result of each multiple comparison test depends on all the data in all the groups, not just the two groups being compared. To see why, read about the use of the pooled SD in the "How It Works" section below.

Multiple comparisons tests account for intertwined comparisons

Once you have compared the mean of Group A with the mean of Group B and the mean of Group A with the mean of Group C, you will already know something about the comparison of Group B with Group C. Consider three groups (with equal n) for which the mean of A is greater than the mean of B, the mean of B is greater than the mean of C, and both comparisons are statistically significant. Without any calculations, you know that the mean of A must be significantly greater than the mean of C. Most multiple comparisons methods account for this interlinking of comparisons.

Multiple comparisons tests can declare the difference between pairs of means to be statistically significant, but they usually do not compute exact P values

When a t test compares two means, it reports an exact P value. You can then decide to compare that P value to a preset significance level in order to decide whether or not that difference is statistically significant.

TUKEY'S MULTIPLE COMPARISONS TEST	MULTIPLICITY-ADJUSTED P VALUE
Recreational runners vs. elite runners	0.9440
Nonrunners vs. elite runners	0.1307
Nonrunners vs. recreational runners	0.0037

Table 40.6. Multiplicity-adjusted P values computed with GraphPad Prism, version 6.

Multiple comparisons tests following ANOVA are different. Statistical significance is determined by accounting for multiple comparisons. The significance level doesn't apply to any one particular comparison, but rather to the entire family of comparisons. Accordingly, exact P values are usually not computed for each comparison. It is common, however, for statistics programs to report statistical significance at several different familywise significance levels (as is done in Table 40.4).

Some programs compute multiplicity-adjusted P values for each comparison

Some programs (including GraphPad Prism 6) can also report *multiplicity-adjusted P values*. Each multiplicity-adjusted P value is for a particular comparison, but it takes into account all the comparisons (Wright, 1992). It is the overall significance level (α) at which that particular comparison would be right at the border of statistical significance (accounting for multiple comparisons). For the sample data, the multiplicity-adjusted P values computed as part of Tukey's multiple comparisons are shown in Table 40.6.

Note that each adjusted P value is computed from the entire data set so thus takes into account the entire family of comparisons. Consequently, the adjusted P value for one particular comparison would have a different value if there were a different number of comparisons or if the data in the other comparisons were changed. The adjusted P value cannot be compared to an individual P value computed by a t test.

OTHER MULTIPLE COMPARISONS TESTS

Statisticians have developed numerous multiple comparisons methods to accompany one-way ANOVA. The example above used Tukey's test, which compares the mean of each group with the mean of every other group. This section explains some (but certainly not all) alternative tests.

Dunnett's test: Compare each mean with the mean of a control group

Dunnett's test compares the mean of each group to the mean of a control group without comparing the other groups among themselves. For example, it would be used in an experiment that tests the effects of six different drugs with a goal of defining which drugs have any effect but not of comparing the drugs with each other.

Because Dunnett's test makes fewer comparisons than Tukey's method, it generates narrower CIs and has more power to detect differences. It is very useful.

The decision to use Dunnett's test (along with a definition of which group is the control group) should be part of the experimental design. It isn't fair to first do Tukey's test to make all comparisons and then switch to Dunnett's test to get more power.

Bonferroni's test: Compare preselected pairs of means

Bonferroni's approach was explained in Chapter 22. Although it can be used to compare all pairs of means, it should not be used in this way, because Tukey's test has more power. Similarly, it should not be used to compare each group against a control, because Dunnett's test has more power for that purpose. Bonferroni's multiple comparisons test should be used when the experiment design requires comparing only selected pairs of means. By making a limited set of comparisons, you get narrower CIs and more statistical power to detect differences.

It is essential that you select those pairs as part of the experimental design, before collecting the data. It is not fair to first look at the data and then decide which pairs you want to compare. By looking at the data first, you have implicitly compared all groups.

Holm's test: Powerful, but no confidence intervals

The *Holm multiple comparisons test* is a powerful and versatile multiple comparisons test. It can be used to compare all pairs of means, compare each group mean to a control mean, or compare preselected pairs of means. It is not restricted to being used as a follow-up to ANOVA but instead can be used in any multiple comparisons context. The test is quite logical and easy to understand.

The Holm multiple comparison method starts with a set of P values and then determines which of these P values are low enough for the corresponding comparison to be declared statistically significant. It doesn't adjust the P values themselves but simply determines the conclusion for each one. The threshold for determining significance depends on the rank of the P value. The threshold for the smallest P value is much smaller than the threshold for the largest P value.

The test is powerful, understandable, and versatile. What's not to like? The Holm test can only be used to decide which comparisons are statistically significant and which are not. Unlike the tests developed by Tukey, Bonferroni, and Dunnett, Holm's test cannot also compute a set of confidence intervals (Westfall et al., 1999, p. 40).

The Holm–Sidak test is a modified version of Holm's test that is slightly more powerful.

Scheffe's test: More general comparisons

Scheffe's multiple comparisons test can make more elaborate comparisons than the other tests. For example, you might want to compare the mean of all treated groups with the mean of a control group. Or you might want to compare the mean of Groups A and B with the mean of Groups C, D, and E. These comparisons are sometimes called *contrasts*.

Scheffe's method can test any number of this kind of comparison. This increased versatility comes with a price, however. To allow for the huge number of possible comparisons, the CIs generated by Scheffe's method are wider than those generated by other methods. Accordingly, the test has less statistical power to detect differences than do the other multiple comparisons tests. The precise comparisons do not need to be defined as part of the experimental design. It is OK to test comparisons that didn't occur to you until you saw the data.

Test for a trend: Is group mean correlated with group order?

The different groups compared by ANOVA often have a natural order—for example, ages, time intervals, or doses. However, one-way ANOVA calculations completely ignore this order. ANOVA analyzes data from different doses or different ages in exactly the same way that it analyzes data from different species or different drugs. If you randomly shuffled the doses or ages, the ANOVA results wouldn't change.

The *test for trend* computes the correlation coefficient between the outcome and group order, along with a P value testing the null hypothesis that there is no trend. If this P value is small, there is a statistically significant trend (correlation) between group order and the outcome.

To learn more, see Altman (1990). Fancier tests for trend can look for nonlinear trends. Look up *polynomial contrasts* in an advanced ANOVA text for more details.

HOW IT WORKS: MULTIPLE COMPARISONS TESTS

Chapter 12 briefly explained how a CI of a mean is computed. The margin of error of the CI of a mean is the product of two values. The first is the SEM, which is computed from the SD and sample size. The second is a critical value from the t distribution, which depends on the desired confidence level (95%) and the number of df $(n - 1)$.

Chapter 30 extended that idea to the CI for the difference between two means. The standard error of the difference between means is computed from both SDs and both sample sizes.

When computing a multiple comparisons test, the standard error for the difference between two means is not computed from the SDs of those two groups, but rather from the pooled SD of *all* the groups. ANOVA software rarely reports this pooled SD, but reports instead the pooled variance, which is labeled "mean square within groups" or "mean square residual." The square root of this value is the pooled SD. The standard error of the difference between two means is computed from that value (which is the same for all comparisons) and the sample size of the two groups being compared (which might not be the same for all comparisons).

The margin of error of the CI is computed by multiplying the standard error of the difference by a critical value that depends on the choice of test, the number of df, the degree of confidence desired, and (importantly) the number of comparisons.

WHEN ARE MULTIPLE COMPARISONS TESTS NOT NEEDED?

Never correct for multiple comparisons?

Correcting for multiple comparisons reduces the risk of a Type I error, but at the cost of increasing the risk of a Type II error (see Table 40.7). Rothman (1990) argues that this trade-off is not worth it and recommends that researchers not correct for multiple comparisons. Instead, it is essential to report all the comparisons you made so that the people reading the research can informally correct for multiple comparisons. This recommendation is sensible, but not mainstream.

Don't correct when some comparisons are more important than others?

Correcting for multiple comparisons can lead to ambiguous situations. Imagine that you are running a research program to investigate the mechanism of action of a drug. You want to know whether it blocks a certain receptor. You set up the proper assay and run some test experiments to make sure everything works. Now imagine two alternative scenarios.

Scenario 1: You run an experiment testing only your one drug against a control. The drug works as you predicted, and the results are statistically significant. Your research moves forward.

Scenario 2: You run an experiment testing not only your drug, but also two additional drugs. There is no reason to expect these drugs to block the receptor you are studying, but it would be very interesting if they did. Setting up the assay and running various controls is a lot of work, but testing three drugs is not a whole lot harder than testing one drug, so you give it a try. Every once in a while this kind of exploratory experiment can move research forward. Not this time. The two extra drugs don't work. Not a big disappointment and not too much time wasted.

The data for the main drug are exactly the same in both scenarios. In the second scenario, two other drugs are tested at the same time. To account for

Risk of Type I error (labeling a chance difference as statistically significant)	Lower
Risk of Type II error (missing a real difference)	Higher
Width of CIs	Wider

Table 40.7. Consequences of correcting for multiple comparisons compared to individual comparisons.

multiple comparisons, the analysis would use Dunnett's method, so the 5% significance level would apply to the family of three comparisons rather than to each individual comparison. With this analysis, the result for the main drug would not be statistically significant. Because more comparisons were made, the method would become more cautious about concluding statistical significance.

In this example, the correction for multiple corrections doesn't really make sense. The experiment was done to test one main hypothesis, and the other two hypotheses were add-ons. When one or a few comparisons are clearly defined in advance as being critical, some statisticians advocate not using any correction for multiple comparisons. This approach is called *planned comparisons*, but that term is ambiguous, because many multiple comparisons are planned in advance.

The statistical principles of this approach are fairly straightforward and do not generate much controversy. However, that doesn't make it easy to decide what to do in this situation. Reasonable statisticians disagree.

COMMON MISTAKES: MULTIPLE COMPARISONS

Mistake: Following ANOVA, using the unpaired t test to compare two groups at a time rather than using multiple comparisons tests

If you perform multiple t tests this way, it is too easy to reach misleading conclusions. Use multiple comparisons designed for this situation.

Mistake: If your experiment has three or more treatment groups, compare the largest mean with the smallest mean using a single unpaired t test

You won't know which group has the largest mean and which has the smallest mean until after you have collected the data. So although you are formally just doing one t test to compare two groups, you are really comparing all the groups. Use one-way ANOVA, followed by multiple comparisons tests.

Mistake: Focusing only on declarations of statistical significance and ignoring confidence intervals

Most multiple comparisons tests can report confidence intervals corrected for multiple comparisons. In many situations, these are more informative (and easier to understand) than declarations of significance.

Mistake: Using multiple comparisons tests to determine the earliest time point (or smallest concentration) at which an effect is significant

If you have collected data over time, or at multiple concentrations of a drug, analyze the data by fitting a line or curve with regression. Some people do one-way ANOVA followed by multiple comparisons to determine the earliest time point (or lowest concentration) at which the difference from control (zero time or zero concentration) is statistically significant. This is rarely a helpful calculation. If you changed sample size, you'd get a different result.

Q & A

What is the difference between a multiple comparisons test and a post hoc test?

The term *multiple comparisons test* applies whenever several comparisons are performed at once with a correction for multiple comparisons. The term *post hoc test* refers to situations in which you can decide which comparisons you want to make after looking at the data. Often, however, the term is used informally as a synonym for multiple comparisons test. *Posttest* is an informal, but ambiguous, term. It can refer to either all multiple comparisons tests or only to post hoc tests.

If one-way ANOVA reports a P value less than 0.05, are multiple comparisons tests sure to find a significant difference between group means?

Not necessarily. The low P value from the ANOVA tells you that the null hypothesis that all data were sampled from one population with one mean is unlikely to be true. However, the difference might be a subtle one. It might be that the mean of Groups A and B is significantly different than the mean of Groups C, D, and E. Scheffe's posttest can find such differences (called contrasts), and if the overall ANOVA is statistically significant, Scheffe's test is sure to find a significant contrast.

The other multiple comparisons tests compare group means. Finding that the overall ANOVA reports a statistically significant result does not guarantee that any of these multiple comparisons will find a statistically significant difference.

If one-way ANOVA reports a P value greater than 0.05, is it possible for a multiple comparisons test to find a statistically significant difference between some group means?

Yes, it is possible.

Are the results of multiple comparisons tests valid if the overall P value for the ANOVA is greater than 0.05?

It depends on which multiple comparisons test you use. Tukey's, Dunnett's, and Bonferroni's tests, mentioned above, are valid even if the overall ANOVA yields a conclusion that there are no statistically significant differences among the group means.

Does it make sense to only focus on multiple comparisons results and ignore the overall ANOVA results?

It depends on the scientific goals. ANOVA tests the overall null hypothesis that all the data come from groups that have identical means. If that is your experimental question—do the data provide convincing evidence that the means are not all identical—then ANOVA is exactly what you want. If the experimental questions are more focused and can be answered by multiple comparisons tests, you can safely ignore the overall ANOVA results and jump right to the results of multiple comparisons.

Note that the multiple comparisons calculations all use the mean square result from the ANOVA table. Consequently, even if you don't care about the value of F or the P value, the multiple comparisons tests still require that the ANOVA table be computed.

Can I assess statistical significance by observing whether two error bars overlap?

If two SE error bars overlap, you can be sure that a multiple comparisons test comparing those two groups will find no statistical significance. However, if two SE error bars do not overlap, you can't tell whether a multiple comparisons test will or will not find a statistically significant difference. If you plot SD, rather than SEM, error bars, the fact that they do (or don't) overlap will not let you reach any conclusion about statistical significance.

Do multiple comparisons tests take into account the order of the groups?

No. With the exception of the test for trend, mentioned above, multiple comparisons tests do not consider the order in which the groups were entered into the program.

Do all CIs between means have the same length?

If all groups have the same number of values, then all the CIs for the difference between means will have identical lengths. If the sample sizes are unequal, then the standard error of the difference between means depends on sample size. The CI for the difference between two means will be wider when sample size is small and narrower when sample size is larger.

Why wasn't the Newman–Keuls test used for the example?

Like Tukey's test, the Newman–Keuls test (also called the Student–Newman–Keuls test) compares each group mean with every other group mean. Some prefer it because it has more power. I prefer Tukey's test, because the Newman–Keuls test does not really control the error rate as it should (Seaman, Levin, & Serlin, 1991) and cannot compute CIs.

CHAPTER SUMMARY

- Multiple comparisons tests follow ANOVA to find out which group means differ from which other means.
- To prevent getting fooled by bogus statistically significant conclusions, the significance level is usually defined to apply to an entire family of comparisons rather than to each individual comparison.
- Most multiple comparisons tests can report both confidence intervals and statements about statistical significance. Some can also report multiplicity-adjusted P values.
- Because multiple comparisons tests correct for multiple comparisons, the results obtained by comparing two groups depends on the data in the other groups and the number of other groups in the analysis.
- To choose a multiple comparison test, you must articulate the goals of the study. Do you want to compare each group mean to every other group mean? Each group mean to a control group mean? Only compare a small set of pairs of group means? Different multiple comparison tests are used for different sets of comparisons.

TERMS INTRODUCED IN THIS CHAPTER

- Contrast (p. 385)
- Dunnett's test (p. 383)
- Holm's test (p. 384)
- Multiplicity-adjusted P value (p. 383)
- Planned comparisons (p. 387)
- Post hoc test (p. 388)
- Post test (p. 388)
- Scheffe's test (p. 384)
- Test for trend (p. 385)
- Tukey's test (or Tukey–Kramer test) (p. 377)

CHAPTER 41

❖❖❖

Nonparametric Methods

Statistics are like a bikini. What they reveal is suggestive, but what they conceal is vital.

AARON LEVENSTEIN

Many of the methods discussed in this book are based on the assumption that the values are sampled from a Gaussian distribution. Another family of methods makes no such assumption about the population distribution. These are called *nonparametric* methods. The nonparametric methods used most commonly work by ignoring the actual data values and instead analyzing only their ranks. Computer-intensive resampling and bootstrapping methods also do not assume a specified distribution, so they are also nonparametric.

NONPARAMETRIC TESTS BASED ON RANKS

The idea of nonparametric tests

The unpaired t test and ANOVA are based on the assumption that the data are sampled from populations with a Gaussian distribution. Similarly, the paired t test assumes that the set of differences (between paired values) are sampled from a Gaussian population. Because these tests are based on an assumption about the distribution of values in the population that can be defined by parameters, they are called *parametric tests*.

Nonparametric tests make few assumptions about the distribution of the populations. The most popular forms of nonparametric tests are based on a really simple idea. Rank the values from low to high and analyze only those ranks, ignoring the actual values. This approach ensures that the test isn't affected much by outliers (see Chapter 25) and doesn't assume any particular distribution.

Comparing two unpaired groups: Mann–Whitney test

The *Mann–Whitney test* is a nonparametric test used to compare two unpaired groups to compute a P value. It works by following these steps:

1. Rank all the values without paying attention to the group from which the value is drawn. In the example from Chapter 30 (comparing the bladder

390

OLD	YOUNG
3.0	10.0
1.0	13.0
11.0	14.0
6.0	16.0
4.5	15.0
8.0	17.0
4.5	9.0
12.0	7.0
2.0	

Table 41.1. Ranks of the data shown in Table 30.1.

The smallest value has a rank of 1, and the largest value has a rank of 17. Note that two values tie for ranks of 4 and 5, so both are assigned a rank of 4.5. These ranks are plotted in Figure 41.1 and are displayed in the same order as the actual values are displayed in Table 30.1.

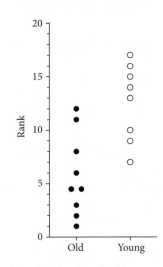

Figure 41.1. Plot of the data (ranks) of Table 41.1.

muscle of old vs. young rats), two values tie for ranks of 4 and 5, so both are assigned a rank of 4.5. Table 41.1 and Figure 41.1 show the ranks.

2. Sum the ranks in each group. In the example data, the sum of the ranks of the old rats is 52 and the sum of the ranks of the young rats is 101. The values for the younger rats tend to be larger and thus tend to have higher ranks.

3. Calculate the mean rank of each group.

4. Compute a P value for the null hypothesis that the distribution of ranks is totally random.

Under the null hypothesis, it would be equally likely for either of the two groups to have the larger mean ranks and more likely to find the two mean ranks close together than far apart. Based on this null hypothesis, the P value is computed by answering the question, If the distribution of ranks between two groups were distributed randomly, what is the probability that the difference between the mean ranks would be this large or even larger? The answer is 0.0061. It certainly is not impossible that random sampling of values from two identical populations would lead to sums of ranks this far apart, but it would be very unlikely. Accordingly, we conclude that the difference between the young and old rats is statistically significant.

Although the Mann–Whitney test makes no assumptions about the distribution of values, it is still based on some assumptions. Like the unpaired t test, the Mann–Whitney test assumes that the samples are randomly sampled from (or representative of) a larger population and that each value is obtained independently. But unlike the t test, the Mann–Whitney test does not assume anything about the distribution of values in the populations from which the data are sampled.

The Mann–Whitney test is equivalent to a test developed by Wilcoxon, so the same test is sometimes called the *Wilcoxon rank-sum test*. Don't mix this test up with the nonparametric test for paired data discussed in the next section.

Comparing two paired groups: Wilcoxon matched-pairs signed-rank test

The *Wilcoxon matched-pairs signed-rank test* (often referred to simply as *Wilcoxon's test*) compares two paired groups. It tests the null hypothesis that there is no difference in the populations and so the differences between matched pairs will be randomly positive or negative.

Like the paired t test (see Chapter 31), Wilcoxon's test can be used in several situations:

- When a variable is measured for each subject before and after an intervention.
- When subjects are recruited as pairs matched for variables such as age, postal code, or diagnosis. One of each pair receives one intervention, and the other receives an alternative treatment.
- When twins or siblings are recruited as pairs. Each receives a different treatment.
- When each run of a laboratory experiment has a control and treated preparation handled in parallel.
- When a part of the body on one side is treated with a control treatment and the corresponding part of the body on the other side is treated with the experimental treatment (e.g., different treatments for right and left eyes).

It works via the following steps:

1. Calculate the difference between each matched pair, keeping track of the sign. A decrease is negative and an increase is positive.
2. Rank the absolute value of the differences, temporarily ignoring the sign.
3. Add up the ranks of all the positive differences and the ranks of all the negative differences. For the example data concerning Darwin's plants (see Chapter 31), the sums of the ranks are 96 and 24.
4. Compute the difference between those two sums (here, 72).

5. Computing a P value answers the question, If the null hypothesis were true, what would be the chance of randomly choosing samples such that the sums of the absolute values of the positive and negative ranks would differ by 72 or more? The answer is 0.0413.

Like the paired t test, Wilcoxon's test assumes that the pairs are randomly selected from (or at least representative of) a larger population and that each pair is selected independently from the others. Unlike the paired t test, Wilcoxon's test does not assume a Gaussian distribution of differences.

Nonparametric correlation

One nonparametric method for quantifying correlation is called *Spearman's rank correlation*. Spearman's rank correlation is based on the same assumptions as ordinary (Pearson) correlations, outlined in Chapter 32, with two exceptions. Rank correlation does not assume Gaussian distributions and does not assume a linear relationship between the variables. However, Spearman's (nonparametric) rank correlation does assume that any underlying relationship between X and Y is monotonic (i.e., either always increasing or always decreasing).

Spearman's correlation separately ranks the X and Y values and then computes the correlation between the two sets of ranks. For the insulin sensitivity example of Chapter 32, the nonparametric correlation coefficient, called r_s, is 0.74. The P value, which tests the null hypothesis that there is no rank correlation in the overall population, is 0.0036.

An easy way to think about how the two kinds of correlation are distinct is to recognize that Pearson correlation quantifies the *linear* relationship between X and Y, while Spearman quantifies the *monotonic* relationship between X and Y.

Nonparametric ANOVA

The nonparametric test analogous to one-way ANOVA is called the *Kruskal–Wallis test*. The nonparametric test analogous to repeated-measures one-way ANOVA is called *Friedman's test*. These tests first rank the data from low to high and then analyze the distribution of the ranks among groups.

THE ADVANTAGES AND DISADVANTAGES OF NONPARAMETRIC TESTS

The advantage of nonparametric tests is clear. They don't require the assumption of sampling from a Gaussian population and so can be used when the validity of that assumption is dubious. And when that assumption is false, nonparametric tests have more power than parametric tests to detect differences.

So why not always use nonparametric tests? There are three reasons.

Nonparametric tests are less powerful when the data are Gaussian

Because nonparametric tests only consider the ranks and not the actual data, they essentially throw away some information, so they are less powerful. If there truly is a difference between Gaussian populations, the P value is likely to be higher with a nonparametric test. How much higher? It depends on sample size.

With large samples, the nonparametric tests are nearly as powerful as the parametric tests when the data are really sampled from Gaussian populations. This is assessed by a value called the *asymptotic relative efficiency*. For example, with large samples drawn from Gaussian populations, the asymptotic relative efficiency of the Mann–Whitney test is 95%. This means that the power of a Mann–Whitney test is equal to the power of a t test with 95% as many data points. The other nonparametric tests have similarly high asymptotic relative efficiencies.

With small samples from Gaussian populations, nonparametric tests have much less power than parametric tests. When the populations are not Gaussian, however, nonparametric tests can be (depending on the distribution) much more powerful than parametric tests (Sawilowsky, 2005).

With tiny samples, nonparametric tests always report P values greater than 0.05 and so have zero power. In order to ever compute a two-tailed P value less than 0.05:

- The Mann–Whitney test requires eight or more values (in total, both groups).
- Wilcoxon's matched pairs test requires six or more data pairs.
- Spearman's correlation requires five or more XY pairs.

Nonparametric results are usually not reported with CIs

This book has emphasized the importance of CIs. Most programs, however, report only P values and not CIs for nonparametric tests.

Some nonparametric tests can be extended to compute CIs, but this requires additional assumptions. For example, the Mann–Whitney test can be extended to provide a CI for the difference between medians (Prism 6 does this). However, this requires assuming that the distributions of the two populations have the same shape, with only the distributions being shifted (and thus having different medians). This assumption is not needed to interpret the P value.

Nonparametric tests are not readily extended to regression models

Chapter 35 showed that t tests can be recast as regression. A t test compares two means. So can simple linear regression. Multiple regression can compare two means after adjusting for differences in other variables. Nonparametric tests cannot be readily extended in this way. There are nonparametric ways to fit curves (look up splines or kernel methods), but these do not estimate values of parameters.

CHOOSING BETWEEN PARAMETRIC AND NONPARAMETRIC TESTS: DOES IT MATTER?

Does it matter whether you choose a parametric or nonparametric test? The answer depends on sample size. There are four cases you must consider (Table 41.2).

DISTRIBUTION	TEST	SMALL SAMPLES	LARGE SAMPLES
Gaussian population	Nonparametric	Misleading. Nonparametric tests have little power with small samples.	Little problem. With large samples, nonparametric tests are nearly as powerful as parametric tests.
Non-Gaussian population	Parametric	Misleading. With small samples, parametric tests are not very robust to violations of the Gaussian assumption.	Little problem. With large samples, parametric tests are robust to violations of the Gaussian assumption.
Not sure	Normality tests	Not very helpful with small samples.	Helpful.

Table 41.2. The problem with small samples.

Using a nonparametric test with a small dataset sampled from a Gaussian population

Nonparametric tests lack statistical power with small samples. The P values will tend to be high.

Using a nonparametric test with a large dataset sampled from a Gaussian population

Nonparametric tests work well with large samples from Gaussian populations. The P values will tend to be a bit too large, but the discrepancy is small. In other words, nonparametric tests are only slightly less powerful than parametric tests with large samples.

Using a parametric test with a small dataset sampled from a non-Gaussian population

The central limit theorem (discussed in Chapter 10) doesn't apply to small samples, so the P value may be inaccurate.

Using a parametric test with a large dataset sampled from a non-Gaussian population

The central limit theorem ensures that parametric tests work well with large samples even if the data are sampled from non-Gaussian populations. In other words, parametric tests are robust to mild deviations from Gaussian distributions, so long as the samples are large. But there are two snags:

- It is impossible to say how large is large enough, because it depends on the nature of the particular non-Gaussian distribution. Unless the population distribution is really weird, you are probably safe choosing a parametric test when there are at least two-dozen data points in each group.
- If the population is far from Gaussian, you may not care about the mean or differences between means. Even if the P value provides an accurate

answer to a question about the difference between means, that question may be scientifically irrelevant.

Summary

Large data sets present no problems. It is usually easy to tell whether the data are likely to have been sampled from a Gaussian population (and normality tests can help), but it doesn't matter much, because in this case, nonparametric tests are so powerful and parametric tests are so robust.

Small data sets present a dilemma. It is often difficult to tell whether the data come from a Gaussian population, but it matters a lot. In this case, nonparametric tests are not powerful and parametric tests are not robust.

SAMPLE SIZE FOR NONPARAMETRIC TESTS

Depending on the nature of the distribution, nonparametric tests might require either more or fewer subjects than the corresponding test that assumes a Gaussian distribution. A general rule of thumb is this: if you plan to use a nonparametric test, compute the sample size required for a parametric test (see Chapter 26) and add 15% (Lehman, 2007).

NONPARAMETRIC TESTS THAT ANALYZE VALUES (NOT RANKS)

There is another approach to analyzing data that makes no assumptions about the shape of the population distribution and also avoids the need to analyze ranks. These tests go by the names *permutation tests*, *randomization tests*, and *bootstrapping*.

The idea of all these tests is to analyze the actual data, not ranks, but not to assume any particular population distribution. All of these tests rely on intensive calculations that require computers and so are called *computer-intensive methods*. Because they don't make any assumptions about population distributions, these tests are nonparametric (even though that term is often used only to refer to those tests that analyze ranks).

To compute a P value, the parametric methods start with an assumption (usually Gaussian distribution) about the population and then use math to figure out the distribution of all possible samples from that population.

The computer-intensive nonparametric methods have a completely different outlook. They make no mathematical assumption about the population beyond the fact that the sample data were selected from it. They work by brute force rather than elegant math.

Randomization or permutation tests work by shuffling. Computers run many simulated "experiments." Each one uses the sample data but changes the labels for the groups. These labels are randomly shuffled among the values. If you are comparing a control and a treated group, each shuffling changes the selection of which values are labeled "control" and which are labeled "treated." Analyzing this set of pseudosamples can lead to valid insights into your data.

With small data sets, these methods can systematically inspect every possible way to shuffle the labels among values. With large data sets, the number of possible rearrangements becomes astronomical, and so these software programs examine a large number (often 1,000 to 10,000) of randomly chosen permutations, which is enough to get valid results.

Another approach called *bootstrapping* or *resampling* has already been mentioned briefly in Chapter 13. The idea of bootstrapping is that all you know for sure about the population is that the sample was obtained from it. Statistical inference requires thinking about what would happen if you picked many samples from that population. Without making any parametric assumptions, this outcome can be approximated by resampling from that single sample. The next two paragraphs give you a taste for how this approach works.

This resampling is done with replacement. Imagine you write each value on a card and place all of the cards in a hat. Mix well, pick a card, and record its value. Then put it back in the hat, mix the cards well again, and pick the next card. Note that some values (cards) may get selected more than once and other values may not be selected at all. Of course the resampling process is actually done with computerized random-number generators.

Resampling produces many pseudosamples, each of which is the same size as the original sample. The samples are different because some values are duplicates and others are omitted. Comparing the actual data with the distribution of bootstrap samples can lead to statistical conclusions expressed as P values and CIs.

This may seem like magic. It sure doesn't seem that useful statistical conclusions will emerge from analysis of sets of pseudosamples created by resampling, or bootstrapping (the two terms are essentially interchangeable). That accounts for the strange name, *bootstrapping*. At first glance, analyzing data this way seems about as (un) helpful as trying to get out of a hole by pulling up on the straps of your boots.

Nevertheless, the theorems have been proven, the simulations have been run, and plenty of real-world experiences have validated these approaches. They work! Many think computer-intensive methods are the future of statistics because they are so versatile. They can be easily adapted for use in new situations, and they don't require making or believing assumptions about the distribution of the population (beyond the assumption that the sample is representative). To learn more, start with a short text by Wilcox (2001) and then a longer book by Manly (2006).

COMMON MISTAKES: NONPARAMETRIC TESTS

Mistake: Using a normality test to automate the decision of when to use a nonparametric test

The decision of when to use a nonparametric test is not straightforward, and reasonable people can disagree about when to use nonparametric tests. Many think that the choice of using a nonparametric test can be automated: First, perform a normality test (see Chapter 24). If the data pass, use a parametric test. If the data fail the normality test, then use a nonparametric test.

This approach is not recommended for the following reasons:

- When analyzing data from a series of experiments, all data should be analyzed the same way (unless there is some reason to think they aren't comparable). In this case, results from a single normality test should not be used to decide whether to use a nonparametric test.
- Data sometimes fail a normality test because the values were sampled from a lognormal distribution (see Chapter 11). In this case, transforming the data to logarithms will create a Gaussian distribution. In other cases, transforming data to reciprocals or using other transformations can often convert a non-Gaussian distribution to a Gaussian distribution.
- Data can fail a normality test because of the presence of an outlier (see Chapter 25). In some cases, it might make sense to analyze the data without the outlier using a conventional parametric test rather than a nonparametric test.
- The decision of whether to use a parametric or nonparametric test is most important with small data sets (because the power of nonparametric tests is so low). But with small data sets, normality tests have little power, so an automatic approach would give you false confidence.

The decision of when to use a parametric test and when to use a nonparametric test really is a difficult one, requiring thinking and perspective. As a result, this decision should not be automated.

Mistake: Referring to "nonparametric" data

The term *nonparametric* refers to a set of statistical tests. It cannot be used to describe data sets.

Mistake: Running both parametric and nonparametric tests and picking the result with the smallest P value

P values can only be interpreted when the test is chosen as part of the experimental design. If you run two (or more) tests and report the results of the one that gives the lowest P value, you cannot interpret that P value at face value.

Q & A

Can data be nonparametric?
> No. *Nonparametric* is an adjective that can only be applied to statistical tests, not to data.

If I am sure my data are not Gaussian, should I use a nonparametric test?
> Not necessarily. It may be possible to transform the data in a way that makes the population Gaussian. Most commonly, a log transform of data from a lognormal distribution (see Chapter 11) will create a Gaussian population.

Are the chi-square test and Fisher's exact test nonparametric?

The distinction between parametric and nonparametric tests doesn't apply to tests that deal with dichotomous outcomes. The term *nonparametric* is sometimes used for these tests, but there is no distinction between parametric and nonparametric tests when analyzing dichotomous data. Most people would not call these tests nonparametric, but some do.

Are the computer-intensive methods (bootstrapping, resampling, permutations) considered to be nonparametric tests?

Those computer-intensive tests don't depend on any assumptions about the distribution of the population, so in that technical sense, they certainly are nonparametric, and that is why I include them in this chapter. But the term *nonparametric* is used inconsistently. Some textbooks of nonparametric tests include computer-intensive methods, and some don't.

Does the Mann–Whitney test compare medians?

Only if you assume that the two populations have identically shaped distributions. The distributions don't have to be Gaussian or even specified, but you do have to assume that the shapes are identical. Given that assumption, the only possible way for two populations to differ is by having different medians (i.e., the distributions are the same shape, but shifted). If you don't make that assumption, then it is not correct to say that a Mann–Whitney test compares medians.

Can nonparametric tests be used when some values are "off scale"?

If some values are too high or too low to be quantified, parametric tests cannot be used, because those values are not known. If they simply are ignored, the test results will be biased because the largest (or smallest) values were thrown out. In contrast, a nonparametric test can work well when a few values are too high (or too low) to measure. Assign values too low to measure an arbitrary very low value and assign values too high to measure an arbitrary very high value. Because the nonparametric test only knows about the relative ranks of the values, it won't matter that you didn't enter those extreme values precisely.

When the decision isn't clear, should I choose a parametric test or a nonparametric test?

When in doubt, some people choose a parametric test because they aren't sure the Gaussian assumption is violated. Others choose a nonparametric test because they aren't sure the Gaussian assumption is met. Ideally, the decision should be based partly on experience with the same kind of data in other experiments. Reasonable people disagree. It really is a hard decision.

Which set of tests gives lower P values, parametric or nonparametric?

It depends on the data.

CHAPTER SUMMARY

- ANOVA, t tests, and many statistical tests assume that you have sampled data from populations that follow a Gaussian bell-shaped distribution.

- Biological data never follow a Gaussian distribution precisely, because a Gaussian distribution extends infinitely in both directions and so includes both infinitely low negative numbers and infinitely high positive numbers! But many kinds of biological data follow a bell-shaped distribution that is approximately Gaussian. Because ANOVA, t tests, and other statistical tests work well even if the distribution is only approximately Gaussian (especially with large samples), these tests are used routinely in many fields of science.

- An alternative approach does not assume that data follow a Gaussian distribution. In this approach, values are ranked from low to high, and the analyses are based on the distribution of ranks. These tests, called nonparametric tests, are appealing because they make fewer assumptions about the distribution of the data.
- Computer-intensive methods (resampling, bootstrapping, permutation tests) also do not rely on any assumption about the distribution of the population and so are also nonparametric.

TERMS INTRODUCED IN THIS CHAPTER

- Asymptotic relative efficiency (p. 394)
- Computer-intensive methods (p. 396)
- Friedman's test (p. 393)
- Kruskall–Wallis test (p. 393)
- Mann–Whitney test (or Wilcoxon rank-sum test) (p. 390)
- Nonparametric tests (p. 390)
- Parametric tests (p. 390)
- Permutation tests (p. 396)
- Randomization tests (p. 396)
- Spearman's rank correlation (p. 393)
- Wilcoxon matched-pairs signed-rank test (or Wilcoxon's test) (p. 392)

❖❖❖

CHAPTER 42

Sensitivity, Specificity, and Receiver-Operating Characteristic Curves

Reports that say that something hasn't happened are always interesting to me, because as we know, there are known knowns; there are things we know we know. We also know there are known unknowns; that is to say we know there are some things we do not know. But there are also unknown unknowns—the ones we don't know we don't know.

DONALD RUMSFELD

This chapter explains how to quantify false positive and false negative results from laboratory tests. Although this topic is rarely found in basic statistics texts, deciding whether a clinical laboratory result is normal or abnormal relies on logic very similar to that used in deciding whether a finding is statistically significant or not. Learning the concepts of sensitivity and specificity presented here is a great way to review the ideas of statistical hypothesis testing and Bayesian logic explained in Chapter 18.

DEFINITIONS OF SENSITIVITY AND SPECIFICITY

This chapter discusses the accuracy of clinical tests that can report two possible results: normal or abnormal. There are two ways such a test can be wrong.

- It can report a *false negative* result—reporting that the result is normal when the patient has the disease for which he or she is being tested. In Table 42.1, false negative results are tabulated in cell C.
- It can report a *false positive* result—reporting that the result is abnormal when the patient really does not have the disease for which he or she is being tested. In Table 42.1, false positive results are tabulated in Cell B.

The accuracy of a diagnostic test is quantified by its sensitivity and specificity.

- The *sensitivity* is the fraction of all those with the disease who get a positive test result. In Table 42.1, sensitivity equals A/(A+C). Sensitivity measures

	DISEASE PRESENT	DISEASE ABSENT	TOTAL
Abnormal (positive) test result	A	B (false positives)	A+B
Normal (negative) test results	C (false negative)	D	C+D
Total	A+C	B+D	A+B+C+D

Table 42.1. The results of many hypothetical lab tests, each analyzed to reach a decision to call the results normal or abnormal.

The top row tabulates results for patients without the disease, and the second row tabulates results for patients with the disease. You can't actually create this kind of table from a group of patients unless you run a "gold standard" test that is 100% accurate.

how well the test identifies those with the disease, that is, how sensitive it is. If a test has high sensitivity, it will pick up nearly everyone with the disease.

- The *specificity* is the fraction of those without the disease who get a negative test result. In Table 42.1, specificity equals D/(B+D). Specificity measures how well the test identifies those who don't have the disease, that is, how specific it is. If a test has very high specificity, it won't mistakenly give a positive result to many people without the disease.

THE PREDICTIVE VALUE OF A TEST

Definitions of positive and negative predictive values

Neither the specificity nor the sensitivity answers the most important questions: If the test is positive (abnormal test result, suggesting the presence of disease), what is the chance that the patient really has the disease? And if the test is negative (normal test result), what is the chance that the patient really doesn't have the disease? The answers to those questions are quantified by the *positive predictive value* and the *negative predictive value*. Based on Table 42.1,

$$\text{Positive Predictive Value} = \frac{\text{True positives}}{\text{All positive results}} = \frac{A}{A + B}$$

$$\text{Negative Predictive Value} = \frac{\text{True negatives}}{\text{All negative results}} = \frac{D}{C + D}$$

The sensitivity and specificity are properties of the test. In contrast, the positive predictive value and negative predictive value are determined by the characteristics of the test and the prevalence of the disease in the population being studied. The lower the prevalence of the disease, the lower the ratio of true positives to false positives. This is best understood by example.

Background to porphyria example

Acute intermittent porphyria is an autosomal dominant disease that is difficult to diagnose clinically. It can be diagnosed by detecting reduced levels of the enzyme porphobilinogen deaminase. However, the levels of the enzyme vary in both the normal population and patients with porphyria, so the test does not lead to an exact diagnosis.

Using the definition that levels less than 98 units are abnormal, 82% of patients with porphyria have an abnormal test result. This means that the sensitivity of the test is 82%. Additionally, 3.7% of normal people have an abnormal test result. This means that the specificity of the test is 100%−3.7%, or 96.3%.

What is the probability that a patient with fewer than 98 units of enzyme activity has porphyria? In other words, what is the positive predictive value of the test? The answer depends on who the patient is. We'll work through two examples.

Predictive values when a test is used for random screening

We'll compute the predictive values in two contexts. In this context, the test was done to screen for the disease, so the people being tested had no particular risk for the disease. Porphyria is a rare disease, with a prevalence of about one in 10,000. Because the people being tested were not selected because of family history or clinical suspicion, you would expect about one in 10,000, or 0.01%, to have the disease.

The test gave a positive result for a patient. Knowing the test result, what is the probability that this patient has the disease? To find out, we must enter numbers into Table 42.1 to create Table 42.2.

1. Assume a population of 1,000,000 . All we care about are ratios of values, so the total population size is arbitrary. A + B + C + D = 1,000,000.
2. Because the prevalence of the disease is 1/10,000, the total number in the disease present column is 0.0001 × 1,000,000, or 100. A + C = 100.
3. Subtract the 100 diseased people from the total of 1,000,000, leaving 999,900 people without the disease. B + D = 999,900.
4. Calculate the number of people with disease present who also test positive. This equals the total number of people with the disease (100) times the sensitivity (0.82). A = 82.
5. Calculate the number of people without the disease who test negative. This equals the total number of people without the disease (999,900) times the specificity (0.963). D = 962,904.
6. Calculate the number of people with the disease who test negative: C = 100 − 82 = 18.
7. Calculate the number of people without the disease who test positive: B = 36,996.
8. Calculate the two row totals. A + B = 37,078. C + D = 962,922.

	DISEASE PRESENT	DISEASE ABSENT	TOTAL
Abnormal test result	82	36,996	37,078
Normal test results	18	962,904	962,922
Total	100	999,900	1,000,000

Table 42.2. Expected results of screening 1 million people with the porphyria test from a population with a prevalence of 0.01%.

Most of the abnormal test results are false positives.

If you screen 1 million people, you will expect to find an abnormal test result in 37,078 people. Only 82 of these cases will have the disease. Therefore, when someone has a positive test, the chance is only 82/37,078=0.22% that he or she has the disease. That is the positive predictive value of the test. Because only about one in 500 of the people with positive test results have the disease, the other 499 of 500 positive tests are false positives.

Of the 962,922 negative tests results, only 18 are false negatives. The predictive value of a negative test is 99.998%.

Predictive value when testing siblings

The disease is autosomal dominant, so there is a 50% chance that each sibling has the gene. If you test many siblings of people who have the disease, you will expect about half to have the disease. Table 42.3 shows the predicted results when you test 1,000 siblings of patients. Positive results are expected in 429 of the people tested. Of these individuals, 410 will actually have the disease and 19 will be false positives. The predictive value of a positive test, therefore, is 410/429, which is about 96%. Only about 4% of the positive tests are false positives.

Analogy to statistical tests

The analogies between interpreting laboratory tests and statistical hypothesis testing are shown in Tables 42.4 and 42.5.

Interpreting a positive or negative test requires knowing who is tested (what the prevalence is). Similarly, as explained in Chapters 18 and 19, the interpretation of statistical significance depends on the scientific context (the prior probability).

	DISEASE PRESENT	DISEASE ABSENT	TOTAL
Abnormal test result	410	19	429
Normal test results	90	481	571
Total	500	500	1,000

Table 42.3. Expected results of testing 1,000 siblings of someone with porphyria.
In this group, the prevalence will be 50% and few of the abnormal test results will be false positives.

LAB TEST	DISEASE PRESENT	DISEASE ABSENT
Decision: Abnormal results		False positive
Decision: Normal result	False negative	
STATISTICAL HYPOTHESIS TEST	**NULL HYPOTHESIS IS FALSE**	**NULL HYPOTHESIS IS TRUE**
Decision: Reject null hypothesis (significant)		Type I error
Decision: Do not reject null hypothesis (not significant)	Type II error	

Table 42.4. False positive and false negative lab results are similar to Type I and Type II errors in statistical hypothesis testing.

	IF…	WHAT IS THE CHANCE THAT…
Lab test: Sensitivity	The patient has disease …	…a lab result will be positive?
Statistical hypothesis test: Power	There is a difference between populations …	…a result will be statistically significant?

	IF…	WHAT IS THE CHANCE THAT…
Lab test: Specificity	The patient does not have the disease…	…a lab result will be negative?
Statistical hypothesis test: 1-alpha	There is no difference between populations …	…a result will not be statistically significant?

Table 42.5. Relationship between sensitivity and power, and between specificity and alpha.

RECEIVER-OPERATING CHARACTERISTIC (ROC) CURVES

The trade-off

When evaluating a diagnostic test, it is often difficult to decide where to set the threshold that separates a clinical diagnosis of normal from one of abnormal.

On one hand, if the threshold is set high (assuming that the test value increases with disease severity), some individuals with low test values or mild forms of the disease will be missed. The sensitivity will be low, but the specificity will be high. Few of the positive tests will be false positives, but many of the negative tests will be false negatives.

On the other hand, if the threshold is set low, most individuals with the disease will be detected, but the test will also mistakenly diagnose many normal individuals as abnormal. The sensitivity will be high, but the specificity will be low. Few of the negative tests will be false negatives, but many of the positive tests will be false positives.

The threshold can be set to have a higher sensitivity or a higher specificity, but not both (until a better diagnostic test is developed).

What an ROC curve plots

A *receiver-operating characteristic* (ROC) *curve* visualizes the trade-off between high sensitivity and high specificity (Figure 42.1). Why the odd name? ROC curves were developed during World War II, within the context of determining whether a blip on a radar screen represented a ship or an extraneous noise. The radar-receiver operators used this method to set the threshold for military action.

The bottom left of the ROC curve is one silly extreme when the test never, ever returns a diagnosis that the person has the disease. At this extreme, every patient is incorrectly diagnosed as healthy (sensitivity = 0%) and every control is correctly diagnosed as healthy (specificity = 100%). At the other extreme, which is in the upper-right corner, the test always returns a diagnosis that the person tested has the disease. Every patient is correctly diagnosed (sensitivity = 100%) and every control is incorrectly diagnosed (specificity = 0%).

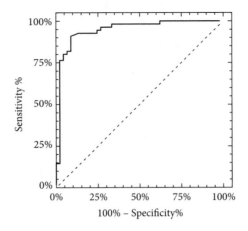

Figure 42.1. An ROC curve.

The solid line shows the trade-off between sensitivity and specificity. Each point along the curve represents a different possible threshold value defining an abnormal test result. The dotted line shows the expected ROC curve for a method that has no predictive value (e.g., flipping a coin).

The dotted diagonal line shows results from a test that has no ability to discriminate between patients and controls. The fraction of patients diagnosed as having the disease (sensitivity) equals the fraction of controls given that diagnosis (1 – specificity).

Where is the best cut-off?

Each point on an ROC curve shows the sensitivity and specificity for one possible threshold value for deciding when a test result is abnormal. Which is the best threshold to use? It depends on the consequences or costs of false negative and false positive diagnoses. If the two are equally bad, then the best threshold is the one that corresponds to the point on the ROC curve that is closest to the upper-left corner of the graph. But in most cases, the consequences of false positives and false negatives are not comparable, so it will be difficult to decide what cut-off makes the most sense. That decision must be made by someone who understands the disease and the test. It is not a decision to be made by a computer program.

BAYES REVISITED

Interpreting clinical laboratory tests requires combining what you know about the clinical context and what you learn from the lab test. This is simply Bayesian logic at work. Bayesian logic has already been discussed in Chapter 18.

Probability versus odds

As you may recall from Chapter 2, chance can be expressed either as a probability or as odds.

- The *probability* that an event will occur is the fraction of times you expect to see that event over the course of many trials.
- The *odds* are the probability that the event will occur divided by the probability that the event will not occur.

A probability is a fraction and always ranges from zero to one. Odds range from zero to infinity. Any probability can be expressed as odds. Any odds can be expressed as a probability. To convert between odds and probability (for situations in which there are only two possible outcomes), use the following equations:

$$Odds = \frac{probability}{1 - probability}$$

$$Probability = \frac{odds}{1 + odds}$$

If the probability is 0.50, or 50%, then the odds are 50:50, or 1. If you repeat the experiment often, you will expect to observe the event (on average) in one of every two trials (probability = 1/2). That means you'll observe the event once for every time it fails to happen (odds = 1:1).

If the probability is 1/3, the odds equal $1/3/(1-1/3) = 1:2 = 0.5$. On average, you'll observe the event once in every three trials (probability = 1/3). That means you'll observe the event once for every two times it fails to happen (odds = 1:2).

The likelihood ratio

The *likelihood ratio* is the probability of obtaining a positive test result in a patient with the disease divided by the probability of obtaining a positive test result in a patient without the disease. The probability of obtaining a positive test result in a patient with the disease is the sensitivity. The probability of obtaining a positive test result in someone without the disease is (1 − specificity). So the likelihood ratio equals sensitivity divided by (1 − specificity).

Using this equation, we can rework the examples. The test used in the intermittent porphyria example has a sensitivity of 82% and a specificity of 96.3%. Thus, the likelihood ratio is $0.82/(1.0-0.963) = 22.2$. A person with the condition is 22.2 times more likely to get a positive test result than a person without the condition.

Bayes as an equation

Bayes's equation for clinical diagnosis can be written in two forms:

$$Posttest\ Odds = Pretest\ odds \cdot \frac{sensitivity}{1 - specificity}$$

$$Posttest\ Odds = Pretest\ odds \cdot Likelihood\ ratio.$$

The posttest odds are the odds that a patient has the disease, taking into account both the test results and your prior knowledge about the patient. The

Who was tested?	PRETEST		POSTTEST	
	Probability	Odds	Odds	Probability
Random screen	0.0001	0.0001	0.0022	0.0022
Sibling	0.50	1.0000	22.2	0.957

Table 42.6. The porphyria calculations computed using Bayes's equation.

The pretest probability is first converted to odds. Then the posttest odds are computed by multiplying the pretest odds by the likelihood ratio, which equals 22.2. Finally, the odds are converted to probabilities.

pretest odds are the odds that the patient has the disease as determined from information you know before running the test.

Table 42.6 reworks the two examples with intermittent porphyria, which has a likelihood ratio of 22.2, using the second equation shown above.

BAYES, GENETIC LINKAGE, AND LOG OF ODDS (LOD) SCORES

When two loci (genes or DNA sequences) are located near each other on the same chromosome, they are said to be linked. If the two loci are very close, crossing over or recombination between the two loci occurs rarely. Thus, alleles of linked loci tend to be inherited together. If the loci are farther apart, recombination (a normal process) occurs more frequently. If the loci are very far apart, they segregate independently, just as if they were on different chromosomes.

Linkage is useful in genetic diagnosis and mapping. Because it is not always possible to detect all abnormal genes directly, geneticists sometimes identify a marker gene (such as those for variable antigens or isozymes) or a variable DNA sequence that is linked to the disease gene. Once you know that the disease gene is linked to a marker, the presence of the marker (which you can identify) can often predict the presence of the disease gene (which you cannot identify directly). This allows detection of genetic diseases prenatally or before they cause clinical problems. It also allows the diagnosis of unaffected heterozygotes (carriers) who can pass the abnormal gene on to their children. This method works best for diseases caused by an abnormality of a single gene.

Before linkage can be useful in diagnosis, you must identify a marker linked to the gene. This is usually done by screening lots of potential markers. How can you tell whether a marker is linked to a disease gene? Geneticists study large families and observe how often the disease and marker are inherited together and how often recombination occurs. If there are few recombination events between the marker and the disease, there are two possible explanations. One possibility is that the two are linked. The other possibility is that the two are not linked, but— just by coincidence—there were few recombination events.

Geneticists don't apply the terms *sensitivity* or *specificity* to linkage studies. But linkage studies, like clinical lab tests, can have false positive and false negative results.

Bayesian logic combines the experimental data with the prior probability of linkage to determine the probability that the gene is truly linked to the disease. To calculate Bayes's equation, we must define the likelihood ratio in the context of linkage. When calculating the predictive values of lab tests, we defined the likelihood ratio as the probability that someone with the disease will have an abnormal test result divided by the probability that someone without the disease will have an abnormal test result. For studies of linkage, therefore, the likelihood ratio is the probability of obtaining the observed data if the genes really are linked divided by the probability of observing those data if the genes really are not linked. The details of the calculations are beyond the scope of this book. When you read papers with linkage studies, you'll rarely see reference to the likelihood ratio. Instead, you'll see the *log of odds (LOD) score*, which is simply the logarithm (base 10) of the likelihood ratio.

An LOD score of 3.0 means that the likelihood ratio equals 1,000 (because $10^3 = 1,000$). This means that the data are 1,000 times more likely to be observed if the marker is linked to the disease than if the marker is not linked. A higher LOD score is stronger evidence for linkage.

To calculate the probability that the marker is linked to the gene requires accounting for the prior probability of linkage using Bayesian logic.

Posttest odds of linkage = Pretest odds of linkage · Likelihood ratio

To calculate Bayes's equation, you must know the prior (or pretest) odds of linkage. Using a common definition of linkage, 2% of randomly selected markers are linked to any particular disease gene. Converting to odds, the pretest odds of linkage is about 0.02. If you pick a marker known to be on the same chromosome as the disease, then the pretest odds of linkage will be higher.

If an LOD score equals 3.0, what is the probability that the marker and the disease are truly linked? The posttest odds equal the pretest odds (0.02) times the likelihood ratio ($10^3 = 1000$), which is 20. Converting to a probability, the posttest probability equals 20/(20+1), which is about 95%. If you observe an LOD score of 3.0, you will conclude that the marker and the gene are linked. When you make that conclusion, there is a 95% chance that you will be correct, leaving a 5% chance that you will be wrong.

If an LOD score equals or exceeds 3.0, geneticists usually conclude that the marker and the disease are linked. If an LOD score is less than or equal to –2.0, geneticists conclude that the marker and the disease are not linked. If the LOD score is between –2.0 and 3.0, geneticists conclude that the evidence is inconclusive.

Q & A

How is the accuracy of a test defined?

"Accuracy" is a fairly ambiguous term. This chapter defines the accuracy of a test in four different ways.

Why is this chapter in a statistics text?

> One reason is that terms like specificity and sensitivity get used in scientific papers, and there is a fuzzy border between epidemiology and biostatistics. The other reason is that interpreting statistical significance works very much like interpreting lab results, and one needs to take into account prior probability (prevalence for lab tests) to properly interpret the results. Review Chapters 18 and 19 and see if they make more sense now that you understand how to interpret clinical lab results.

CHAPTER SUMMARY

- When you report a test result as positive or negative, you can be wrong in two ways: false positives and false negatives.
- The sensitivity of a test is the fraction of all those with a disease who correctly get a positive result.
- The specificity is the fraction of those without the disease who correctly get a negative test result.
- The positive predictive value of a test answers the question, If the test is positive (abnormal test result, suggesting the presence of disease), what is the chance that the patient really has the disease? The answer depends in part on the prevalence of the disease in the population you are testing.
- The negative predictive value of a test answers the question, If the test is negative (normal test result, suggesting the absence of disease), what is the chance that the patient really does not have the disease? The answer depends in part on the prevalence of the disease in the population you are testing.
- A receiver-operating characteristic (ROC) curve visualizes the trade-off between high sensitivity and high specificity depending on where you designate the cut-off between normal and abnormal test results.
- The likelihood ratio is the probability of obtaining a positive test result in a patient with the disease divided by the probability of obtaining a positive test result in a patient without the disease.
- When geneticists ask whether a marker is linked to a genetic abnormality, they use the term *LOD score* to refer to the logarithm (base 10) of the likelihood ratio.

TERMS INTRODUCED IN THIS CHAPTER

- False negative (p. 401)
- False positive (p. 401)
- Likelihood ratio (p. 407)
- Log of odds (LOD) score (p. 409)
- Negative predictive value (p. 402)
- Positive predictive value (p. 402)
- Receiver-operating characteristic (ROC) curve (p. 405)
- Sensitivity (p. 401)
- Specificity (p. 402)

CHAPTER 43

❖❖❖

Meta-analysis

Let my data set change your mindset!
HANS ROSLING

When many studies have been performed on a particular topic, it is natural to want to combine the results. This chapter explains the methods and pitfalls of meta-analysis.

INTRODUCING META-ANALYSIS

What is meta-analysis?

Meta-analysis is used to combine evidence from multiple studies, usually clinical trials testing the effectiveness of therapies or tests. The pooled evidence from multiple studies can give much more precise and reliable answers than any single study can.

A large part of most meta-analyses is a descriptive review that points out the strengths and weaknesses of each study. Additionally, a meta-analysis pools the data to report an overall effect size (perhaps a relative risk or odds ratio), an overall confidence interval, and an overall P value. Meta-analysis is often used for clinical trials.

Why meta-analysis is difficult

Investigators planning a meta-analysis face many problems, summarized in Table 43.1. Meta-analysis is not simple.

The promise of meta-analysis

If a huge, well-designed clinical trial has been conducted, there is little need for a meta-analysis. The results from the large study will be compelling. Meta-analyses are needed when the individual trials have conflicting, ambiguous, or unconvincing conclusions. The hope is that combining the trials in a meta-analysis will lead to a persuasive conclusion. Many researchers are convinced that meta-analyses are hugely important and helpful (Cumming, 2011; Goldacre, 2013). Others doubt that meta-analysis can consistently overcome the huge challenges listed in Table 43.1 and produce consistently useful results (Dallal, 2012; Bailar, 1997).

CHALLENGE	SOLUTION
Investigators may use inconsistent criteria to define a disease, and so may include different groups of patients in different studies.	Define which patient groups are included.
Studies asking the same clinical question may use different outcome variables. For example, studies of cardiovascular disease often tabulate the number of patients who have experienced new cardiac events. Some studies may only count people who have documented myocardial infarctions (heart attacks) or who have died. Others may include patients with new angina or chest pain.	Define which clinical outcomes are included.
Some studies, especially those that don't show the desired effect, are not published. Omitting these studies from the meta-analysis causes publication bias (discussed below).	Seek unpublished studies; review registries of study protocols.
Some studies are of higher quality than others.	Define what makes a study of high enough quality to be included.
Some relevant studies may be published in a language other than English.	Arrange for translation.
Published studies may not include enough information to perform a proper meta-analysis.	Estimate data from figures; obtain unpublished details from the investigators or web archives.
Published data may be internally inconsistent. For example, Garcia-Berthou and Alcaraz (2004) found that 11% of reported P values were not consistent with the reported statistical ratios (t, F, etc.) and degrees of freedom.	Resolve inconsistencies without bias; request details from original investigators.
Data from some patients may be included in multiple publications. Including redundant analyses in a meta-analysis would be misleading.	When multiple studies are published by the same investigators, ask them to clarify which patients were included in more than one publication.

Table 43.1. The challenges of meta-analysis.

PUBLICATION BIAS

The problem

Scientists like to publish persuasive findings. When the data show a large, statistically significant effect, it is easy to keep up the enthusiasm necessary to write up a paper, create the figures, and submit the resulting manuscript to a journal. If the effect is large, the study is likely to be accepted for publication, as editors prefer to publish papers that report results that are statistically significant.

If the study shows a small effect, especially if it is not statistically significant, it is much harder to stay enthusiastic through this process and stick it out until the work is published. Studies that show large effects tend to get submitted for publication, while studies that show small effects may get left in a file drawer or computer file. And even if the scientist did the work to prepare the manuscript for publication, editors tend to reject papers that show negative findings.

Many people have documented that this really happens. Studies that find large, statistically significant effects are much more likely to get published, while studies showing small effects tend to be abandoned by the scientists or rejected by journals. This tendency is called *publication bias*.

Meta-analyses really can only be interpreted at face value when they summarize all the studies that have been performed. A good meta-analyst, therefore, has to try hard to identify and obtain unpublished studies.

An example of publication bias

Turner and colleagues (2008) demonstrated this kind of selectivity in industry-sponsored investigations of the efficacy of antidepressant drugs. Between 1987 and 2004, the FDA reviewed 74 such studies and categorized them as positive, negative, or questionable. The FDA reviewers found that 38 studies showed a positive result (i.e., the antidepressant worked). All but one of these studies was published. The FDA reviewers found that the remaining 36 studies had negative or questionable results. Of these, 22 were not published, 11 were published with a "spin" that made the results seem somewhat positive, and only 3 were published with clearly presented negative findings.

Think about that. The investigators were forced to dig through the FDA files to find all the studies. Only about half of these studies (38 of 74) showed a positive result. Almost all (49 of 52) of the published studies, however, concluded that the drugs worked. What a difference! Selective publication leads to misleading conclusions and reduces the value of a meta-analysis. Goldacre (2013) gives more examples of this bias in the first chapter ("Missing Data") of his compelling book *Bad Pharma*.

How a meta-analysis can deal with publication bias

Good meta-analyses deal with publication bias in two ways.

First, the analyst tries hard to include unpublished studies. Sometimes, when these studies are first proposed, they are entered into a registry of study protocols (e.g., clintrials.gov). Such a registry can help an analyst track down the investigator and obtain unpublished results. But, by definition, unpublished studies are hard to learn about, and you can never be sure that you have found them all. Even if you know about some unpublished studies, the investigators may not have completed the study, may not have kept good records, or may choose not to share the data.

Second, a meta-analysis can try to detect publication bias by looking at the pattern of data. Look up *funnel plots* to see how this is done.

RESULTS FROM A META-ANALYSIS

Part of a meta-analysis is a narrative report that summarizes the studies that have been done and evaluates their strengths and weaknesses.

The rest of a meta-analysis is quantitative. The results of each study are summarized by one value, called the *effect size,* along with its confidence interval.

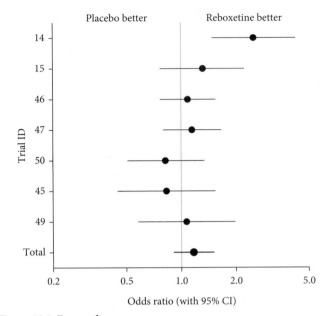

Figure 43.1. Forest plot.

This forest plot, adapted from Eyding et al. (2010), shows the results from seven studies of the effectiveness of reboxetine as a treatment for severe depression. The horizontal axis is the odds ratio. A value of 1.0 means no effect; a value greater than 1.0 means the drug works better than placebo; and a value less than 1.0 means the drug works worse than placebo. The top 7 lines are results of individual studies, identified by labels in the Y-axis (with a key in the meta-analysis publication). For each study, the plot shows the odds ratio (circle) and 95% confidence interval (line). At the very bottom, the graph shows the total (pooled) effect computed from all seven studies, also shown as an odds ratio with its 95% confidence interval.

The effect size is usually a relative risk or odds ratio, but it could also be some other measure of treatment effect. In some cases, different studies use different designs, and so the meta-analyst has to do some conversions so that the effect sizes are comparable.

Combining all the studies, the meta-analysis computes the pooled treatment effect, its confidence interval, and a pooled P value.

The results for the individual studies and the pooled results are plotted on a graph known as a *forest plot* or *blobbogram*. Figure 43.1 is an example. It shows part of a meta-analysis performed by Eyding et al. (2010) on the effectiveness of the drug reboxetine, a selective inhibitor of norepinephrine re-uptake, marketed in Europe as an antidepressant. The researchers tabulated two different effects for this drug: the odds ratio for remission and the odds ratio for any response. They computed these odds ratios comparing reboxetine against either placebo or other antidepressants.

Figure 43.1 shows the odds ratio for remission comparing reboxetine to placebo. An odds ratio of 1.0 means no effect, an odds ratio greater than 1.0 means that reboxetine was more effective than placebo, and an odds ratio less than 1.0 means that reboxetine was less effective than placebo. The graph shows 95% confidence intervals for seven studies identified with numbers (which can be used to look up details presented in the published meta-analysis). The bottom symbol shows the overall odds ratio and its 95% confidence interval.

In six of the seven studies, the 95% confidence interval includes 1.0. With 95% confidence, you cannot conclude from these six studies that reboxetine works better than placebo. Accordingly, a P value would be greater than 0.05. In only one of the studies (the one shown on top of the graph and identified as Study 14) does the 95% confidence not include 1.0. The data from this study, but not the others, would lead you to conclude with 95% confidence that reboxetine worked better than placebo and that the effect is statistically significant (with P < 0.05). As explained in Chapter 22, even if the drug were entirely ineffective, it is not too surprising that one of six studies would find a statistically significant effect just by chance.

The bottom part of the graph shows the total, or pooled, effect, as computed by the meta-analysis. The 95% confidence interval ranges from 0.98 to 1.56. The confidence interval contains 1.0, so the P value must be greater than 0.05. From this confidence interval alone, it might be hard to know what to conclude. The overall odds ratio is 1.24. That is a small effect, but not a tiny one. The 95% confidence interval includes 1.0 (no effect), but just barely. The paper also compares reboxetine to other antidepressants and concludes that reboxetine is less effective than the others.

ASSUMPTIONS OF META-ANALYSIS

The calculations used in a meta-analysis are complicated and beyond the scope of this book. But you should know that there are two general methods that are used, each of which is based on different assumptions:

- Fixed effects. This model assumes that all the subjects in all the studies were really sampled from one large population. Thus, all the studies are estimating the same effect, and the only difference between studies is due to random selection of subjects.
- Random effects. This model assumes that each study population is unique. The difference among study results is due both to differences between the

Q & A

How does the meta-analysis combine the individual P values from each study?
> It doesn't! A meta-analysis pools weighted effect sizes, not P values. It calculates a pooled P value from the pooled effect size.

Why not just average the P values?
> A meta-analysis accounts for the relative sizes of each study. This is done by computing a weighted average of the effect size. Averaging P values would be misleading.

How is a meta-analysis different from a narrative review article?

> In a narrative review, the author discusses all the studies and explains their strengths and weaknesses. This is a very subjective process. In contrast, a meta-analysis is quantitative. It combines the evidence and calculates a pooled effect size (and its confidence interval).

Are meta-analyses objective?

> The calculations are very objective and use standard methods. But before the calculations can be performed, the investigator must decide which studies to include and which to reject. This decision can include some subjective criteria.

What is the minimum number of studies that can be combined in a meta-analysis?

> Two. Meta-analyses don't have to be huge. It can be helpful to combine the results from as few as two studies.

populations and to random selection of subjects. This model is more realistic and is used more frequently.

CHAPTER SUMMARY

- A meta-analysis pools multiple studies and reports an overall effect size (perhaps a relative risk or odds ratio), an overall confidence interval, and an overall P value.
- Studies that show positive results are far more likely to be published than studies that reach negative or ambiguous conclusions. Selective publication (publication bias) makes it impossible to properly interpret the published literature and can make the results of meta-analyses suspect. A good meta-analyst has to work hard to find relevant unpublished data.
- The main results of a meta-analysis are shown in a forest plot, also known as a blobbogram. This kind of graph shows the effect size of each study with its confidence interval, as well as the pooled (overall) effect with its confidence interval.

TERMS INTRODUCED IN THIS CHAPTER

- Effect size (p. 413)
- Forest plot (or blobbogram) (p. 414)
- Meta-analysis (p. 411)
- Publication bias (p. 413)

PART I

Putting It All Together

❖❖❖

CHAPTER 44

The Key Concepts of Statistics

> If you know twelve concepts about a given topic you will look
> like an expert to people who only know two or three.
>
> SCOTT ADAMS

W hen learning statistics, it is easy to get bogged down in the details and lose track of the big picture. This short chapter reviews the most important concepts in statistical inference that every scientist should know.

Many statistical terms are also ordinary words

A large part of the challenge of learning any new field is mastering its vocabulary. This is a special challenge with statistics, because so many of the terms are ordinary words given special meanings. It is easy to think that a term has an ordinary meaning when it actually is being used in a very special way. This causes lots of confusion and miscommunication about statistics.

Statistics helps you make general conclusions from limited data

The whole point of statistics is to extrapolate from limited data to make a general conclusion. *"Descriptive statistics"* simply describe data without reaching any general conclusions. But the challenging and difficult aspects of statistics are all about making inferences—reaching general conclusions from limited data.

Statistical conclusions are always presented in terms of probability

Myles Hollander has said, "Statistics means never having to say you are certain" (quoted in Samaniego, 2008). Every statistical conclusion must include words like "probable," "most likely," or "almost certainly." Be wary if you ever encounter statistical conclusions that seem 100% definitive.

All statistical tests are based on assumptions

Every statistical inference is based on a list of assumptions. Review that list before interpreting any statistical results.

The standard deviation and standard error of the mean are quite different

As explained in Chapter 14, the standard deviation (SD) and standard error of the mean (SEM) are quite distinct. The standard deviation quantifies the average variation among a set of values, while the SEM estimates how precisely the sample mean captures the population mean. The SEM is computed by dividing the SD by the square root of n (the sample size), so it is always smaller than the SD. The SEM is used to compute the confidence interval (summarized below) and is sometimes used to plot error bars on graphs.

Confidence intervals quantify precision

Almost all results—proportions, relative risks, odds ratios, means, differences between means, slopes, rate constants, and so on—can and should be reported with a confidence interval. Whenever you see a value computed from data, ask for its confidence interval.

Every P value tests a null hypothesis

The logic of P values seems strange at first, and they are commonly misinterpreted. The key to interpreting a P value is to first identify the null hypothesis that the P value refers to. Every P value is associated with a null hypothesis, so you cannot begin to understand a P value until you can precisely state which null hypothesis is being tested.

The sample size has a huge effect on the P value

Table 44.1 demonstrates that sample size has a huge impact on the P value. Each P value in the table was calculated using an unpaired t test (see Chapter 30) comparing two groups. One group has a mean of 10 and an SD of 5, and the other group has a mean of 15 and an SD of 5. The only thing that differs between the difference comparisons is sample size. When n (in each group) is three, the P value is 0.29. When n equals 50, the P value is 0.000003. This huge spread is due only to differences in sample size, as the samples have exactly the same means and SD in every case.

"Statistically significant" does not mean the effect is large or scientifically important

The term *significant* has a special meaning in statistics. It means that, by chance alone, a difference (or association, or correlation, etc.) as large (or larger than) the one you observed would happen less than 5% of the time (or some other stated value). That's it. A tiny effect can be statistically significant and yet scientifically or clinically trivial. Spector and Vesell (2006b) give an example. Use of the drug montelukast for treatment of allergic rhinitis reduces symptoms, and the results are statistically significant. However, the drug reduces allergy symptoms by only 7%, so it may not be clinically useful.

The conclusion that a finding is statistically significant does not tell you that the effect is large, does not tell you that the effect is important, and does not tell you that it is worth following up.

SAMPLE SIZE (PER GROUP)	P VALUE (UNPAIRED T TEST)
3	0.287864
5	0.152502
7	0.085946
10	0.038250
15	0.010607
20	0.003073
25	0.000912
50	0.000003

Table 44.1. The impact of sample size on P value.

The P value in each row was computed by an unpaired t test comparing one group with mean = 10 and SD = 5 with another group with mean = 15 and SD = 5. The first column shows the sample size for each of the two groups.

"Not significantly different" does not mean the effect is absent, small, or scientifically irrelevant

If a difference is not statistically significant, you can conclude that the observed results are not inconsistent with the null hypothesis. Note the double negative. You *cannot* conclude that the null hypothesis is true. It is quite possible that the null hypothesis is false and that there really is a small difference between the populations. All you can say is that the data are not strong or consistent enough to persuade you to reject the null hypothesis.

The concept of statistical significance is designed to help you make a decision based on one result

It is helpful to define a result as being either statistically significant or not statistically significant when you plan to make a crisp decision based on a single result. If you don't plan to use this one result to make a clear decision, the concept of statistical significance is not necessary and may not be helpful.

If you decide to classify all results as "statistically significant" and "not statistically significant," then some of your results will inevitably be very close to the deciding line. There really is not much difference between $P = 0.045$ and $P = 0.055$! If you are using the concepts of statistical hypothesis testing and have set the significance level to 5% (as is conventional), then the first result is statistically significant and the second one is not.

Setting the dividing line at 0.05 is just a convention. Figure 44.1 is a cartoon that shows how some scientists seem to mistakenly think that $P = 0.05$ has magic qualities.

Multiple comparisons make it hard to interpret statistical results

When many hypotheses are tested at once, the problem of multiple comparisons makes it very easy to be fooled. If 5% of tests will be statistically significant by chance, you will expect lots of statistically significant results if you test hundreds or thousands of hypotheses. Special methods can be used to reduce the problem

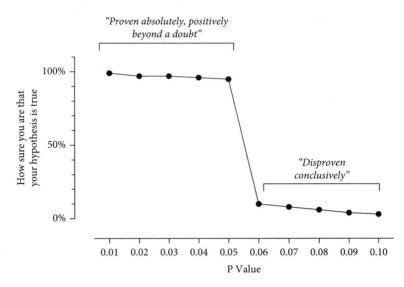

Figure 44.1. "P = 0.05" does not have magical properties.

This figure exaggerates what many researchers seem to believe—that P values less than 0.05 definitely prove that the experimental hypothesis is true and P values greater than 0.05 definitely prove that the experimental hypothesis is wrong. Of course, this belief is wrong. There is very little difference between 0.04 and 0.06. Adapted from Goldstein (2006).

of finding false but statistically significant results, but these methods also make it harder to find true effects.

Multiple comparisons can be insidious. To correctly interpret statistical analyses, all analyses must be planned (before collecting data), and all planned analyses must be conducted and reported. However, these simple rules are widely broken.

Correlation does not prove causation

If two variables are correlated (or associated) and that difference is statistically significant, you should not conclude that changes in the variable labeled X causes changes in the variable labeled Y. It is possible that changes in Y cause changes in X, that changes in X cause some other change that indirectly changes Y (or vice versa), that both X and Y are influenced by a third variable, or that the correlation is spurious and due to a coincidence of random sampling. Correlation does not prove causation.

Published statistics tend to be optimistic

By the time you read a paper, a great deal of selection has occurred. When experiments are successful, scientists tend to continue the project. Lots of other projects get abandoned. When the project is done, scientists are more likely to write up projects that lead to remarkable results or keep analyzing the data in various ways to extract a statistically significant conclusion. Finally, journals are more likely to publish "positive" studies. If the null hypothesis were true, you would expect a statistically significant result in 5% of experiments. But those 5% are more likely to get published than the other 95%. This is called *publication bias* and was discussed in Chapter 43.

What about the published findings that are not false positives? As pointed out in Chapter 26, these studies tend to inflate the size of the difference or effect. The explanation is simple. If many studies were performed, you'd expect the average of the effects detected in these studies to be close to the true effect. By chance, some studies will happen to find larger effects and some studies will happen to find smaller effects. However, studies with small effects tend not to get published. On average, therefore, the studies that do get published tend to have effect sizes that overestimate the true effect (Ioannidis, 2008). This is called the *winner's curse* (Zollner and Pritchard, 2007). This term was coined by economists to describe the winner of an auction who tends to overpay.

Decisions about how to analyze data should be made in advance
Analyzing data requires many decisions: Collect how many data points? Run a parametric or nonparametric test? Eliminate outliers or not? Transform the data first? Normalize to external control values? Adjust for covariates? Use weighting factors in regression? All these decisions (and more) should be made as part of the experimental design. When decisions about statistical analysis are made only after inspecting the data, it is too easy for statistical analysis to become a high-tech Ouija board—a method to produce preordained results rather than an objective method of analyzing data.

Statistics is only part of interpreting data
Statistical calculations provide limited help in overcoming the following common issues in interpreting data:

- The population you really care about is more diverse than the population from which your data were sampled.
- The subjects in the study were not randomly sampled from a larger population.
- The measured variable is a proxy for another variable you really care about.
- The measurements may be made or recorded incorrectly, and assays may not always measure exactly the right thing.
- Scientific (or clinical) conclusions require examining multiple outcomes, not just one.

Understanding the statistical calculations is only part of evaluating scientific data. Interpreting data also requires a great deal of judgment.

Statistical calculations are helpful, but should augment (not replace) looking at graphs of the data
Always look at graphs (or tables) first to understand the data. Then look at statistical calculations to sharpen your understanding.

TERM INTRODUCED IN THIS CHAPTER
- Winner's curse (p. 423)

CHAPTER 45

❖❖❖

Statistical Traps To Avoid

> When the data don't make sense, it's usually because you have
> an erroneous preconception about how the system works.
>
> ERNEST BEUTLER

To avoid being misled by statistical results when reading scientific papers or using statistical software, beware of the traps listed in this chapter. Be sure to review Chapter 44 before reading this one.

TRAP #1: FOCUSING ON P VALUES AND STATISTICAL SIGNIFICANCE RATHER THAN EFFECT SIZE

P values and conclusions about statistical significance can be useful, but there is more to statistics than P values and asterisks. Focus instead on the effect size (the size of the difference or association, etc.). Is the difference or association or correlation (effect) big enough to care about? Is that effect determined with sufficient precision (as assessed by the confidence interval)? With huge sample sizes, tiny effects might be statistically significant yet irrelevant. With small sample sizes, you may not have determined the effect with sufficient precision to make any useful conclusion.

Gelman (2013) suggests a clever approach to avoid getting distracted by statements of statistical significance. He suggests imagining that the sample size is huge ("one zillion"). In that case, essentially every effect will be statistically significant, but most will be trivial, so "you'll need to ignore conclusions about statistical significance and instead think about what you really care about".

TRAP #2: TESTING HYPOTHESES SUGGESTED BY THE DATA

When studies are designed to answer a focused question, the results are often straightforward to interpret. But some studies analyze data for many variables, in many groups of subjects, and have no real hypothesis when they begin. Such studies can be a useful way to generate hypotheses as part of *exploratory data analysis*. New data, however, are needed to test those hypotheses.

The trap occurs when a scientist analyzes the data many ways (which might include many variables and many subgroups), discovers an intriguing relationship, and then publishes the results so it appears that the hypothesis was stated before the data collection began As Chapter 23 noted, Kerr (1998) coined the acronym *HARK*: hypothesizing after the results are known. Kriegeskorte and colleagues (2009) call this *double dipping*.

It is impossible to evaluate the results from such studies unless you know exactly how many hypotheses were actually tested. You will be misled if the results are published as if only one hypothesis was tested. A XKCD cartoon points out the folly of this approach (Figure 45.1).

Also beware of studies that don't truly test a hypothesis. Some investigators believe their hypothesis so strongly (and may have stated the hypothesis so vaguely) that no conceivable data would lead them to reject that hypothesis. No matter what the data showed, the investigator would find a way to conclude that the hypothesis is correct. Such hypotheses "are more *vampirical* than *empirical*— unable to be killed by mere evidence" (Gelman & Weakliem, 2009; Freese, 2008).

TRAP #3: ANALYZING WITHOUT A PLAN—"P-HACKING"

Analyzing data requires many decisions. How large should the samples be? Which statistical test should you use? What should you do with outliers? Should you transform the data first? Should the data be normalized to external control values? Should you run a multivariable regression to control for differences in confounding variables? All of these decisions (and more) should be made at the time you design an experiment.

If you make analysis decisions only after you see the data, there is a danger that you will pick and choose among analyses to get the results you want, and thus fool yourself. We already discussed the problem of ad hoc sample size selection in Chapter 26, and the folly of p-hacking in Chapter 23.

Gotzsche (2006) presented evidence of p-hacking (without calling it that). He tabulated the actual distribution of P values in journals and found that "P values were much more common immediately below .05 than would be expected based on the number of P values occurring in other ranges." Masicampo and Lalande (2012) found similar results (see Figure 45.2). In addition to being caused by p-hacking, this excess of P values just below 0.05 could be due to ad hoc sample size selection (see Chapter 26) and publication bias (see Chapter 43).

TRAP #4: MAKING A CONCLUSION ABOUT CAUSATION WHEN THE DATA ONLY SHOW CORRELATION

Messerli (2012) wondered why some countries produce many more Nobel Prize winners than others, and to answer this question, he plotted the data shown in Figure 45.3. The Y-axis plots the total number of Nobel Prizes ever won by citizens of each country. The X-axis plots chocolate consumption in a recent year (different years for different countries, based on the availability of data).

Figure 45.1. Nonsense conclusions via HARKing (hypothesizing after the results are known).

Source: xkcd.com.

Figure 45.2. Too many published P values are just a tiny bit less than 0.05.

Masicampo and Lalande(2012) collected P values from three respected and peer-reviewed journals in psychology and then tabulated their distribution. The figure shows a "peculiar prevalence" of P values just below 0.05. This figure was made from the raw list of 3,627 P values kindly sent to me by Masicampo, matching a graph published by Wasserman (2012).

Both X and Y values are standardized to the country's current population. The correlation is amazingly strong, with r = 0.79. A P value testing the null hypothesis of no real correlation is tiny, less than 0.0001.

Of course, these data don't prove that eating chocolate helps people win a Nobel Prize. Nor does it prove that increasing chocolate imports into a country will increase the number of Nobel Prizes that residents of that country will win.

When two variables are correlated or associated, it is possible that changes in one of the variables causes the other to change. But it is also likely that both variables are related to a third variable that influences both. There are many variables that differ among the countries shown in the graph, and some of those probably correlate with both chocolate consumption and number of Nobel Prizes.

This point is often summarized as "Correlation does not *imply* causation," but it is more accurate to say that correlation does not *prove* causation.

When each data point represents a different year, it is even easier to find silly correlations. For example, Figure 45.4 shows a very strong negative correlation between the total number of pirates in the world and one measure of global average temperature. But correlation (pirates would say *carrrrelation?*) does not prove causation. It is quite unlikely that the lack of pirates caused global warming or that global warming caused the number of pirates to decrease. More likely, this graph simply shows that both temperature and the number of pirates have changed over time. The variable time is said to be *confounded* by the other two variables.

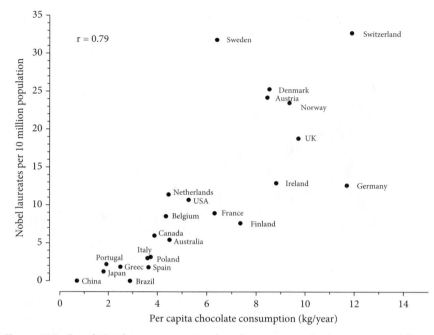

Figure 45.3. Correlation between average chocolate consumption by country and the number of Nobel Prize winners from that country.

Data are from Messerili (2012). The Y-axis plots the total number of Nobel Prizes won by citizens of each country. The X-axis plots chocolate consumption in a recent year (different years for different countries, based on the availability of data). Both X and Y values are normalized to the country's current population. The correlation is amazingly strong, but this doesn't prove that eating chocolate will help you win a Nobel Prize.

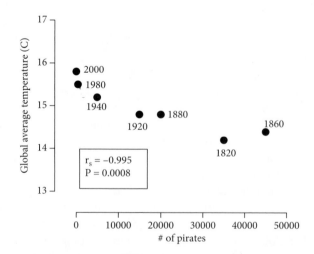

Figure 45.4. Correlation between the number of pirates worldwide and average world temperature.

Adapted from Henderson (2005). Correlation does not prove causation!

Figure 45.5. Correlation does not imply causation.

Sources: Top: xkcd.com. Bottom: DILBERT © 2011 Scott Adams. Used By permission of UNIVERSAL UCLICK. All rights reserved.

The cartoons in Figure 45.5 drive home the point that correlation does not prove causation.

TRAP #5: OVERINTERPRETING STUDIES THAT MEASURE A PROXY OR SURROGATE OUTCOME

Measuring important outcomes (e.g., survival) can be time consuming and expensive. It is often far more practical to measure *proxy*, or *surrogate*, *variables*. However, an intervention that improves the proxy variable may not improve the more important variable.

Chapter 15 already provided an example. For many years, people who had myocardial infarctions (heart attacks) were treated with anti-arrhythmic drugs. These drugs prevented extra heartbeats (premature ventricular contractions) and thus were thought to reduce the incidence of sudden death due to arrhythmia. The logic was clear. Abnormal results (extra beats) on an electrocardiogram (ECG) were known to be associated with sudden death, and treatment with anti-arrhythmic drugs was known to reduce the number of extra beats. So it made sense that taking those drugs would extend life. The evidence was compelling enough that the FDA approved use of these drugs for this purpose. But a randomized

TREATMENT	SURROGATE OR PROXY VARIABLE	IMPORTANT VARIABLE(S)
Two anti-arrhythmic drugs	• Fewer premature heart beats • Conclusion: Good treatment	• More deaths • Conclusion: Deadly treatment
Torcetrapib	• Increase in HDL ("good" cholesterol) • Decrease in LDL ("bad" cholesterol) • Conclusion: Good treatment	• More deaths • More heart attacks • Conclusion: Deadly treatment

Table 45.1. Results using proxy or surrogate variables can produce incorrect conclusions.

study to directly test the hypothesis that anti-arrhythmic drugs would reduce sudden death showed just the opposite. Patients taking two specific anti-arrhythmic drugs had (fewer extra beats, the proxy variable) but were more likely to die (CAST Investigators, 1989). Fisher and VanBelle (1993) summarize the background and results of this trial.

Another example is the attempt to prevent heart attacks by using drugs to raise HDL levels. Low levels of HDL ("good cholesterol") are associated with an increased risk of atherosclerosis and heart disease. Pfizer Corporation developed torcetrapib, a drug that elevates HDL, with great hope that it would prevent heart disease. Barter and colleagues (2007) gave the drug to thousands of patients with a high risk of cardiovascular disease. LDL ("bad cholesterol") decreased 25% and "good" (HDL) cholesterol increased 72%. The CIs were narrow, and the P values were tiny (<0.001). If the goal were to improve cholesterol levels, the drug was a huge success. Unfortunately, however, treatment with torcetrapib also increased the number of heart attacks by 21% and increased the number of deaths by 58%.

The take-home message is clear: treatments that improve results of lab tests may not improve health or survival (see Table 45.1). Svennson (2013) lists 14 additional examples.

TRAP #6: OVERINTERPRETING DATA FROM AN OBSERVATIONAL STUDY

Munger and colleagues (2013) wondered whether deficiency of vitamin D (25-hydroxyvitamin D, abbreviated 25(OH)D) predisposed people to developing type I diabetes. The researchers compared a group of diabetics with a group of people without diabetes who were similar in other ways and then compared the vitamin D levels in blood samples drawn before disease onset. They found that people with average 25(OH)D levels greater than 100 nmol/L had a much lower risk of developing diabetes than those whose average 25(OH)D levels were less than 75 nmol/L. The risk ratio was 0.56, with a 95% CI ranging from 0.35 to 0.90 (P = 0.03).

Intriguing data! Do these findings mean that taking vitamin D supplements will prevent diabetes? No. The association of low vitamin D levels and onset of diabetes could be explained many ways. Sun exposure increases vitamin D

| INTERVENTION | INCIDENCE OF | RESULTS FROM . . . | |
		OBSERVATIONAL STUDIES	EXPERIMENT
Hormone replacement therapy after menopause	Cardiovascular events	Decrease	Increase
Megadose vitamin E	Cardiovascular events	Decrease	No change
Low-fat diet	Cardiovascular events and cancer	Decrease	No change
Calcium supplementation	Fractures and cancer	Decrease	No change
Vitamins to reduce homocysteine	Cardiovascular events	Decrease	No change

Table 45.2. Five hypotheses suggested by observational studies proven not to be true by experiment.
Adapted from Spector and Vesell (2006a). "Cardiovascular events" include myocardial infarction, sudden death, and stroke.

levels. Perhaps sunlight exposure also creates other hormones (which we may not have yet identified) that decrease the risk of diabetes. Perhaps the people who are exposed more to the sun (and thus have higher vitamin D levels) also exercise more, and it is the exercise that helps prevent diabetes. Perhaps the people with higher vitamin D levels drink more fortified milk, and it is the calcium in the milk that helps prevent diabetes. The only way to find out for sure whether ingestion of vitamin D prevents diabetes is to conduct an experiment comparing people who are given vitamin D supplements with those who are not.

The take-home message is to be skeptical about conclusions derived from observational data. Although observational studies are much easier to conduct than experiments, data from experiments are more definitive. With observational studies, it is difficult to deal with confounding variables, and so it is nearly impossible to convincingly untangle cause and effect. Compared to most experiments, observational studies often require more complicated analyses and yield less certain results.

To emphasize this point, Spector and Vesell (2006a) reviewed five hypotheses suggested by observational studies that turned out not to be valid when tested with clinical experiments (see Table 45.2).

TRAP #7: MAKING CONCLUSIONS ABOUT INDIVIDUALS WHEN THE DATA WERE ONLY COLLECTED FOR GROUPS

Figure 45.6 shows the relationship between the median income in 2011 and the fraction of voters who voted for Romney, the Republican candidate in the 2012 U.S. presidential election (this example was adapted from Statwing, 2012). There are 50 dots on the graph, one for each state. The relationship is striking. In states with higher income, a smaller percentage of voters tended to vote for Romney.

Does that mean that people with higher income were less likely to vote for Romney? No! The data in Figure 45.6 only let you make conclusions about

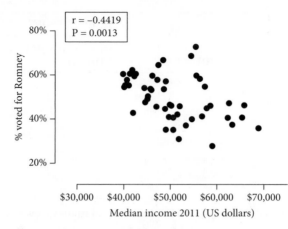

Figure 45.6. Median income and votes for Romney by state.
Each symbol represents one of the 50 states (excluding the District of Columbia). The X-axis shows median income in 2011 as determined by the U.S. Census (2011), and the Y-axis shows actual voting data from the Federal Election Commission (2012).

Figure 45.7. Income and votes for Romney by individual.
These are national polling data collected about two months before the election (Gallup, 2012).

states. If you want to make a conclusion about individuals, you must analyze data from individuals. Figure 45.7 shows results from polling individuals. There is a strong relationship between income and voting, but it goes in a direction *opposite* to the state data. People with higher income were *more* likely to vote for Romney, even though states with higher average income tended to have fewer Romney voters.

What accounts for the discrepancy? There are lots of differences between states. A correlation of statewide data does not tell you about the individuals in those states (Gelman & Feller, 2012). Mistakenly using relationships among groups to make inferences about individuals is called the *ecological fallacy*. Another example is the data on Nobel Prizes and chocolate presented earlier in this chapter.

TRAP #8: FOCUSING ONLY ON MEANS WITHOUT ASKING ABOUT VARIABILITY OR UNUSUAL VALUES

"Math Scores Show No Gap for Girls, Study Finds," reported the *New York Times* several years ago (Lewin, 2008). Not quite (Briggs, 2008a). In addition to showing essentially no difference in the average math scores of boys and girls, the study showed that there was a difference in variation (Hyde et al., 2008). The scores of boys were more variable, and twice as many boys as girls were in the top 1%. To fully understand these data, you must consider more than just the averages. Kane and Mertz (2012) continued this line of inquiry using data sets from many countries and showed that boys' math scores are not universally more varied than girls' math scores.

Here is another example. Steven Jay Gould wondered about changes in professional baseball from 1870 to 1970 (Gould, 1997). Why have there been no players with batting averages over 0.400 since 1941, even though the mean batting average has been consistent (at about 0.260)? It's not enough to look at the mean. Gould discovered that the SD of players' batting averages has dropped considerably (by almost 50%), especially in the first half of that period. This decrease in variability, he concludes, occurred because players, coaching, refereeing, and equipment have all become more consistent. Since the mean hasn't changed and the SD is much smaller than it was in earlier times, a batting average greater than 0.400 is now incredibly rare. Gould wasn't able to understand the changes in baseball until he investigated changes in variability (standard deviation).

Variability in biological or clinical studies often reflects real biological diversity (rather than experimental error). Appreciate this diversity! Don't get mesmerized by comparing averages. Pay attention to variation and to the extreme values. Nobel Prizes have been won from studies of individuals whose values were far from the mean.

TRAP #9: COMPARING STATISTICALLY SIGNIFICANT WITH NOT STATISTICALLY SIGNIFICANT

Figure 45.8 is similar to many you'll see in scientific journals. It compares a response in wild-type and mutant animals. The results might be summarized in a journal as, "The drug-induced increase in uptake in the wild-type mice was not statistically significant (P = 0.068), but the increase in the mutant mice was statistically significant (P = 0.004)."

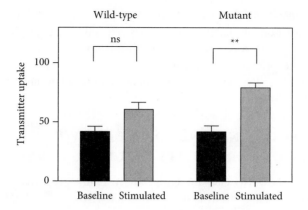

Figure 45.8. The difference between statistically significant and not statistically significant is not itself statistically significant.

The data are simulated. The bars show the mean and SEM of triplicate values. Unpaired t tests were used to compare baseline and stimulated results, with separate t tests done for the wild-type and mutant animals. A seemingly obvious conclusion—that there is a statistically significant difference between the drug-stimulated transmitter uptake in mutant and wild-type mice—is incorrect. If you want to compare one change with another change, you need to run statistical tests with that purpose in mind. Here, the difference between wild-type and mutant drug-induced transmitter uptake is not statistically significant.

What can you conclude from that statement? A seemingly obvious conclusion is that there is a statistically significant difference between the drug-stimulated uptake in mutant mice compared to wild-type mice. But this would be an incorrect conclusion (Nieuwenhuis, Forstmann, & Wagenmakers, 2011). As summarized by Gelman and Stern (2006), "the difference between 'significant' and 'not significant' is not itself statistically significant."

If you want to make a conclusion about the difference between two differences, you need to use a statistical method designed to test that exact question. Since the baseline values are very similar for the two groups of animals, the stimulated values in the wild-type animals can be directly compared to the stimulated values in the mutant animals with an unpaired t test. The difference between means is 18.7, with a 95% confidence interval ranging from −1.8 to 39.1 (P = 0.064). A fancier approach (beyond what is covered in this book) is to use two-way ANOVA to compute a P value that tests the null hypothesis of no interaction (the null hypothesis is that the difference between baseline and stimulated is identical in wild-type and mutant animals). The conclusion would be the same. The difference in drug-stimulated transmitter uptake between mutant and wild-type animals is not significantly significant.

This is not a trivial issue that only applies, say, when one P value is 0.04 and the other is 0.06. Bland and Altman (2011) have done simulations showing how easy it is to be led astray by basing one's conclusion on the difference between

	ACCEPTED	DENIED	% ACCEPTED
Men	3,738	4,704	44.3
Women	1,494	2,827	34.6

Table 45.3. Admissions to the graduate programs in Berkeley, 1973. Pooled data.

At first glance, the data seem to provide evidence of sexism, but that is only because the data pool admission rates from multiple graduate programs, each of which makes its own admissions decisions.

EVIDENCE FOR DISCRIMINATION?	NUMBER OF PROGRAMS
No statistically significant difference in admission decisions	75
Males preferred (P < 0.05)	4
Females preferred (P < 0.05)	6
Total number of programs	85

Table 45.4. Admissions to the graduate programs at Berkeley, 1973, by program.

a statistically significant change in one group and a not statistically significant change in the other. In their simulations, there was no difference in the actual mean increase in the two populations. Yet in 38% of their simulations, the increase was statistically significant in one group and not in the other.

TRAP #10: MISSING IMPORTANT FINDINGS BECAUSE DATA COMBINE POPULATIONS

Admission was offered to 44.3% of the men but only 34.6% of the women who applied to a graduate school (see Table 45.3; Bickel, Hammel, & O'Connell, 1975). The ratio is 1.28, with a 95% CI ranging from 1.22 to 1.34. The P value is less than 0.0001.

While these data seem to be convincing proof of sexism, in fact the data are very misleading. The problem is that the data pool results from many graduate programs. That would make sense if students applied to many graduate programs and if admissions decisions were centralized. But that is not the way graduate admissions work. Students choose the programs to which they apply, and each program makes its own admissions decisions. Therefore, it is essential to analyze the data for each program individually. When the investigators did this, they found that the admission rates of men and women were about the same. The difference was not statistically significant in 75 of the 85 programs (see Table 45.4). In four programs, the difference was statistically significant, and women were less likely to be admitted. In six programs, the difference was statistically significant, and women were more likely to be admitted. Overall, the authors found no evidence of sexism. So what's going on? Why do the pooled data suggest sexism?

Some graduate programs accept a high fraction of applicants and others accept a low fraction of applicants. The two departments most popular among women admitted only 34% and 24% of the applicants, respectively, whereas the two departments most popular among men admitted 62% and 63% of the applicants, respectively (Freedman, 2007). A smaller fraction of women were admitted overall for a simple reason: women tended to apply to more selective programs than did men.

This is a classic example of *Simpson's paradox*. Analysis of pooled data can lead to misleading results.

The same problem probably happens in some medical studies. Imagine how little progress you'd make if you tested whether a new drug cured cancer by pooling patients with all kinds of cancers. Cancers are a group of many diseases that respond to different drugs. Combining all those diagnoses into one study would lead to frustration and inconclusive results. Most likely, many medical conditions (including, perhaps, septic shock, breast cancers and autism) really are combinations of distinct disorders. Until we figure out how to identify the individual diseases, studies of therapies are likely to be ambiguous.

TRAP #11: INVALID MULTIPLE REGRESSION ANALYSES AS A RESULT OF AN OMITTED VARIABLE

This example (extended from one presented by Freedman, 2007) is a bit silly but makes an important point. Pretend that your goal is to find a model that predicts the area of a rectangle from its perimeter.

Figure 45.9 shows that rectangles with larger perimeters also tend to have larger areas, with two outliers. Figure 45.10 fits the remaining points (after removing the "outliers") to possible models. The straight-line model (Figure 45.10, left side) might be adequate, but the sigmoid-shaped model (Figure 45.10, right side) fits the data better.

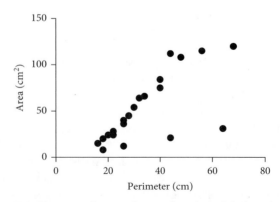

Figure 45.9. The area of rectangles as a function of their perimeter.

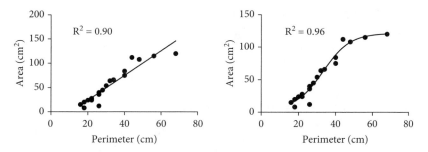

Figure 45.10. Can we fit a model to predict area from perimeter?
We removed two outliers from Figure 45.9 and fit the curve to a straight line (left) or a sigmoid-shaped curve (right).

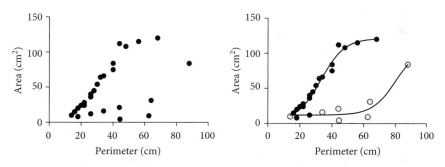

Figure 45.11. Two kinds of rectangles, maybe?
Now we have collected data from more rectangles and realize those two outliers were not really so unusual (left). Instead, it seems that there are two different kinds of rectangles requiring two different models (right).

Figure 45.11 adds data from more rectangles. Now it seems that those two outliers were really not so unusual. Instead, it seems that there are two distinct categories of rectangles. The right side of Figure 45.11 tentatively identifies the two types of rectangles with open and closed circles and fits each to a different model.

While this process may seem like real science, it is not. Two rectangles with the same perimeter can have vastly different areas depending on their shapes. It simply is not possible to predict the area of a rectangle from its perimeter. The area must be computed from both height and width (or, equivalently, from perimeter and either height or width, or from perimeter and the ratio of height/ width). An important variable (either height, width, or their ratio) encoding the shape of the rectangle was missing from our analysis. Understanding these data required simple thinking to identify the missing variable, rather than fancy statistical analyses. A missing variable that influences both the dependent and independent variables is called a *lurking variable*.

TRAP #12: NOT RECOGNIZING HOW COMMON FALSE POSITIVE FINDINGS ARE

Chapter 22 pointed out that statistically significant results can occur by chance, even when the null hypothesis is true. This is called making a Type I error. Chapter 18 showed you how to compute the fraction of all statistically significant findings that are Type I errors based on three values:

- Definition of statistical significance, α. If this is set to a tiny value (say 0.001 rather than the usual 0.05), it is much more likely that statistically significant findings will be real.
- Power. When the experiment has low power, a smaller fraction of statistically significant results will be true.
- The scientific context. Consider the two extreme cases. If you do a series of positive control experiments in which you know for sure that the effect is real, then every single statistically significant result will definitely be real. There can be no false positives. The other extreme would be running a series of negative controls, in which you know for sure that the null hypothesis is true. Among these experiments, every statistically significant result must be a Type I error, a false positive. Actual experiments fall between these two extremes. Statistically significant results from focused experiments based on solid science and prior data are more likely to be true. Statistically significant results from unfocused experiments are more likely to be false positives.

Table 45.5 shows one scenario that uses the standard definition of statistical significance ($P < 0.05$) and a standard goal for power (80%) and assumes a scientific context in which there is a real difference (or effect) in 6% of the experiments and the null hypothesis is actually true in the other 94%.

The numbers in the table are easy to compute. Of 1,000 experiments, there is a real effect (given our assumption) in 60, or 6%. Of those 60, we will detect a statistically significant effect in 48, or 80% (the power). In the other 94% of

	STATISTICALLY SIGNIFICANT: REJECT NULL HYPOTHESIS	NOT STATISTICALLY SIGNIFICANT: DO NOT REJECT NULL HYPOTHESIS	TOTAL
No real effect (null hypothesis true)	47	893	940
Effect is real	48	12	60
Total	95	905	1,000

Table 45.5. Results of 1,000 comparisons with 80% power, a 5% significance level, and a prior probability of 6%.

Of the 60 comparisons for which there really is an effect, 80% (the power), or 48, will be statistically significant. Of the remaining 940 experiments for which there is really no effect, 5%, or 47, will be statistically significant. Thus, half (47/95) of the statistically significant results are false.

FACTOR	EXPLANATION
Small sample size	Small studies have less power than large ones. Statistically significant results from studies with lower power are more likely to be false than studies with lots of power.
Reported effect is small	Studies have less power to detect small effects than large ones. Statistically significant results from studies with lower power are more likely to be false than studies with lots of power.
Many comparisons; only report the ones that are statistically significant	Chapter 23 has explained how easy it is to be fooled by multiple comparisons.
Flexible data analysis without a plan	Studies that analyze data flexibly, without a rigid analysis protocol established before the data were collected, pose challenges. Chapter 23 already warned about this problem.
Biased investigators with financial motives	If the investigators have a financial motive for the results to be statistically significant, they will try harder to extract a statistically significant conclusion from the data.
Biased investigators, who are opinionated	Investigators with strong opinions about how the result is supposed to come out may tend to accept and publish results that go the "right way" and reanalyze or ignore data that go the "wrong way." These investigators are more likely to publish work that turns out to be wrong than are open-minded investigators.
Hot field	When a field is hot, investigators may rush to publish statistically significant findings and tend to spend less time running all the needed controls.

Table 45.6. Factors that increase the chance that a published statistically significant finding will turn out not to be true (inspired by Ioannidis, 2005).

experiments, the null hypothesis is true. Given the standard definition of statistical significance, random sampling will lead to a statistically significant conclusion in 47 (or 5%) of the experiments. Thus, $47/(47 + 48) = 50\%$ of the statistically significant results in this scenario will be false positives.

Many experiments are designed to have less than 80% power, and the prior probability of a true effect could possibly be less than 6%. Thus, it is quite possible to report statistical significance in a situation in which the chance is much greater than 50% that the finding is a false positive.

Ioannidis (2005) argues that *most* published statistically significant research findings are false and lists factors that increase the chance that a published result is false (see Table 45.6). Empirical research confirms this point. Two drug companies have tabulated systematic efforts to reproduce published preclinical findings and have discovered that only a small fraction of published findings can be reproduced (Prinz, Schlange, & Asadullah, 2011; Begley & Ellis, 2012).

CHAPTER SUMMARY

• Don't get distracted by P values and conclusions about statistical significance without also thinking about the size of difference or correlation, and its precision (as assessed by its confidence interval).

- Beware of HARKing (hypothesizing after results are known). Testing a hypothesis proposed as part of the experimental design is quite distinct from "testing" a hypothesis suggested by the data.
- Beware of p-hacking. You cannot interpret P values at face value if the investigator analyzed the data many different ways.
- Correlation does not prove causation.
- Beware of surrogate or proxy variables. A treatment can improve laboratory results while harming patients.
- If you want to know if an intervention makes a difference, you need to intervene. Observational studies can be useful but rarely are definitive.
- Don't fall for the ecological fallacy. If the data were collected for groups, the conclusion can only be applied to the set of groups, not to the individuals within the groups.
- Ask about more than averages. Be on the lookout for differences in variability.
- The difference between "significant" and "not significant" is not itself statistically significant.
- If the study ignores a crucial variable, it doesn't really matter how fancy the analyses are.
- False positive conclusions are much more common than most people realize.

TERMS INTRODUCED IN THIS CHAPTER

- Double dipping (p. 425)
- Ecological fallacy (p. 433)
- Exploratory data analysis (p. 424)
- HARK, Hypothesizing After the Results are Known (p. 425)
- Lurking variable (p. 437)
- Proxy variable (or surrogate variable) (p. 429)
- Simpson's paradox (p. 436)
- Vampirical hypotheses (p. 425)

CHAPTER 46

❖❖❖

Capstone Example

I must confess to always having viewed studying statistics as similar to a screening colonoscopy; I knew that it was important and good for me, but there was little that was pleasant or fun about it.

ANTHONY N. DEMARIA

This chapter was inspired by Bill Greco, who is its narrator and coauthor. One definition of capstone is "a final touch," and this chapter reviews many of the statistical principles presented throughout this book. It also demonstrates the usefulness and versatility of simulations. Although based on a true situation, many of the elements of the storyline, peoples names, and drug names have been changed to protect the innocent.

THE CASE OF THE EIGHT NAKED IC$_{50}$s

Then

It was Friday, May 2, 1975. It was a chilly spring day in Buffalo, New York. Gerald Ford was the president of the United States. There were no PCs on the desks of scientists. The Internet hadn't been invented yet. I was a new graduate student who had just received an A in a beginning biostatistics course.

An interim chairman, Dr. Jeremy Bentham, asked me for help. Two months ago, he had submitted a manuscript that compared the potency of two drugs to block the proliferation of cultured cancer cells. The key results are shown in Table 46.1, which tabulates the amount of drug needed to inhibit cell growth by 50% (the IC$_{50}$).

The average IC$_{50}$ for trimetrexate (TMQ) was 21 nM and that for methotrexate (MTX) was 2.4 nM. A smaller IC$_{50}$ corresponds to a more potent drug, because less drug is needed to inhibit growth. MTX was 8.8-fold (21 nM/2.4 nM) more potent than TMQ. MTX was the standard dihydrofolate reductase (DHFR) inhibitor used in the clinic; TMQ was a new lipophilic DHFR inhibitor with "better" cellular uptake properties.

The investigator was sure of the conclusion based on his experience with these types of agents and the specific cell growth inhibition assay. He was very confident that in this KB cell line, TMQ was about 10-fold less potent than MTX.

TMQ (nM)	MTX (nM)
5.5	0.83
12	1.1
19	1.9
47	5.8

Table 46.1. Sample IC50 data for this example.

Human KB cells were grown in monolayer culture in glass T-25 flasks with RPMI 1640 medium and 10% fetal calf serum for seven days. Each IC_{50} was calculated from the concentration-effect curve from a separate experiment, consisting of five control flasks (no added drug) and three flasks at each of seven different logarithmically spaced concentrations of TMQ or MTX. Cell growth was assessed by measuring total protein content with the Lowry assay.

Each of the eight experiments had taken one week to conduct. All eight experiments were conducted with the same passage of the KB cell line over the six-month period preceding the submission of the manuscript. The IC_{50}s for TMQ and MTX of 19 and 1.9 nM, respectively, were determined from a joint experiment with shared controls. The other six IC_{50} determinations were made at different times and preceded this last paired experiment. The overall theme of the manuscript was a comparison of the two DHFR inhibitors against the single KB human cell line regarding antiproliferative potency; cellular uptake; metabolism; and inhibition of the target enzyme, DHFR.

As part of the usual process of peer review, the journal's editor sent the manuscript to several anonymous reviewers. The paper was rejected largely because of the following comment by one reviewer: "This reviewer performed an unpaired Student's t test on the data [see Table 46.1], and found that there is no significant difference between the mean IC_{50} of TMQ and MTX (P = 0.092) against KB cells. The data do not support your conclusion."

Not believing the reviewer, Dr. Bentham asked me to check the calculations.

Using the formulas and tables in my biostatistics text and a scientific calculator, I performed a standard unpaired two-tailed Student's t test (with the usual assumption of equal error variance for the two populations). I found that t = 2.01 and P = 0.092, in total agreement with the reviewer. Because P > 0.05, I informed the investigator that the reviewer was correct: the difference in potency was not statistically significant.

Because the P value was 0.092, the investigator asked, doesn't that mean there is a (1.000 − 0.092) = 0.908, or 90.8%, chance that the findings are real? I explained to him that a P value cannot be interpreted that way. The proper interpretation of the P value is that even if the two drugs really have the same potency (the null hypothesis), there is a 9.2% chance of observing this large a discrepancy (or larger still) by chance. If I had had a crystal ball, perhaps I could have shown him Chapter 19 of this text!

I took the analysis one step further and computed the CI of the difference. The difference in observed IC_{50}s was −18.5 nM (the order of subtraction was arbitrary, so the difference could have been +18.5 nM; it is important only to be consistent). The 95% CI for the difference was from −41.0 to +4.06 nM. I pointed out that the lower confidence limit was negative (MTX was more potent) and the upper confidence limit was positive (TMQ was more potent), so the 95% CI included the possibility of no difference in potency (equal IC_{50} values; the null hypothesis). This was consistent with the P value (see Chapter 17). When a P value is greater than 0.05, the 95% CI must include the value specified by the null hypothesis (zero, in this case).

The investigator became enraged. He thought the data in the table would convince any competent pharmacologist that TMQ is, beyond any reasonable doubt, less potent than MTX. Statistics weren't needed, he thought, spouting off half-remembered famous quotes: "When the data speaks for itself, don't interrupt." "There are liars; there are damn liars; and then there are statisticians." "If you need statistics, you've done the wrong experiment."

The investigator sent the manuscript unaltered to another journal, which accepted it.

Now

It is Friday, January 23, 2009. It is a chilly winter day in Buffalo, New York. Barack Obama has just become the president of the United States. There are small, powerful PCs on the desks of virtually all scientists. Powerful statistical PC software is available, at reasonable or no cost, to all scientists. The Internet is a major part of professional and personal life. I am the instructor of an applied biostatistics course. I use *Intuitive Biostatistics* as the main textbook in the course. I just retrieved the published paper on TMQ and MTX. As I stare at the IC_{50} table reproduced here as Table 46.1, it stares back at me.

The mismatch of statistical reasoning and common sense gnawed at me. I knew the statistical calculations were correct. But I also felt the senior and experienced investigator's scientific intuition and conclusion were also probably correct. Something was wrong, and this chapter is the culmination of decades of angst. I must resolve this 34-year-old quandary. How should I have analyzed these data back in 1975? How could I analyze these data now in 2009?

Before reading on, think about how you would analyze these data.

LOOK BEHIND THE DATA

If I were asked to consult on this project today, I would not begin with Table 46.1. Instead, I would go back to the raw data. Each of the values in Table 46.1 came from analyzing an experiment that measured cell number at one time point in the presence of several different concentrations of drug. The IC_{50} is defined as the concentration of drug that inhibits cell growth to a value halfway between the maximum and minimum values. This raises many questions:

- What are the details of the cell proliferation assay? Which details provide the opportunity for experimental artifacts? Were cells actually counted? Or did they measure some proxy variable (perhaps protein content) that is usually proportional to cell count? Over what period of time was cell growth measured? Has the relationship between cell growth and time been well characterized in previous experiments? Does cell growth slow down because of cell-to-cell contact inhibition (or depletion of nutrients) as the flasks fill with cells? Are the different drug treatments randomly assigned to the flasks? Is it possible that cell growth depends on where each flask is placed in the incubator?
- How was 100% cell growth defined—as growth with no drug? What if growth was actually a bit faster in the presence of a tiny concentration of drug? Should the 100% point be defined by averaging the controls or by fitting a curve and letting the curve-fitting program extrapolate to define the top plateau?
- How was 0% cell growth defined? Is 0% defined to be no cell growth at all or cell growth with the largest concentration of drug used? What if an even larger concentration had been used? Or was 0% defined by a curve-fitting program that extrapolated the curve to infinite drug concentration? The answer to this seemingly simple technical question can have a large influence on the estimation of drug potency. The determination of the IC_{50} can be no more accurate than the definitions of 0 and 100%.
- How was the IC_{50} determined? By hand with a ruler? By hand with a French curve? By linearizing the raw data and then using linear regression? By fitting the four-parameter concentration-effect model (see Chapter 36) to the raw data using nonlinear regression? If so, with what weighting scheme?

Here, we will skip all those questions and assume that the values in Table 46.1 are reliable. But do note that many problems in data analysis actually occur while planning or performing the experiment or while wrangling the data into a form that can be entered into a statistics program.

STATISTICAL SIGNIFICANCE BY CHEATING

One way to solve the problem is to report a one-tailed P value, rather than a two-tailed P value. It isn't too hard to justify that decision. All I have to do is state that the experimental hypothesis was to confirm that TMQ is less potent than MTX. The one-tailed P value is half the two-tailed P value, so it is 0.046. That is less than 0.05, so now the difference is statistically significant. Mission accomplished. That was easy!

Of course, this is cheating. To justify a one-tailed P value, the decision to report a one-tailed P value would need to have been made before the data were collected. It wasn't. At that time, it would also be necessary to decide that any data showing that TMQ is more potent than MTX would be attributed to chance and

Figure 46.1. The data shown in Table 46.1.

not worth pursuing. In reality, if the results had gone the other way, with MTX being substantially less potent, the investigator probably would have found an explanation and reported the findings.

USING A t TEST THAT DOESN'T ASSUME EQUAL SDs

One of the assumptions for the unpaired t test—that the data are sampled from populations with equal SDs—seems dubious with these data. Look at Figure 46.1. The TMQ data are way more scattered. This can be quantified by comparing the two SDs. The sample SD for the TMQ data was 18.3 nM, whereas the sample SD for the MTX data was 2.31 nM.

Chapter 30 provides a formal way to test the assumption of equal variances Calculate F as the square of the ratio of standard deviations, $(18.3/2.31)^2 = 62.7$. If the null hypothesis that the two populations have the same standard deviation were true (so the observed discrepancy is the result of chance) what is the chance that F would be that high or higher? The answer depends on the df of the numerator and denominator, which are both are $n - 1$, or 3. Calculate P using the free GraphPad QuickCalc (Appendix A) or the Excel formula = FDIST(62.7,3,3). The P value is 0.0033. That is the probability that the first group would have a standard deviation so much larger than the second group by chance alone. Doubling that value to also allow for the chance that the second group would have had the larger standard deviation, produces a P value of 0.0066. With such a low P value, it makes sense to reject the null hypothesis of equal standard deviations.

Now what? A modified t test allows for unequal SDs. With this modified test, the t value is still 2.01, but the P value is now 0.136, with df = 3. The P value is higher, but perhaps more appropriate, because it does not make an assumption that is probably false. However, according to Moser and Stevens (1992), the decision to use this test should be made as part of the experimental design and should not be based on the results of the F test to compare SDs.

This, perhaps, is a more appropriate way to analyze the data, but it doesn't really help. It doesn't bridge the gap between the result that is scientifically obvious and the result computed by formal statistics. In addition, the P value moved in the "wrong" direction! I know that it is cheating to just run a bunch of rival statistical tests and then pick the one that yields the smallest P value, but the increase in P value with the modified t test surely suggests that something is fundamentally wrong with my thinking!

UNPAIRED t TEST AS LINEAR OR NONLINEAR REGRESSION

Hidden in Chapters 35 and 36 is an intriguing regression approach that might be appropriate for these data. I defined X to equal zero for the TMQ group and one for the MTX group and defined Y to be the individual IC_{50} values. I then used linear regression to fit a straight line to the data, recast as the X,Y pairs shown in Table 46.2. The slope reported by linear regression equals the difference in Y values divided by the difference in X values. The X values for the two drugs differ by 1.0, so the slope quantifies the difference in mean IC_{50}s between TMQ and MTX.

Linear regression calculated that the best-fit slope equals –18.5 nM, with a standard error of 9.21 nM. The 95% CI for the true slope (difference in mean IC_{50}) is –41.0 to 4.06 nM. The associated P value (testing the null hypothesis that the population slope equals zero) is 0.092. These are exactly the same as the results calculated by the simple unpaired t test. I expected this, because Chapter 35 emphasized that the two approaches are equivalent. However, the regression approach is more versatile, because it lets me compensate for nonuniform variance by appropriately weighting the data. Data with more scatter can be given less weight, and data with less scatter can be given more weight. Unfortunately, most linear regression software does not permit this type of weighting, so I switched to nonlinear regression software (which can also fit straight lines).

To apply this weighted regression approach, I fit a straight line to the X,Y data using weighted nonlinear regression (see Chapter 36). I assumed that both sets of data were drawn from Gaussian populations but that the standard deviations of these populations are not equal but instead are proportional to their mean. To fit this model, I asked the regression program to minimize the square of the relative distance of each point from the fitted line (weighting with the reciprocal

X: DRUG	Y: IC_{50} (nM)
0	5.5
0	12
0	19
0	47
1	0.83
1	1.1
1	1.9
1	5.8

Table 46.2. Data as entered into a regression program.
The X value was defined to be zero for the drug TMQ and one for the drug MTX.

of the square of the predicted Y). The slope of this line is the difference between IC_{50} values, which equals –18.5 nM, with a 95% CI ranging from –42.1 to 5.12 nM. One end of the CI is negative, which corresponds to the IC_{50} of MTX being smaller. But the other end of the CI is positive, which means the IC_{50} of TMQ is smaller. Like before, this method concludes that the data are consistent with no difference in IC_{50}. Because the 95% CI includes zero, the P value must be higher than 0.05. In fact, it is 0.104.

In summary, this alternative logical approach lengthens the CI and increases the P value! I don't seem to be making any real progress.

NONPARAMETRIC MANN–WHITNEY TEST

The t test assumes that the data were sampled from populations with Gaussian distributions. With only four values in each group, it is impossible to seriously assess this assumption. Normality tests (see Chapter 24) don't work with such tiny samples.

In fact, I find that normality tests are rarely helpful when used as part of an analysis of a particular experiment, but they can be quite useful when characterizing an assay. I prefer to collect many values (dozens to hundreds) while characterizing (developing) an experimental assay. I then examine the distribution of these values using various tests (including normality tests) to decide how these kinds of data should be analyzed. That decision then applies to all experiments that use the assay. In this example, we don't have the large number of measurements we'd need to tell us whether the Gaussian assumption is reasonable for this kind of data.

Rather than assuming the data were sampled from Gaussian populations, we can use the nonparametric Mann–Whitney test instead. With this test, P = 0.057. Although the P value is closer to 0.05, it is still greater than 0.05. Sure, using 0.05 as a threshold for defining statistical significance is arbitrary, but it is widely accepted. It is also tempting to just erase that last digit and call it 0.05—but that would be cheating. Rounded to two decimal places, the P value is 0.06.

Although you could argue that 0.057 is close enough to "tentatively suggest" that the potencies are different for the two drugs, this conclusion is somehow unsatisfying. In my gut, I really know that TMQ is less potent than MTX.

JUST REPORT THE LAST CONFIRMATORY EXPERIMENT?

I considered just presenting the data from the last paired experiment, in which TMQ and MTX were studied simultaneously, with a set of shared control flasks. The individual IC_{50}s for TMQ and MTX were 19 and 1.9, respectively, with a potency ratio of 10. Perhaps I might consider the first six individual experiments exploratory and the last paired experiment confirmatory.

This last paired experiment's potency ratio is close to the overall potency ratio of the mean IC_{50}s of 8.7. Perhaps it would be best to eliminate the table and just list in the text the individual IC_{50}s for TMQ and MTX, as well as the potency ratio for this last confirmatory experiment, with a statement that this experiment was representative of six previous individual exploratory experiments with the two agents.

This seems to be a scientifically ethical approach to the dilemma. However, because I like to think of myself as a stubborn, honest, and thorough scientist, I suspect that there is a better solution—one that involves calculating, rather than avoiding, statistics. I want to honestly showcase and analyze all of the experimental work.

Studying TMQ and MTX in paired experiments with common controls in order to assess their potency ratio is a good idea. Pairing (see Chapter 31) is usually recommended because it reduces variability and so increases power. This example combines data from paired experiments with data from unpaired experiments, which makes the analysis awkward.

INCREASE SAMPLE SIZE?

Of course, we'd like to get the results we want without collecting more data. But maybe more data are required. Equation 26.2 computes the approximate sample size per group (n) needed to detect a specified difference (w) given an anticipated SD (s) with specified power and significance level.

$$n = 2C\left(\frac{s}{w}\right)^2$$

To compute a sample size, all we need are values for C, s, and w.

Look up the value of C in Table 26.1. For 80% power and a statistical significance level (α) of 5%, C = 7.8.

For s, we must enter the anticipated SD. In many situations, we may have lots of prior data to use to estimate this. Here, we have only the data at hand. Our two samples have different standard deviations. The sample SD for the TMQ data was 18.3 nM, whereas the sample SD for MTX was 2.31 nM. One thought might be to average the two and enter s = 10.3. Actually, with the same sample size in both groups, it is better to use the square root of the average of the variances (the square of the SDs). With this procedure, s is approximately 13 nM.

For w, we must enter the smallest difference between mean potencies that we'd find scientifically interesting. The choice of this number is a bit arbitrary, but to keep the calculations easy, let's set it to 18 nM, which is a bit smaller than the observed difference of 18.6 nM.

$$n = 2C\left(\frac{s}{w}\right)^2 = 2 \cdot 7.8 \cdot \left(\frac{13}{18}\right)^2 = 8.1 \approx 8 \text{ per group}$$

So we need to run eight experiments per drug, and each of those experiments requires 26 flasks of cells measured at multiple time points. This would be an outrageously large amount of work to answer a relatively trivial scientific question for which I am quite sure that I already know the answer.

COMPARING THE LOGARITHMS OF IC$_{50}$ VALUES

Lognormal data

IC$_{50}$ values tend to follow lognormal distributions (see Chapter 11) rather than Gaussian distributions. This leads to larger SDs when the mean is larger, as we saw. It can also lead to false identification of outliers (see Chapter 25).

Analyzing lognormal data is simple. I first calculated the (base 10) logarithms of the eight IC$_{50}$s. The results are shown in Table 46.3 and Figure 46.2. The mean log(IC$_{50}$) for TMQ is 1.19 and that for MTX is 0.251. The variability of the two sets of log(IC$_{50}$) values is very similar, with nearly identical sample SDs (0.389 and 0.373). Thus, there is no problem accepting the assumption of the standard two-sample unpaired t test that the two samples come from populations with the same SDs.

Reversing the logarithmic transform (i.e., taking the antilog; see Appendix E) converts the values back to their original units. Taking the antilog is the same as taking 10 to that power, and taking the antilog of a mean of logarithms computes the geometric mean (see Chapter 11). The corresponding geometric mean for the IC$_{50}$ for TMQ is 15.6 nM and that for MTX is 1.78 nM.

TMQ	MTX
0.740	−0.081
1.079	0.041
1.279	0.279
1.672	0.763

Table 46.3. The logarithms of the IC$_{50}$ values of TMQ and MTX.
These are the logarithms of the values (in nM) shown in Table 46.1.

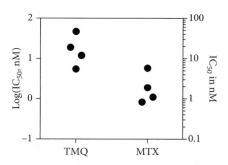

Figure 46.2. The logarithms of the data shown in Figure 46.1.
The left axis shows the logarithms, and the right axis shows the antilogarithms. Comparing the two lets you review the meaning of logarithms. The logarithm of 100 is 2, because $10^2 = 100$. The logarithm of 1.0 is 0.0, because $10^0 = 1.0$. The logarithm of any value between 0 and 1 is negative.

Unpaired t test of log-transformed data

The calculated t ratio is 3.49 and the corresponding P value is 0.013, df = 6. MTX is more potent than TMQ and that discrepancy is statistically significant (according to the conventional definition).

The difference between the $\log(IC_{50})$ s is 0.942. Transform this to its antilog to turn it into a potency ratio (remember that the difference between the logarithms of two values equals the logarithm of the ratio of those two values). The antilog is 10 to that power, or $10^{0.942}$, which equals 8.75. In other words, MTX is almost nine times more potent than TMQ. It require one ninth as much MTX to have the same effect as TMQ.

95% CI

The 95% CI for the difference in mean $\log(IC_{50})$ ranges from 0.282 to 1.60. Transform each limit to its antilog (10 to that power) to determine the 95% confidence limits of the population potency ratio, 1.91 to 39.8. This interval does not encompass 1.0 (the value of the null hypothesis of equal drug potency), so it is consistent with a P value of less than 0.05. With 95% confidence, we can say that MTX is somewhere between twice as potent as TMQ and 40 times as potent. Analyzed this way, the intuition of the investigator is proven to be correct.

Why this is the optimal method

This is probably the best simple way to analyze the data. The t test of the untransformed IC_{50} values is inferior to the t test of the log-transformed IC_{50} values for three important reasons.

- The log transform equalized the two SDs and thus eliminated the assumption violation of unequal variances.
- As suggested in Chapter 11, the log transformation changed the distribution of IC_{50}s from a lognormal distribution to a Gaussian distribution, another assumption of the t test. With only two sets of data, each with only four values, we really don't know much about the distribution of the population of IC_{50}s. But extensive past experience with concentration-effect experiments tells us to expect a lognormal distribution.
- The t test asks about differences between means. The t test on log-transformed data essentially asks about ratios. Pharmacologists universally ask about ratios of drug potencies, not differences. Transforming to logarithms helps match the statistical method to the scientific question. The t test of the log-transformed IC_{50} values is a nice statistical translation of the scientific question, What is the relative potency of TMQ and MTX?

If I had access to the original raw concentration-effect data, I would likely try to fit a variant of the four-parameter concentration-effect model showcased in Chapter 36 to all of the data at once using nonlinear regression. I would take into account differences among the experiments run at different times and would estimate the overall potency ratio with an accompanying 95% CI. I would take advantage of the true sample size of 203 flasks from eight experiments in my analysis. However, in this case, I don't have the original raw data; I just have the eight naked IC_{50}s shown in Table 46.1.

SAMPLE SIZE CALCULATIONS REVISITED

Now that we realize the data follow a lognormal distribution, let's go back and redo the sample size calculations. On a log scale, we found that the SD of both groups was 0.38, so we enter into the sample size formula a conservative 0.40 as s. How large a difference in $\log(IC_{50})$ are we looking for? Let's say we are looking for an eightfold difference in potency, so the $\log(IC_{50})$ values differ by $\log(8)$, which equals about 0.9. Set w = 0.9.

$$n = 2C\left(\frac{s}{w}\right)^2 = 2 \cdot 7.8 \cdot \left(\frac{0.4}{0.9}\right)^2 = 3.1 \approx 3 \text{ per group}$$

This equation is an approximation that works well with large n. Here, n is tiny and that equation is not accurate. A more complicated equation (by a computer program designed for that purpose) suggests a sample size of four per group. So once the analysis was done properly, we can see that the original sample size was indeed sufficient.

This might be stated in a paper as follows: a sample size of four in each group was computed to have 80% power to detect as statistically significant ($P < 0.05$) an eightfold difference in potency, assuming that $\log(IC_{50})$ values follow a Gaussian distribution with a standard deviation of 0.40.

IS IT OK TO SWITCH ANALYSIS METHODS?

This whole exercise to search for a better statistical analysis than was suggested by the reviewer has left me frustrated.

Data analysis should not be a process of fishing for statistical significance. It is not ethical to conduct multiple statistical procedures on the same data set for the purpose of finding the statistical approach that yields the lowest P value. This whole procedure of analyzing the data this way, then that way, and then another way seems to be a form of scientific fraud.

At the same time, the logarithm method makes scientific sense. As explained in the previous section, this is the optimal method for comparing IC_{50} values. The method matches the scientific context and goals, whereas the other methods do not. Of course, the decision to analyze the data in this way should have been planned before the data were collected. But it wasn't. Is it OK to switch methods after the fact?

Here's a reality check: this happens all the time in basic research. Many scientists never even think about how to analyze the data until after the experiments are done. But just because something is done commonly doesn't mean it is correct. Chapter 23 pointed out how easy it is to be fooled by multiple analyses.

Certainly, clinical trials that are reviewed by the FDA require that the analysis method be precisely specified in advance.

In this case, I have a strong scientific justification for switching to the proper method. Also, I am responding, albeit indirectly, to comments made by an expert reviewer of the manuscript. Does this rationale make it OK? I think so, but I will let you, the reader, decide for yourself. It certainly would be OK if these were preliminary data and if this whole exercise was done as a way to make decisions about how I will analyze data from future experiments.

THE USEFULNESS OF SIMULATIONS

Analyzing log(IC$_{50}$) values has more power

For this example, analyzing the log(IC$_{50}$) values produced a smaller P value than analyzing the IC$_{50}$ values and seems to better match the scientific goals of the study. If the data really do follow a lognormal distribution, how much is lost by analyzing the IC$_{50}$ values rather than the log(IC$_{50}$) values? Our one example gives a hint of the answer, but only a hint. To find out a fuller answer, I performed Monte Carlo simulations.

Table 46.4 shows the simulated results. The first row in Table 46.4 is the result of 1 million simulated experiments. For each, four values were randomly generated to simulate an experimental measurement of potency for one drug, and four other values were randomly generated to simulate the potency of the other drug. All in all, 8 million values were randomly generated to create the first row of Table 46.4.

The data for one drug were simulated with an IC$_{50}$ of 2 nM, so the logarithm of 2 nM, 0.301, was entered as the mean into the simulation program. The data for the other drug were simulated using an IC$_{50}$ of 20 nM, so the logarithm of 20 nM,

N PER GROUP	RELATIVE POTENCIES OF TWO DRUGS	DATA ANALYZED	HOW OFTEN IS P < 0.05?
4	Differ by factor of 10	log(IC$_{50}$)	83.6%
4	Differ by factor of 10	IC$_{50}$	45.6%
7	Differ by factor of 10	log(IC$_{50}$)	99.0%
7	Differ by factor of 10	IC$_{50}$	79.2%

Table 46.4. Simulations in which the drugs have different potencies.
Each row is the result of 1 million simulations. The log(IC$_{50}$) values were simulated using a random number method that generates a Gaussian distribution, and the IC$_{50}$ values were computed by taking the antilog (10 to that power). Therefore, the log(IC$_{50}$) values are sampled from a Gaussian distribution, and the IC$_{50}$ values are sampled from a lognormal distribution. The true population mean IC$_{50}$s of the two groups in each row differ by a factor of 10 (2 and 20 nM), and the samples included either four or seven IC$_{50}$s per group. The differences in mean log(IC$_{50}$) (or IC$_{50}$) were compared using the unpaired t test, with the assumption of uniform variances.

1.301, was entered into the simulation program. Each log(IC$_{50}$)—all 8 million of them—was chosen using a random number method that simulates sampling from a Gaussian distribution, assuming a mean of 0.301 or 1.301 and a standard deviation of 0.40.

These simulations mimic the data presented at the beginning of this chapter. The potency of the two drugs differs by a factor of 10, the log(IC$_{50}$) values vary randomly according to a Gaussian distribution, the SD of the logarithms is 0.4, and n = 4 for each group. The log(IC$_{50}$) values from each simulated experiment were compared with an unpaired t test, and the P value was tabulated.

The last column of Table 46.4 shows the percentage of the simulated experiments in which the P value is less than 0.05, the usual definition of statistical significance. In other words, that last column shows the power of the experimental design (see Chapter 20). For this design, the power is 83.6%. In the remaining simulated experiments (16.4% of them), the P values are greater than 0.05 and so result in the conclusion that the results were not statistically significant. This happens when the means happen to be close together or the scatter among the values in that simulated experiment happens to be large.

The second row of Table 46.4 shows analyses of IC$_{50}$ values. Each IC$_{50}$ was computed by taking the antilog (10 to that power) of a log(IC$_{50}$) generated as explained for the first row. The first row shows the results of a simulation of sampling log(IC$_{50}$) values from a Gaussian distribution, and the second row shows the results of a simulation of sampling IC$_{50}$ values from a lognormal distribution. The IC$_{50}$ values from each simulated experiment were compared with an unpaired t test, and the P value was tabulated.

In only 45.6% of these experiments was the P value less than 0.05. Because the violation of all three of its assumptions (lognormal distribution, unequal variance, differences examined rather than ratios) discourages the use of the standard t test on untransformed IC$_{50}$ values, this approach has much less statistical power (45.6% vs. 83.6%).

Analyzing log(IC$_{50}$) better controls the Type I error rate

Table 46.5 shows simulations identical to those shown in Table 46.4, except that the simulations were done using the same mean IC$_{50}$ for both drugs. In other words, for these simulations, the null hypothesis was true—the drugs had equal potency.

The first row shows the analyses of the log(IC$_{50}$) values. For both drugs, these values were drawn from the same Gaussian distribution, so the assumptions of a t test are met. If the simulations work as they should, 5% of these P values should be less than 0.05, and that is precisely what I observed. No surprise here, but this is a good way to check that the simulations worked properly.

The second row shows the analyses of the IC$_{50}$ values. Only 3.23% of these simulated experiments resulted in a P value less than 0.05, although the null hypothesis is true. The data violate one of the assumptions of a t test (Gaussian distribution) and so the definition of statistical significance is off. When $P < 0.05$ is defined to be statistically significant, you can expect that 5% of experiments

N PER GROUP	RELATIVE POTENCIES OF TWO DRUGS	DATA ANALYZED	HOW OFTEN IS P < 0.05?
4	Equal	$\log(IC_{50})$	5.00%
4	Equal	IC_{50}	3.24%
7	Equal	$\log(IC_{50})$	4.99%
7	Equal	IC_{50}	3.42%

Table 46.5. Simulations in which the null hypothesis is true (drugs have equal potencies).

The simulations were done as explained in Table 46.4, except that these simulations used equal potencies for the two drugs (here, 20 nM, but nearly identical results were obtained when using 2 nM).

under the null hypothesis will end up being statistically significant. The Type I error is defined to be 5%. Here, only 3.24% of the experiments under the null hypothesis resulted in a statistically significant result. The Type I error is only 3.24%. Less Type I error might seem to be a good thing, but it is accompanied by a decrease in power, which is definitely not desirable.

Increasing sample size increases power

The third and fourth rows of Table 46.4 show the results of increasing the sample size to seven per group. The power rose to 99.0% for the t test of the $\log(IC_{50})$s. Is this extra effort worth it? It depends on the context of the work, the cost of the experiments (in money and effort), and the consequences of missing a real difference. For this experiment, I think 80% power is plenty and that it is not necessary to spend the time and money it would take to increase the sample size to seven experiments.

Increasing the sample size to n = 7 increases the power to 79.2% for the t test of the IC_{50}s. This simulation result is similar to our previous sample size calculation with the simple formula.

The third and fourth rows of Table 46.5 show that an increase to n = 7 barely changes the Type I error rates.

Simulating experiments is not hard

These simulations took me an hour or two to set up and less than 20 minutes to compute using a computer purchased in 2008 for about US$1,000. In contrast, it would have taken many hours to complete such simulations on many "mainframe" computers in 1975.

Many software packages can perform these simulations (e.g., GraphPad Prism, Minitab, Excel, R, etc.). I used 1 million simulations per experiment to be compulsive, but I would have gotten virtually the same results with 10,000 or even 1,000 simulated experiments. If you plan to design experiments, it is well worth your time to learn how to run these kinds of simulations. You can save a lot of time in the lab by using simulations to test the design of your experiments.

CHAPTER SUMMARY

- Lognormal distributions are common in biology. Transforming data to logarithms may seem arcane, but it is actually an important first step for analyzing many kinds of data.

- This chapter was designed to review as many topics as possible, not to be as realistic as possible. It rambles a bit. However, it realistically shows that the search for the best method of analyzing data can be complicated. Ideally, this search should occur before collecting the data or after collecting preliminary data but before doing the experiments that will be reported. If you analyze final data in many different ways in a quest for statistical significance, you are likely to be misled.

- Statistical methods are all based on assumptions. The holy grail of biostatistics is to find methods that match the scientific problem and for which the assumptions are likely to be true (or at least not badly violated). The solution to this chapters problem (using logarithms) was not optimal because it yielded the smallest P value but rather because it melded the statistical analysis with the scientific background and goal. It was a nice translation of the scientific question to a statistical model.

- Simulating data is a powerful tool that can reveal deep insights into data analysis approaches. Learning to simulate data is not hard, and it is a skill well worth acquiring if you plan to design experiments and analyze scientific data.

- Statistical textbooks and software aren't always sufficient to help you analyze data properly. Data analysis is an art form that requires experience as well as courses, textbooks, and software. Until you have that experience, collaborate or consult with others who do. Learning how to analyze data is a journey, not a destination.

CHAPTER 47

❖❖❖

Review Problems

I may not have gone where I intended to go, but I think I have ended up where I intended to be.

DOUGLAS ADAMS

These problems do not cover all the material in the book but instead focus on fundamental areas of statistics that are widely misunderstood. Chapter 48 repeats the questions and includes answers.

A. PROBLEMS ABOUT CIS OF PROPORTIONS, SURVIVAL CURVES, AND COUNTS

A1. Of the first 100 patients to undergo a new operation, 6 die. Can you calculate the 95% CI for the probability of dying with this procedure? If so, calculate the interval. If not, what information do you need? What assumptions must you make?

A2. A new drug is tested in 100 patients and lowers blood pressure by an average of 6%. Can you calculate the 95% CI for the fractional lowering of blood pressure by this drug? If so, calculate the interval. If not, what information do you need? What is the CI for the fractional lowering of blood pressure? What assumptions must you make?

A3. You use a hemocytometer to determine the viability of cells stained with trypan blue. You count 94 unstained cells (viable) and 6 stained cells (indicating that they are not viable). Can you calculate the 95% CI for the fraction of stained (dead) cells? If so, calculate the interval. If not, what information do you need? What assumptions must you make?

A4. In 1989, 20 of 125 second-year medical students in San Diego failed to pass their biostatistics course (until they came back for an oral exam). Can you calculate the 95% CI for the probability of passing the course? If so, calculate the interval. If not, what information do you need? What assumptions must you make?

A5. Ross Perot won 19% of the vote in the 1992 U.S. presidential election. Can you calculate the 95% CI for the fraction of voters who voted for him? If so, calculate the interval. If not, what information do you need? What assumptions must you make?

456

A6.   In your city (population = 1 million) last year, a rare disease had an incidence rate of 25 in 10,000 people. Can you calculate the 95% CI for the incidence rate? If so, calculate the interval. If not, what information do you need? What assumptions must you make?

A7.   Is it possible to calculate the median survival time when some of the subjects are still alive?

A8.   Why are survival curves drawn as staircase curves?

A9.   A survival curve includes many subjects who were censored because they dropped out of the study when they felt too ill to come to the clinic appointments. Is the survival curve likely to overestimate or underestimate survival of the population?

A10. A study began on January 1, 1991, and ended on December 31, 1994. How will each of these subjects appear on the graph?
      a. Entered March 1, 1991. Died March 31, 1992.
      b. Entered April 1, 1991. Left study March 1, 1992.
      c. Entered January 1, 1991. Still alive when study ended.
      d. Entered January 1, 1992. Still alive when study ended.
      e. Entered January 1, 1993. Still alive when study ended.
      f. Entered January 1, 1994. Died December 30, 1994.
      g. Entered July 1, 1992. Died of car crash March 1, 1993.

A11. A survival study started with 100 subjects. Before the end of the fifth year, no patients had been censored and 25 had died. Five patients were censored in the sixth year, and five died in the seventh year. What is the percentage survival at the end of the seventh year?

A12. You use a hemocytometer to count white blood cells. When you look in the microscope you see lots of squares, and 25 squares enclose 0.1 μL. You count the number of cells in nine squares and find 50 white blood cells. Can you calculate the 95% CI for the number of cells per microliter? If so, calculate the interval. If not, what information do you need? What assumptions must you make?

A13. In 1988, a paper in *Nature* (Davenas et al., 1988) caused a major stir in the popular and scientific press. The authors claimed that antibodies diluted even to 10^{-120} of the starting concentration stimulated basophils to degranulate. With that much dilution, the probability that even a single molecule of antibody would remain in the tube is infinitesimal. The investigators hypothesized that the water somehow "remembered" that it had seen antibody. These results purported to give credence to homeopathy, the theory that extremely low concentrations of drugs are therapeutic.
      The assay to test such a hypothesis is simple. Add a test solution to basophils and incubate. Then stain. Count the number of stained cells in a certain volume. Compare the number of cells in control and treated tubes. A low number indicates that many cells have degranulated, because cells that haven't degranulated won't take up the stain.
      The authors present the "mean and standard error of basophil number actually counted in triplicate." In the first experiment, the three control

tubes are reported as 81.3 ± 1.2, 81.6 ± 1.4, and 80.0 ± 1.5. The results from a second experiment conducted on another day are very similar. The tubes treated with the diluted antibody give much lower numbers of stained cells, indicating that the diluted antibody has caused degranulation. For this problem, think only about the control tubes.
a. Why are these control values surprising?
b. How could these results have been obtained?

B. PROBLEMS ABOUT SD, SEM, CIS, AND LOGNORMAL DISTRIBUTIONS

B1. Estimate the SD of the age of the students in a typical medical school class (just a rough estimate will suffice).

B2. Estimate the SD of the number of hours slept each night by adults.

B3. The following cholesterol levels were measured in 10 people (mg/dl):

260, 150, 160, 200, 210, 240, 220, 225, 210, 240

Calculate the mean, median, variance, SD, and CV. Make sure you include units.

B4. Add an eleventh value (931) to the numbers in Question B3 and recalculate all values.

B5. The level of an enzyme in blood was measured in 20 patients, and the results were reported as 79.6 ± 7.3 units/ml (mean ± SD). No other information is given about the distribution of the results. Estimate what the probability distribution might have looked like.

B6. The level of an enzyme in blood was measured in 20 patients and the results were reported as 9.6 ± 7.3 units/ml (mean ± SD). No other information is given about the distribution of the results. Estimate what the probability distribution might have looked like.

B7. The Weschler IQ scale was created so that the mean is 100 and the standard deviation is 15. What fraction of the population has an IQ greater than 130?

B8. An enzyme level was measured in cultured cells. The experiment was repeated on three days, and on each day, the measurement was made in triplicate. The experimental conditions and cell cultures were identical on each day; the only purpose of the repeated experiments was to determine the value more precisely. The results are shown in Table 47.1 as enzyme activity in units per minute per milligram of membrane protein. Summarize these data as you would for publication. The readers have no interest in the individual results for each day, just in one overall

	REPLICATE 1	REPLICATE 2	REPLICATE 3
Monday	234	220	229
Tuesday	269	967	275
Wednesday	254	249	246

Table 47.1. Data for Problem B8.

mean with an indication of the scatter. Give the results as mean, error value, and n. Justify your decisions.

B9.   A paper reports that cell membranes have 1,203 ± 64 (mean ± SEM) fmol of receptor per milligram of membrane protein. These data come from nine trials of an experiment.

    a. Calculate the 95% CI. Explain what it means in plain language.

    b. Calculate the 90% CI.

    c. Calculate the CV.

B10. You measure blood pressure in 10 randomly selected subjects and calculate that the mean is 125 and the SD is 7.5 mmHg. Calculate the SEM and 95% CI. Then you measure blood pressure in 100 subjects randomly selected from the same population. What values do you expect to find for the SD and SEM?

B11. Does it ever make sense to calculate a CI from a population SD (with a denominator of n) as opposed to the sample SD (with a denominator of [n − 1])?

B12. Data were measured in 16 subjects, and the 95% CI of the mean ranged from 97 to 132. Calculate the 99% CI.

B13. The standard error of the difference between two means is larger than the SE of either mean. Why?

B14. Pullan et al. (1994) investigated the use of transdermal nicotine for treating ulcerative colitis. The plasma nicotine level at baseline was 0.5 ± 1.1 ng/ml (mean ± SD; n = 35). After six weeks of treatment, the plasma level was 8.2 ± 7.1 ng/ml (n = 30). Is it more likely that the nicotine levels are sampled from a Gaussian or a lognormal distribution?

C. PROBLEMS ABOUT P VALUES AND STATISTICAL SIGNIFICANCE

C1.   You conduct an experiment and calculate that the two-tailed P value is 0.08. What is the one-tailed P value? What assumptions are you making?

C2.   The FDA requires two-tailed statistical tests for Phase III new drug registration trials, defining $P < 0.05$ as statistically significant. What would happen if they required one-tailed P values instead, still using the threshold of 0.05 to define significance?

C3.   All members of a class of 111 pharmacy practice majors volunteered for a study in which a single blood sample was taken from each student at a single session. All blood samples were analyzed by the same technician for the concentration of a natural human biochemical in plasma, 17β-estradiol. Gender was recorded for each volunteer.

    The investigator hypothesized that the 43 males in the class had lower levels of 17β-estradiol than did the 68 females. For the males, the mean level = 17.2 pg/ml and the median level = 15.5 pg/ml. For the females, the mean level = 238 pg/ml and the median level = 146 pg/ml. The investigator then compared the two groups with the Mann–Whitney nonparametric test. The P value was tiny, less than 0.0001. To what question is this P value the answer?

C4. A student wants to determine whether treatment of cells with a particular hormone increases the number of a particular kind of receptor. She and her advisor agree that an increase of less than 100 receptors per cell is too small to care about. Based on the SD of results she has observed in similar studies, she calculates the necessary sample size to have 90% power to detect an increase of 100 receptors per cell. She performs the experiment that number of times, pools the data, and obtains a P value of 0.04.

The student thinks that the experiment makes a lot of sense and believes that the prior probability that her hypothesis was true was 60%. Her advisor is more skeptical and believes that the prior probability was only 5%.

a. Combining the prior probability and the P value, what is the chance that these results are caused by chance? Answer from both the student's perspective and that of the advisor.

b. Explain why two people can interpret the same data differently.

c. How would the advisor's perspective be different if the P value were 0.001 (and the power were still 90%)?

C5. All members of a class of 111 third-year pharmacy practice majors volunteered for a study in which a single blood sample was taken from each student. All blood samples were analyzed by the same technician for the concentration of a natural human biochemical in plasma, Compound X, in nanograms per milliliter. No additional patient identifiers or covariates were recorded.

The investigator, after looking over the data from the 111 volunteers (see Figure 47.1, left), hypothesized that the subjects came from two distinct populations, one with low values of Compound X and one with high values of Compound X. He divided the data set at the overall median into two groups, a low group (5–44 ng/ml) and a high group (47–1,466 ng/ml). There were no values between 44 and 47. For all of the subjects, n = 111; mean = 159; SD = 250; and median = 44. For the low group, n = 56; mean = 20.0; SD = 9.45; and median = 19.8. For the high group, n = 55; mean = 300; SD = 295; and median = 220. The investigator compared the two groups (shown in Figure 47.1, right) using the nonparametric Mann–Whitney test (see Chapter 41). The P value was tiny, less than 0.0001. Interpret what this P value means and what you can conclude from the study.

C6. Assume that you are reviewing a manuscript prior to publication. The authors tested many drugs to see whether they increased enzyme activity. All P values were calculated using a paired t test. Each day they also performed a negative control, using just a vehicle with no drug. After discussing significant changes caused by some drugs, they state, "In contrast, the vehicle control has consistently given a nonsignificant response (P > 0.05) in each of 250 experiments." Comment on these negative results. Why are they surprising?

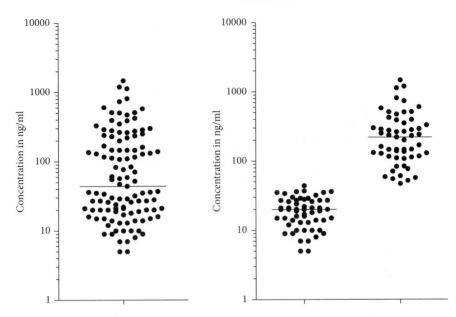

Figure 47.1. Compound X in all students (left panel) and separately for values below and above the median (right panel).

C7. The serum level of a hormone (the Y factor) was measured to be 93 ± 1.5 units/ml (mean ± SEM) in 100 nonpregnant women and 110 ± 2.3 units/ml in 100 women in the first trimester of pregnancy. The difference between the means is 110 − 93 = 17, with a 95% CI extending from 11.6 to 22.4 units/ml.

a. Interpret that CI.

b. Estimate the P value.

c. What assumptions are you making to answer a and b?

d. Would this be a useful test for diagnosing pregnancy?

C8. Cohen et al. (1993) investigated the use of active cardiopulmonary resuscitation (CPR). In standard CPR, the resuscitator compresses the victim's chest to force the heart to pump blood to the brain (and elsewhere) and then lets go to let the chest expand. Active CPR is done with a suction device. This enables the resuscitator to both pull up on the chest to expand it and press down to compress it. These investigators randomly assigned cardiac arrest patients to receive either standard or active CPR. Eighteen of 29 patients treated with active CPR were resuscitated. In contrast, 10 of 33 patients treated with standard CPR were resuscitated.

a. Using Fisher's test, the two-sided P value is 0.0207. Explain what this means in plain language.

b. What would a Type I error mean in this context?

c. What would a Type II error mean in this context?

Figure 47.2. Data demonstrating the winner's curse.
This figure is similar to Figure 2 of Ioannidis (2008) and was created using raw data provided by Dr. Ioannidis.

C9. Which of the following are inconsistent?
  a. Mean difference = 10. The 95% CI is –20 to 40. P = 0.45.
  b. Mean difference = 10. The 95% CI is –5 to 15. P = 0.02.
  c. Relative risk = 1.56. The 95% CI is 1.23 to 1.89. P = 0.013.
  d. Relative risk = 1.56. The 95% CI is 0.81 to 2.12. P = 0.04.
  e. Relative risk = 2.03. The 95% CI is 1.01 to 3.13. P < 0.001
C10. In response to many case reports of connective tissue diseases after breast implants, the FDA called for a moratorium on breast implants in 1992. Gabriel and colleagues (1994) did a prospective study to determine whether there really was an association between breast implants and connective tissue (and other) diseases. They studied 749 women who had received a breast implant and twice that many control subjects. They analyzed their data using survival analysis to account for different times between implant and disease and to correct for differences in age. You can analyze the key findings more simply. They found that 5 of the cases and 10 of the controls developed connective tissue disease.
  a. What null hypothesis would a P value test?
  b. What is that P value (hint: no calculations or computer needed)?
C11. Figure 47.2 shows data from 461 meta-analyses that all showed a statistically significant effect (P < 0.05). A meta-analysis pools data from several (or many) studies to obtain one P value and CI that take into account all the data. Each symbol in the graph represents one such study. The graph shows that studies with larger samples tend to show a smaller effect size than do smaller (presumably less accurate) studies. Ioannidis (2008) presents that as evidence for the winner's curse explained at the end of Chapter 44.
  Figure 47.2 shows Spearman's correlation coefficient and the corresponding P value.

    a. Interpret the meaning of the P value.

    b. The graph is plotted using a logarithmic scale for both axes, but it isn't clear whether the correlation was computed using the raw data or the log-transformed data. Can you tell from the figure? Does it matter? Would the two P values be different or identical?

C12. Two gentleman, John and Frank, decided to fly in a hot air balloon across the Atlantic Ocean, from North Carolina to Europe. After a frightening five-week journey, the pair finally spotted land. Frank thought that they were now over Ireland. They saw a man in a tweed suit walking along a path. John shouted down to the man, "Where are we?" The man stopped walking, looked up, thought for five seconds, and replied, "You're in a balloon." Frank then said to John, "That man must be a statistician." John asked, "How did you know?" What was Frank's answer?

D. PROBLEMS ABOUT SAMPLE SIZE AND POWER

D1. Which will require more subjects, detecting a difference between two proportions when $p_1 = 0.10$ and $p_2 = 0.20$, or detecting a difference between two proportions when $p_1 = 0.20$ and $p_2 = 0.40$? Assume you are looking for the same difference with the same power and are using samples of equal size.

D2. You read the following in a journal: "Before starting the study, we calculated that with a power of 80% and a significance level of 5%, 130 patients would be required in each group to demonstrate a 15-percentage-point reduction in mortality (from the expected rate of 33 down to 18%)." Explain in plain language what this means.

D3. Lymphocytes contain β-adrenergic receptors. Epinephrine binds to these receptors and modulates immune responses. It is possible to count the average number of receptors on human lymphocytes using a small blood sample. You wish to test the hypothesis that people with asthma have fewer receptors. By reading various papers, you learn that there are about 1,000 receptors per cell and that the CV in a normal population is about 25%.

    a. How many asthmatic subjects do you need to determine the receptor number to ± 100 receptors per cell with 95% confidence?

    b. You want to compare a group of normal subjects with a group of asthmatics. How many subjects do you need in each group to have 80% power to detect a mean change of 10% of the receptors using $\alpha = 0.05$ (two-tailed)?

    c. How many subjects do you need in each group to have 95% power to detect a mean difference in receptor number of 5% with $\alpha = 0.01$?

D4. You are preparing a grant for a study that will test whether a new drug treatment lowers blood pressure substantially. From previous experience, you know that 15 rats in each group is a large enough sample size and that the SD of measurements in different rats is about 10 mmHg. Prepare a convincing power analysis to justify that sample size.

D5. The overall incidence of a disease is 1 in 10,000. You think that a risk factor increases the risk. How many subjects do you need if you want to have 80% power to detect a relative risk as small as 1.1 in a prospective study?

D6. You are planning a clinical study. You know that the mean heart rate for the placebo-treated control group will be about 65 beats per minute with a standard deviation of about 8 beats per minute. You don't expect the heart rate to change with treatment (a LDL cholesterol-lowering agent). The study will set the Type I error rate, α, to the standard value of 0.05. Someone reviewing the protocol asks you what the power of the study is to detect a tiny change in pulse (1 beat per minute). Estimate that power without doing any calculations.

E. PROBLEMS ABOUT CORRELATION AND REGRESSION

E1. Figure 47.3 shows the relationship between the population of countries (or territories) and their area. Each dot is a separate country. All 237 countries are shown. The graph on the left panel shows linear axes. The graph on the right panel shows both axes using a logarithmic axis.
a. Interpret the P value.
b. Were the two variables properly assigned to the X-axis and Y-axis?
c. Would the correlation coefficient or P value be different if the definitions of X and Y were swapped?
d. What are the advantages of using logarithmic axes?

E2. Figure 47.4 shows the same data as Figure 47.3 but is fit with a linear regression program.
a. What are the units of the slope? What does it mean?
b. Interpret the P value.
c. Imagine that the graph of data remained the same but the definitions of the independent (X) and dependent (Y) variables were switched.

Figure 47.3. Correlation of population and area of different countries.

Figure 47.4. Linear regression of population versus area of different countries.

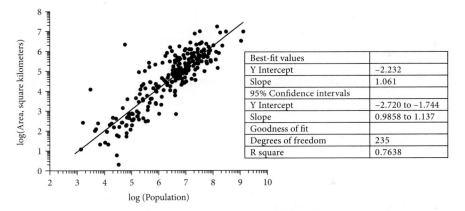

Best-fit values	
Y Intercept	−2.232
Slope	1.061
95% Confidence intervals	
Y Intercept	−2.720 to −1.744
Slope	0.9858 to 1.137
Goodness of fit	
Degrees of freedom	235
R square	0.7638

Figure 47.5. Each dot is a different country, with a line fit by linear regression.

Now imagine that linear regression is used to find the line that best predicts population (horizontal axis) from area (vertical axis). Would the best-fit line be the same or different?

d. On logarithmic axes, why is the "line" curved?

e. On the logarithmic graph, why does the "line" stray so far from many points?

f. Figure 47.5 was also fit by linear regression. Why does it show a different line and different R^2 than those shown in Figure 47.4?

E3a. Figure 47.6 (left) shows a log(dose)–response curve, fit with nonlinear regression. The shaded region shows the 95% CI for the EC_{50} (the dose that provokes a response halfway between the lowest and highest possible responses). Why is this range so wide?

E3b. Figure 47.6 (right) shows the same data, also with a log(dose)–response curve fit by nonlinear regression. Why is the shaded area denoting the 95% CI of the $logEC_{50}$ so much narrower?

E3c. One of the points shows a response greater than 100%. Should it be removed because that is impossible?

Bottom	−1.587
Top	140.4
LogEC50	−6.684
Hill slope	0.6616
EC50	2.071e-007
95% Confidence Intervals	
Bottom	−4.519 to 1.345
Top	71.90 to 209.0
LogEC50	−7.349 to −6.019
Hill slope	0.4425 to 0.8807
EC50	4.476e-008 to 9.577e-007

Bottom	= 0.0
Top	= 100.0
LogEC50	−7.114
Hill slope	0.8408
EC50	7.699e-008
95% Confidence intervals	
LogEC50	−7.232 to −6.996
Hill slope	0.7658 to 0.9157
EC50	5.867e-008 to 1.010e-007

Figure 47.6. Dose–response data fit with nonlinear regression.
Why is the CI for the EC_{50} (shaded) so much narrower on the right?

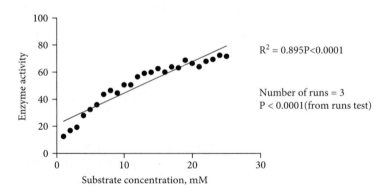

$R^2 = 0.895 P < 0.0001$

Number of runs = 3
$P < 0.0001$ (from runs test)

Figure 47.7. Linear regression and runs test.

E4a. Figure 47.7 shows data and a line fit by linear regression. Interpret the R^2 and P value shown on the graph.

E4b. Figure 47.7 also shows results from a runs test. This is not a test covered in this book. Read about it in other books or on the Web and interpret what the P value means.

E5. Arad and colleagues (2000) performed a large study to see whether measuring calcium in coronary arteries can help predict heart attacks. They studied 1,177 people who were free of cardiac symptoms. They measured calcium in the coronary arteries using a noninvasive method called electron beam–computed tomography. The results are reported as a coronary artery calcium score. The investigators followed all these

VARIABLE	ODDS RATIO	95% CI
Coronary calcium score > 160	14.3	4.9 to 42.3
Age > 55 years old	3.3	1.3 to 8.4
Elevated cholesterol	4.0	1.3 to 12.2
Hypertension	2.6	1.1 to 6.1
Diabetes	4.8	1.6 to 13.9

Table 47.2. Results of logistic regression to predict cardiac events (death, nonfatal myocardial infarction, need for a revascularization procedure).

Data are from a portion of Table 3 from Arad et al. (2000).

people for three to four years and recorded cardiac events (3 people died, 15 had a nonfatal myocardial infarction, and 21 required a revascularization procedure to relieve chest pain). The question is whether the calcium score predicts the occurrence of these cardiac events better than the standard cardiac risk factors (age, high cholesterol, high blood pressure, diabetes). Complete data were available for 787 of the subjects. The data were analyzed by logistic regression, and the results are shown in Table 47.2.

a. Odds ratios compare the risk in one group with the risk in another. When the logistic regression model computes odds ratios, what group is the comparison group?

b. Which of the five variables make a statistically significant ($P < 0.05$) contribution to the model? Which has the highest P value?

c. Write out the logistic regression model.

d. What is the predicted odds ratio for someone who is older than 55 and has a coronary artery calcium score greater than 160?

e. Did this study have enough subjects?

f. Most of the cardiac events (21 of 39) were people requiring angioplasty or coronary artery surgery to restore blood flow in the coronary arteries. In some cases, the decision to perform these procedures might have been a difficult one. How would the results be affected if the decision to perform the procedures was partially influenced by knowing that the coronary calcium score was high?

g. Explain in plain language what the odds ratio of 14.3 means.

E6. The "Oscar curse" is a phenomenon in which the career of an actor or actress plummets just after he or she wins an Academy Award (Oscar) for best or best supporting actor or actress. Some examples of actors and actresses who are said to have been hexed by the Oscar curse include Halle Berry, Mira Sorvino, Cuba Gooding Jr, Marisa Tomei, Roberto Benigni, F. Murray Abraham, Adrien Brody, and Marlee Matlin. From Chapter 1, which statistical/probability concept is exemplified by the Oscar curse?

CHAPTER 48

Answers to Review Problems

The questions from Chapter 47 are all repeated here, along with answers. Reading this chapter can help you review statistics, even if you don't want to actually work the problems.

A. PROBLEMS ABOUT CIs OF PROPORTIONS, SURVIVAL CURVES, AND COUNTS

A1. **Of the first 100 patients to undergo a new operation, 6 die. Can you calculate the 95% CI for the probability of dying with this procedure? If so, calculate the interval. If not, what information do you need? What assumptions must you make?**

Yes, it is possible to calculate a CI. This is a straightforward calculation of the 95% CI from a proportion. The proportion is 6/100. Using the modified Wald method:

First compute p', the center of the CI.

$$p' \approx \frac{s+2}{n+4} \approx \frac{6+2}{100+4} \approx 0.0769 \approx 7.69\%$$

The margin of error, W, is computed as follows:

$$W \approx 2\sqrt{\frac{.0769(1-0.0769)}{n+4}} \approx 0.0523 = 5.23$$

The 95% CI extends from p' − W to p' + W, or from 2.46 to 12.92%.

To interpret these results, you must define the population. If these were real data, you'd want to see how the investigators selected the patients. Are they particularly sick? Do they represent all patients with a certain condition, or only a small subset? If you can't figure out how the investigators selected their patients, then the data are not very useful. If you can define the population from which they were selected, you must assume that the 100 patients are representative of these patients and that each was selected independently. If you accept these assumptions, then you can be 95% confident that the confidence interval includes the overall death rate.

A2. **A new drug is tested in 100 patients and lowers blood pressure by an average of 6%. Can you calculate the 95% CI for the fractional lowering of blood pressure by this drug? If so, calculate the interval. If not, what information do you need? What is the CI for the fractional lowering of blood pressure? What assumptions must you make?**

Although this looks similar to the previous question, it is actually quite different. In the previous question, the percentage was really a proportion. In this question, the percentage is really change in a measured value. The CI cannot be computed from the data presented. You'd need to know the average blood pressure before treatment and its standard deviation.

Beware of percentages. Percentages can express a proportion, a relative difference in proportions, a relative difference in a measured value, or many other things. Use the methods of Chapter 4 only with proportions.

A3. **You use a hemocytometer to determine the viability of cells stained with trypan blue. You count 94 unstained cells (viable) and 6 stained cells (indicating that they are not viable). Can you calculate the 95% CI for the fraction of stained (dead) cells? If so, calculate the interval. If not, what information do you need? What assumptions must you make?**

This is essentially the same problem as Problem A1. The population is the entire tube of cells, and you have assessed the viability of a sample of these cells. Assuming that you mixed the tube well, it is reasonable to assume that your sample is randomly selected from the population. You can calculate a 95% CI for the proportion of dead cells in the entire population. You can say that you are 95% confident that the fraction of stained (dead) cells ranges from 2.5 to 12.9% by the modified Wald method.

A4. **In 1989, 20 of 125 second-year medical students in San Diego failed to pass their biostatistics course (until they came back for an oral exam). Can you calculate the 95% CI for the probability of passing the course? If so, calculate the interval. If not, what information do you need? What assumptions must you make?**

It is easy to plug the numbers into a computer program to calculate a CI. But what would it mean? There were only 125 students in 1989, so we don't really have a sample. It would make sense to calculate a CI only if you imagined the 125 students in 1989 to be representative of students in other years (or other cities) and if there would be no differences in course content, student preparation, exam difficulty, or passing threshold. You would also have to assume that students in future years hadn't read this question, because it would give them an incentive to study harder. In fact, the students in 1990 read this question. They got the message that the course required work, and they studied hard. All passed the course the first time they took the exam.

Because those assumptions are unlikely to be true, I don't think it would be meaningful to calculate a CI. If you are persuaded that the assumptions are met, you can calculate that the 95% CI for passing ranges from 76 to 90%.

A5. **Ross Perot won 19% of the vote in the 1992 U.S. presidential election. Can you calculate the 95% CI for the fraction of voters who voted for him? If so, calculate the interval. If not, what information do you need? What assumptions must you make?**

It wouldn't be meaningful to calculate a CI. We know the exact fraction of voters who voted for each candidate. This is not a poll before the election; these are results from the election. Because we have not sampled from a population, a CI would be pointless and meaningless.

A6. **In your city (population = 1 million) last year, a rare disease had an incidence rate of 25 in 10,000 people. Can you calculate the 95% CI for the incidence rate? If so, calculate the interval. If not, what information do you need? What assumptions must you make?**

First, ask yourself whether it makes sense to calculate a CI. You know the incidence of the disease in the entire population last year, so there is no need for a CI for the incidence of the disease last year. If you assume that the incidence of the disease is not systematically changing from year to year, it may be reasonable to consider last year's population to be a sample from the populations of the next few years. So you can calculate a CI for the overall incidence rate for the next few years.

The problem is presented with the number of disease cases per 10,000 people, but the data were collected from a population of 1,000,000. So the sample proportion is 2,500 of a population of 1,000,000. Applying the modified Wald method for 95% confidence (see Problem A1):

$$p' = \frac{S+2}{n+4} = \frac{2500+2}{1000000+4} = 0.002502$$

$$W \approx 2\sqrt{\frac{p'(1-p')}{n+4}} \approx 2\sqrt{\frac{0.002502(1-0.002502)}{1000000+4}} \approx .0001 \approx 0.01\%$$

The 95% CI ranges from approximately 0.24 to 0.26%. Assuming that the incidence of the disease is not changing, you can be 95% sure that the true average incidence rate for the next few years will be somewhere within that range.

A7. **Is it possible to calculate the median survival time when some of the subjects are still alive?**

If you listed all the survival times in order, the median survival time would be the middle survival time. If subjects are still alive, then you don't know their survival time. But you do know that their survival time will be longer than the time they have been followed. So you can still find the median time, even if some subjects are still alive, provided that at least half have died. Median survival time is the time at which half the subjects are still alive and half are dead.

A8. **Why are survival curves drawn as staircase curves?**

The curves plot the survival of the subjects in your sample. Whenever a subject dies, the percentage survival drops. The percentage survival then remains at a steady plateau until the next patient dies.

You could imagine drawing the curve differently if your goal were to draw your best guess of survival in the population. You might then draw a sloping line from one patient's death to the next. Then the curve would not look like a staircase. But the convention is to draw survival curves showing the survival of your sample, so staircase curves are used.

A9. **A survival curve includes many subjects who were censored because they dropped out of the study when they felt too ill to come to the clinic appointments. Is the survival curve likely to overestimate or underestimate survival of the population?**

Because the censored patients are sicker than the average subjects, they tend to die sooner than the noncensored patients. The remaining patients tend to die later than the overall average, and so the data on the remaining patients overestimates the survival of the entire population.

A survival curve can only be interpreted when you assume that the survival of the censored subjects would, on average, equal the survival of the uncensored subjects. If many subjects are censored, this is an important assumption. If the study is large and only a few subjects were censored, then it doesn't matter as much.

A10. **A study began on January 1, 1991, and ended on December 31, 1994. How will each of these subjects appear on the graph?**
   a. Entered March 1, 1991. Died March 31, 1992.
   b. Entered April 1, 1991. Left study March 1, 1992.
   c. Entered January 1, 1991. Still alive when study ended.
   d. Entered January 1, 1992. Still alive when study ended.
   e. Entered January 1, 1993. Still alive when study ended.
   f. Entered January 1, 1994. Died December 30, 1994.
   g. Entered July 1, 1992. Died of car crash March 1, 1993.

   a. This subject is shown as a downward step at time = 13 months.
   b. This subject was censored at 11 months. He or she will appear as a blip at time = 11 months.
   c. The subject was alive for the entire 48 months of the study. He or she was censored at time = 48 months. Because the graph ends at 48 months, a blip may or may not be visible.
   d. This subject was alive for 36 months after starting the trial. He or she was censored at 36 months and will appear as a blip at time = 36 months.
   e. This subject was alive for 24 months at the time the study ended. He or she will appear as a censored subject at time = 24 months.
   f. This subject died 12 months after starting the trial. He or she will appear as a downward step at time = 12 months.

g. This is ambiguous. After eight months, the subject died of a car crash. This might count as a death or as a censored subject, depending on the decision of the investigator. Ideally, the experimental protocol should have specified how to handle this situation before the data were collected. Some studies only count deaths from the disease being studied, but most studies count deaths from all causes. Some studies analyze the data both ways. Why would you include a death from a car crash in a study of cancer? The problem is that it is impossible to know for sure that a death is really unrelated to the cancer (and its therapy). Maybe the car crash in this example would have been avoided if the study medication hadn't slowed the reaction time or if the cancer hadn't limited peripheral vision.

A11. A survival study started with 100 subjects. Before the end of the fifth year, no patients had been censored and 25 had died. Five patients were censored in the sixth year, and five died in the seventh year. What is the percentage survival at the end of the seventh year?

This book did not give the equations for calculating a survival curve, but it is pretty easy to figure out. At the end of the fifth year, 25 patients had died and 75 were still alive, so the fraction survival is 75%. During the sixth year, five subjects were censored. This does not change percentage survival, which remains at 75%. Five subjects died in the next year. At the start of that year, 70 subjects were alive $(100 - 25 - 5)$. At the end of the seventh year, 65 subjects were alive. For subjects alive at the start of the seventh year, the probability of surviving that year is 65/70, or 92.86%. The probability of surviving up to the beginning of the seventh year, as we already calculated, is 75%. So the probability of surviving from time zero to the end of the seventh year is the product of those two probabilities, $75\% \times 92.86\% = 69.6\%$.

A12. You use a hemocytometer to count white blood cells. When you look in the microscope you see lots of squares, and 25 squares enclose 0.1 μL. You count the number of cells in nine squares and find 50 white blood cells. Can you calculate the 95% CI for the number of cells per microliter? If so, calculate the interval. If not, what information do you need? What assumptions must you make?

You counted 50 cells in nine squares. To calculate the CI, we must work with the number actually counted, 50. From Table 6.1, you can see that the 95% CI for the average number of cells in nine squares ranges from 37.1 to 65.9. Because there are 250 squares/μL and 9 were counted, multiply by $(250/9) = 27.77$ to obtain cells per microliter. The 95% CI ranges from 1,030 to 1,830 cells/μL.

A13. In 1988, a paper in Nature (Davenas et al., 1988) caused a major stir in the popular and scientific press. The authors claimed that antibodies diluted   even to 10^{-120} of the starting concentration stimulated basophils to degranulate. With that much dilution, the probability that even a single molecule of antibody would remain in the tube is infinitesimal. The investigators hypothesized that the

water somehow "remembered" that it had seen antibody. These results purported to give credence to homeopathy, the theory that extremely low concentrations of drugs are therapeutic.

The assay to test such a hypothesis is simple. Add a test solution to basophils and incubate. Then stain. Count the number of stained cells in a certain volume. Compare the number of cells in control and treated tubes. A low number indicates that many cells have degranulated, because cells that haven't degranulated won't take up the stain.

The authors present the "mean and standard error of basophil number actually counted in triplicate." In the first experiment, the three control tubes are reported as 81.3 ± 1.2, 81.6 ± 1.4, and 80.0 ± 1.5. The results from a second experiment conducted on another day are very similar. The tubes treated with the diluted antibody give much lower numbers of stained cells, indicating that the diluted antibody has caused degranulation. For this problem, think only about the control tubes.

a. Why are these control values surprising?
b. How could these results have been obtained?

a. The authors counted only the cells that still had granules. In the results shown, the average number of cells in the volume counted was about 80. If this is true, then the Poisson distribution tells us what we expect to see in different experiments. From Table 6.1, the 95% CI ranges from 63 to 100. If you repeated the experiment many times, you'd expect the number of cells counted to be within this range 95% of the time and outside the range 5% of the time. The results reported by the authors have far less variation than that.

b. One possibility is that the investigators were lucky and just happened to have so little variation. The chance of having so little variation is very small, so this is quite unlikely. If the cells are well mixed and randomly sampled, the Poisson distribution tells us the distribution of values we expect to see. This is the "best" you can do, assuming perfect experimental technique. Any sloppiness in the experiment will lead to more variability. No matter how carefully you do the experiment, you can't get less scatter than that predicted by the Poisson distribution unless you are extremely lucky.

So how then could the results have been obtained? There is no way to be sure, but here is one possibility. It is difficult to count cells in a reliable way. Counting requires some judgment as you ask yourself these kinds of questions: How do you deal with cells on the border? How do you distinguish between real cells and debris? When two cells are stuck together, does that count as one or two? If you know the results you are "supposed" to get, your answers might be swayed to get the count you expect to get.

What do you do when the number of stained cells in a treated sample exceeds the number in the control sample? By chance, and assuming that the diluted antibodies do nothing, you'd expect half the samples to have more stained cells than the average and half to have less. If the experimenter threw out the data from the samples for which the number of stained cells exceeded the control average (because those data seemed to be impossible), the data will be biased and will give a smaller average. In this case, the results would be impossible to interpret.

For an assay like this one (counting cells manually), it is essential that the person counting the cells not know which tube is which so that he or she can't let the expected results bias the count. Even better, use a machine that counts in an unbiased way. The machines can also count more cells, thus narrowing the CI.

You could also improve the experimental methodology in other ways. Why not count all cells, not just the ones that haven't degranulated? Why not use a more precise assay, such as measuring the amount of histamine released from the cells instead of the number of cells that have released histamine?

B. PROBLEMS ABOUT SD, SEM, CIs, AND LOGNORMAL DISTRIBUTIONS

B1.   **Estimate the SD of the age of the students in a typical medical school class (just a rough estimate will suffice).**

Without data, you can't determine the SD exactly. But you can estimate it. The mean age of first-year medical students (in the United States) is probably about 23. Lots of students are one year older or younger. A few might be a lot younger or a lot older. If you assume that the age distribution is approximately Gaussian, you will include two-thirds of the students in a range of ages that extends 1 SD on either side of the mean. Therefore, the SD is probably about one year. Obviously, it will vary from class to class. It is likely that the distribution will be skewed (asymmetric) by the presence of a few students much older than the others, which may make the SD a bit larger.

The point of this question is to remind you that the SD is not just a number that falls out of an equation. Rather, the SD describes the scatter of data. If you have a rough idea of how the values are scattered, you can estimate the SD without any calculations.

B2.   **Estimate the SD of the number of hours slept each night by adults.**

Without data, you can't know the SD exactly, but you can estimate it. On average, adults sleep about eight hours each night. Some sleep more, some less. If you go 1 SD from the mean, you'll include two-thirds of the population, assuming that sleep time is roughly Gaussian. I'd estimate that the SD is about one hour. If this were true, then about two-thirds of adults would sleep between seven and nine hours per night, and 95% would sleep between six and ten hours per night.

Again, the point of this question is to remind you that the SD is not just a number that falls out of an equation. Rather, the SD describes the scatter of data. If you have a rough idea of how the values are scattered, you can estimate the SD without any calculations.

B3. **The following cholesterol levels were measured in 10 people (mg/dl):**
260, 150, 160, 200, 210, 240, 220, 225, 210, 240
Calculate the mean, median, variance, SD, and CV. Make sure you include units.

Mean = 211.5 mg/dl
Median = 215 mg/dl
Variance = 1200.28 $(mg/dl)^2$
SD = 34.645 mg/dl
CV = 34.645/211.5 = 16.4%

B4. **Add an eleventh value (931) to the numbers in Question B3 and recalculate all values.**

Mean = 276.9 mg/dl
Median = 220 mg/dl
Variance = 48,142 $(mg/dl)^2$
SD = 219.4 mg/dl
CV = 219.4/276.9 = 79.23%

One value much larger than the rest increases the mean, SD, CV, and variance considerably. It has little impact on the median.

B5. **The level of an enzyme in blood was measured in 20 patients, and the results were reported as 79.6 ± 7.3 units/ml (means ± SD). No other information is given about the distribution of the results. Estimate what the probability distribution might have looked like.**

With no other information to go on, it is reasonable to speculate that the enzyme activity might be roughly Gaussian. If so, then about two-thirds of the values should be within 1 SD of the mean, within the range of about 72 to 87, and about 95% of the values should be within 2 SD of the mean (65 to 94).

B6. **The level of an enzyme in blood was measured in 20 patients and the results were reported as 9.6 ± 7.3 units/ml (means ± SD). No other information is given about the distribution of the results. Estimate what the probability distribution might have looked like.**

Again, you might start with the assumption that the distribution is roughly Gaussian. If that were true, then 95% of the population would be within a range defined by the mean ± 2 SD. However, this range would include negative numbers, and enzyme activity cannot be negative. Therefore, the distribution cannot possibly be Gaussian with the stated mean and SD. The distribution probably is asymmetrical with a "tail" to the right and may be lognormal. It could also possibly be bimodal.

B7. **The Weschler IQ scale was created so that the mean is 100 and the standard deviation is 15. What fraction of the population has an IQ greater than 130?**

The problem asks what fraction of a Gaussian distribution has a value greater than two standard deviations above the mean. Approximately 95% of a Gaussian population is within a range defined by the mean ± 2 SD, so 5% is outside that range. A Gaussian distribution is symmetrical, so 2.5% of a population has a value greater than 2 SD above the mean.

B8.   **An enzyme level was measured in cultured cells. The experiment was repeated on three days, and on each day, the measurement was made in triplicate. The experimental conditions and cell cultures were identical on each day; the only purpose of the repeated experiments was to determine the value more precisely. The results are shown in Table 47.1 as enzyme activity in units per minute per milligram of membrane protein. Summarize these data as you would for publication. The readers have no interest in the individual results for each day, just in one overall mean with an indication of the scatter. Give the results as mean, error value, and n. Justify your decisions.**

There are two difficulties with this problem. First, you must decide what to do with Tuesday's second replicate. It is wildly out of whack with the others. Deciding what to do with outlying values is a difficult problem, and different scientists have different policies. In this case, I think it is clear that the value can't possibly be correct. Not only is it far, far away from the other two replicates, but it is far, far away from the six replicates measured on other days. Because the experiment was performed with cultured cells, there is no biological variability from day to day. The only contribution to scatter consists of experimental errors. It seems plain to me that the 967 value is clearly incorrect. First, I would look back at the lab notebook and see whether the experimenter noticed anything funny about that tube during the experiment. If there was a known problem with that replicate, then I would feel very comfortable throwing it out. Even otherwise, I would delete it from the analysis, while recording in a lab notebook exactly what I had done. Including that value would throw off the calculations, so the result wouldn't be helpful.

If the rows represented different patients (rather than different experiments on the same cell cultures), then my answer would be different. With different patients, biological variability plays a role. Maybe the large number is correct and the two small numbers are wrong. Maybe the second patient is really different from the first and third. If possible, I'd do the experiment with Patient 2 a second time to find out.

The second problem is how to pool the eight remaining values. It is not really fair to consider these as eight independent measurements. The replicates in one day are closer to one another than they are to measurements made on other days. This is expected, because there are more contributions to variability between days than between replicates. So calculating the mean and SD or SEM of all eight values is not valid.

How to compute the CI of a grand mean (often called a *consensus mean*) is a surprisingly complicated question and has multiple answers, especially because the sample size (after excluding the outlier) is not the same for all three days. A simple approach is to calculate the mean value for each day, then calculate the mean and SEM of the three means, and finally calculate a 95% CI. This makes sense if you think of the population as the means of all possible experiments, with n = 3. We can be 95% sure that the true mean lies somewhere within the CI.

The three means are 227.7, 272.0, and 249.7. The grand mean is 249.8, with an SEM of 12.8 (n = 3). The 95% CI for the grand mean ranges from 194.7 to 304.9. If you were to repeat a similar experiment many times, the 95% CI would contain the true mean in 95% of experiments. This could be reported as follows: "The experiment was performed three times in triplicate. After excluding one wildly high value, the mean for each experiment was calculated. The grand mean is 249.8, with an SEM of 12.8 (n = 3). The 95% CI ranges from 194.7 to 304.9."

B9. **A paper reports that cell membranes have 1,203 ± 64 (mean ± SEM) fmol of receptor per milligram of membrane protein. These data come from nine trials of an experiment.**
   a. **Calculate the 95% CI. Explain what it means in plain language.**
   b. **Calculate the 90% CI.**
   c. **Calculate the CV.**

   a. With n = 9, df = 8 and t* = 2.3060 (see Appendix D). So the 95% CI ranges from 1,055 to 1,351 fmol/mg protein. Assuming that each of the nine experiments was conducted independently, you can be 95% sure that the true mean (if you repeated the experimental trial many times) lies within this range.
   b. With df = 8, t* for 90% CI is 1.8595 (see Appendix D). The 90% CI ranges from 1,084 to 1,322.
   c. The coefficient of variation is the SD divided by the mean. The SD equals the SEM times the square root of sample size, 64×3, or 192. So the CV equals 192/1,203, or 16.0%. Because the mean and SD have the same units, the CV is a unitless ratio.

B10. **You measure blood pressure in 10 randomly selected subjects and calculate that the mean is 125 and the SD is 7.5 mmHg. Calculate the SEM and 95% CI. Then you measure blood pressure in 100 subjects randomly selected from the same population. What values do you expect to find for the SD and SEM?**

The SEM is the SD divided by the square root of the sample size, so it equals $7.5/\sqrt{10}$, or 2.37. With n = 10 and df = 9, t* = 2.26. The 95% CI extends 2.26 SEMs on either side of the mean: 119.6 to 130.4 mmHg.

When you increase the sample size, you don't expect the SD to change. The SD measures the scatter of the data, and the scatter doesn't change by sampling more subjects. You'll know the SD more accurately, so the value will probably change. But the SD is just as likely to increase as to decrease when you add more subjects. Your best guess is that the SD will still equal 7.5 mmHg.

The SEM quantifies how well you have determined the population mean. When you increase the sample size, your sample mean will probably be closer to the population mean. So the SEM gets smaller as the sample size gets larger. With n = 100, your best guess for the SEM is $7.5/\sqrt{100}$, or 0.75 mmHg.

Many students mistakenly think that the SD ought to get smaller as the sample size gets larger. Not so. The SD quantifies variability or scatter, and the scatter is real. The values in the population are not all the same. Collecting a larger sample helps you know the SD better, but the size of the SD does not change predictably as the sample size increases. The SD quantifies an attribute of the population, and that attribute (scatter) doesn't change when you sample more subjects.

If you are still confused, think of how the mean will change when you sample more subjects. If you collect a larger sample, you'll know the population mean better and your sample mean from n = 100 won't exactly equal the sample mean from n = 10. But it is just as likely that the mean will increase as it is that it will decrease. The sample mean quantifies an attribute of the population (its average), and the population doesn't change by sampling more subjects.

The standard error is different. It does not quantify an attribute of the population. It quantifies sampling error, and its value is determined (in part) by the sample size. Collecting bigger samples reduces the value of the standard error of the mean.

B11. **Does it ever make sense to calculate a CI from a population SD (with a denominator of n), as opposed to the sample SD (with a denominator of [n – 1])?**

No! If you know the population SD, then you must know all the values in the population. If this is true, then you know the population mean exactly, and so there is no point in calculating a CI. CIs are only useful when you are unsure about the true population mean, so you want to calculate a CI to quantify your uncertainty.

B12. **Data were measured in 16 subjects, and the 95% CI of the mean ranged from 97 to 132. Calculate the 99% CI.**

Because the CI is symmetrical around the mean, the mean must equal the average of 97 and 132, or 114.5. With 16 subjects, the 95% CI extends in each direction from the mean a distance equal to 2.13 times the SEM. Therefore, the SEM must equal the difference between 132 and 97, divided by twice 2.13. So the SEM is 35/(2 × 2.13), or about 8.22. The SD equals the SEM times the square root of the sample size, which is 4 × 8.22, or 32.88. To calculate a 99% CI with df = 15, t* = 2.95 (see Appendix D). Therefore, the margin of error of the 99% CI equals 2.95 times the SEM, which is 2.95 × 8.22, or 24.24. The 99% CI extends from 114.5 – 24.24 to 114.5 + 24.24, or from 90.3 to 138.7. Note that the 99% CI is wider than the 95% CI.

B13. **The standard error of the difference between two means is larger than the SE of either mean. Why?**

The SE of a difference quantifies how well you know the difference between two population means. It depends on how well you know each mean as reflected in the two SEMs. Uncertainty in each of the two means contributes to uncertainty in the difference.

B14. **Pullan et al. (1994) investigated the use of transdermal nicotine for treating ulcerative colitis. The plasma nicotine level at baseline was 0.5 ± 1.1 ng/ml (mean ± SD; n = 35). After six weeks of treatment, the plasma level was 8.2 ± 7.1 ng/ml (n = 30). Is it more likely that the nicotine levels are sampled from a Gaussian or a lognormal distribution?**

If the populations were Gaussian, about 95% of the values in the population would be within 2 SDs of the mean and the distribution would be symmetrical around the mean. With the values given in this problem, this would be possible only if nicotine levels could be negative, which is, of course, impossible. Also, the SD is much larger for the group with the larger mean. For these two reasons, the distribution is more likely to be lognormal than Gaussian.

C. PROBLEMS ABOUT P VALUES AND STATISTICAL SIGNIFICANCE

C1. **You conduct an experiment and calculate that the two-tailed P value is 0.08. What is the one-tailed P value? What assumptions are you making?**

You can only calculate a one-tailed P value if the direction of the expected change was predicted before the data were collected. If the change occurred in the predicted direction, then the one-tailed P value will be half the two-tailed P value, or 0.04.

If the change occurred in the opposite direction, the situation is trickier. The one-tailed P value answers the question, If the null hypothesis were true, what is the chance of randomly picking subjects such that the difference goes in the direction of the experimental hypothesis with a magnitude as large or larger than that observed? When the result goes in the direction opposite to the experimental hypothesis, the one-tailed P value equals 1.0 minus half the two-tailed value. So the one-tailed P value is 0.96.

It really doesn't help to calculate a one-tailed P value when the observed difference is opposite in direction to the experimental hypothesis. You should conclude that the P value is very high and that the difference was caused by chance.

C2. **The FDA requires two-tailed statistical tests for Phase III new drug registration trials, defining P < 0.05 as statistically significant. What would happen if they required one-tailed P values instead, still using the threshold of 0.05 to define significance?**

First, a reality check. When deciding whether to approve a drug, there are many considerations. This problem just considers one—the

P value for the difference in treatments for the primary endpoint in the Phase III registration clinical trial. In some cases, this critical P value can have a strong influence on the decision. For this example, we will assume that this one P value is the only information used to decide whether to approve a new drug.

The definition of a 5% significance level (two-tailed) is simple. If the new drug really does not work better than whatever it is being compared with (placebo or a standard treatment), there is a 5% chance that random chance will result in a difference large enough to be labeled statistically significant. Half of the time (half of 5%, or 2.5%), this difference will be in the right direction (indicating the new drug worked well) and the drug will be approved. Half of the time, the difference will be in the wrong direction (indicating the new drug worked less well than the standard drug). Although the results are statistically significant, the drug won't be approved, because the data show the drug is working significantly *more poorly* than the treatment with which it is being compared. Effectively, the current policy uses a 2.5% significance level for a one-tailed P value to decide on drug approval.

What happens if the policy changes to use a one-tailed P value with a 5% threshold for significance? If there is no true underlying superiority of the new treatment, the chance of concluding that a new treatment is statistically significantly better than an older treatment would double from 2.5 to 5%.

By making it easier to declare results statistically significant, this change would also increase power (assuming studies are done with the same sample size). Thus, this alternate policy change would also likely increase the approval rate for drugs that are truly better than the standard. Some of these drugs would have two-tailed P values between 0.05 and 0.10. By switching to a one-tailed criterion, these P values would be between 0.025 and 0.05, and so the drugs would be approved.

The FDA must find the right balance of approving truly better new drugs while protecting the public from inferior drugs. Switching to a one-tailed P value would change this balance, allowing more drugs to be approved. More good drugs would be approved, but so would many more worthless drugs.

C3. **All members of a class of 111 pharmacy practice majors volunteered for a study in which a single blood sample was taken from each student at a single session. All blood samples were analyzed by the same technician for the concentration of a natural human biochemical in plasma, 17β-estradiol. Gender was recorded for each volunteer.**

The investigator hypothesized that the 43 males in the class had lower levels of 17β-estradiol than did the 68 females. For the males, the mean level = 17.2 pg/ml and the median level = 15.5 pg/ml. For the females, the mean level = 238 pg/ml and the median level = 146 pg/ml. The investigator then compared the two groups with the Mann–Whitney nonparametric test. The P value was tiny, less than 0.0001. To what question is this P value the answer?

Statistical inference is used to analyze data from a sample to make inferences about a much larger, or infinite, population. The Mann–Whitney test and the t test are statistical procedures for making this kind of generalization or inference from sample to population. In this problem, the investigator seems to be only interested in this particular group of 43 male and 68 female subjects and does not seem to have any more general question in mind. As such, the P value is totally meaningless. It can make sense to summarize the data by stating the medians, as was done in this problem. It might also help to show the minimum and maximum values, or the interquartile range, or the SD. It would be even better to show a graph of all the values or a graph of their frequency distribution. But P values (and CIs) only make sense when the goal is to extrapolate from sample to population.

C4. **A student wants to determine whether treatment of cells with a particular hormone increases the number of a particular kind of receptor. She and her advisor agree that an increase of less than 100 receptors per cell is too small to care about. Based on the SD of results she has observed in similar studies, she calculates the necessary sample size to have 90% power to detect an increase of 100 receptors per cell. She performs the experiment that number of times, pools the data, and obtains a P value of 0.04.**

The student thinks that the experiment makes a lot of sense and believes that the prior probability that her hypothesis is true was 60%. Her advisor is more skeptical and believes that the prior probability was only 5%.

a. **Combining the prior probability and the P value, what is the chance that these results are caused by chance? Answer from both the student's perspective and that of the advisor.**

b. **Explain why two people can interpret the same data differently.**

c. **How would the advisor's perspective be different if the P value were 0.001 (and the power were still 90%)?**

a. The student's perspective is illustrated in Table 48.1, which shows the results of 1,000 hypothetical experiments.

Of 556 experiments with a P value less than 0.04, receptor number really increases in 540. So the chance that the receptor number really increases is 540/556 = 97.1%, leaving a 2.9% chance that the results are caused by coincidence.

The advisor's perspective is different, as illustrated in Table 48.2. Of all the hypothetical experiments with $P < 0.04$, 45/83 = 54.2% occurred when receptor number really increased. The other 45.8% of the low P values were Type I errors, caused by random sampling. So the new

PRIOR PROBABILITY = 0.60	RECEPTOR NUMBER REALLY INCREASES BY MORE THAN 100 SITES/CELL	NULL HYPOTHESIS IS TRUE: RECEPTOR NUMBER DOESN'T CHANGE	TOTAL
P < 0.04	540	16	556
P > 0.04	60	384	444
Total	600	400	1,000

Table 48.1. Student's perspective.

PRIOR PROBABILITY = 0.05	RECEPTOR NUMBER REALLY INCREASES BY MORE THAN 100 SITES/CELL	NULL HYPOTHESIS IS TRUE: RECEPTOR NUMBER DOESN'T CHANGE	TOTAL
P < 0.04	45	38	83
P > 0.04	5	912	917
Total	50	950	1,000

Table 48.2. Advisor's perspective.

evidence is not convincing to the advisor, because she still thinks there is only about a 50% chance that the effect is real.

b. The results could have been obtained in two ways.

First possibility: The treatment really does not change receptor number, and the change we observed is simply the result of coincidence. The statistical calculations tell you how unlikely such a coincidence is. For this example, you would see a difference as large as or larger than the difference you observed in 4% of experiments, even if the null hypothesis were true.

Second possibility: The hormone really does increase receptor number.

In interpreting the results, you must decide which of these possibilities is more likely. The student in this example thought that the second possibility was probably true even before collecting any data. She is far happier believing this hypothesis is true rather than believing in a coincidence that happens only 1 in every 25 experiments.

The advisor has reasons to think that this experiment shouldn't work. Perhaps she knows that the cells used do not have receptors for the hormone used, so it seems very unlikely that the hormone treatment could alter receptor number. The advisor must choose between believing that an "impossible" experiment worked and believing that a 4% coincidence has occurred. It is a toss up, and the conclusion is not clear.

Interpreting experimental results requires integrating the results from one particular experiment (summarized with a P value) with your prior opinions about the experiment. If different people have different prior opinions, they may reach different conclusions when the P value is only moderately low.

PRIOR PROBABILITY = 0.05	RECEPTOR NUMBER REALLY INCREASES BY MORE THAN 100 SITES/CELL	NULL HYPOTHESIS IS TRUE: RECEPTOR NUMBER DOESN'T CHANGE	TOTAL
P < 0.001	45	1	46
P > 0.001	5	949	954
Total	50	950	1,000

Table 48.3. Advisor's perspective if the P value was 0.001.

    c. Now the advisor must choose between a result that makes no sense to her or a coincidence that happens only 1 time in 1,000. She believes the result rather than such an unlikely coincidence. See Table 48.3. Although she probably makes this judgment intuitively, you can explain it with Table 48.3. Given the assumptions of this problem, only 1 in 46 results like this would be caused by chance, whereas 45 of 46 would reflect a true difference. (To maintain the power with a smaller threshold value of P requires increasing the sample size.)

C5. **All members of a class of 111 third-year pharmacy practice majors volunteered for a study in which a single blood sample was taken from each student. All blood samples were analyzed by the same technician for the concentration of a natural human biochemical in plasma, Compound X, in nanograms per milliliter. No additional patient identifiers or covariates were recorded.**

    The investigator, after looking over the data from the 111 volunteers (see Figure 47.1, left), hypothesized that the subjects came from two distinct populations, one with low values of Compound X and one with high values of Compound X. He divided the data set at the overall median into two groups, a low group (5–44 ng/ml) and a high group (47–1,466 ng/ml). There were no values between 44 and 47. For all of the subjects, n = 111; mean = 159; SD = 250; and median = 44. For the low group, n = 56; mean = 20.0; SD = 9.45; and median = 19.8. For the high group, n = 55; mean = 300; SD = 295; and median = 220. The investigator compared the two groups (shown in Figure 47.1, right) using the nonparametric Mann–Whitney test (see Chapter 41). The P value was tiny, less than 0.0001. Interpret what this P value means and what you can conclude from the study.

    A P value can only be interpreted when you have random samples from populations and you want to make inferences about those populations. Here the investigator created the two groups based on the concentration of Compound X. Of course, the two means and medians are farther apart than you'd expect to see by chance. There is zero possibility that chance had anything to do with it. The investigator decided to put the higher values in one group and the lower values in the other. Essentially, this investigator has used statistics to attempt to prove that high numbers are larger than low numbers!

The P value is completely meaningless and provides no help at all in interpreting the data. No conclusion can be reached from these data.

C6. **Assume that you are reviewing a manuscript prior to publication. The authors tested many drugs to see whether they increased enzyme activity. All P values were calculated using a paired t test. Each day, they also performed a negative control, using just a vehicle with no drug. After discussing significant changes caused by some drugs, they state, "In contrast, the vehicle control has consistently given a nonsignificant response (P > 0.05) in each of 250 experiments." Comment on these negative results. Why are they surprising?**

In the negative control experiments, the null hypothesis is true. Therefore, you expect 1 of 20 control experiments to have a P value less than 0.05. In 250 experiments, therefore, you expect to see about $0.05 \times 250 = 12.5$ significant responses. You'd be very surprised to see zero significant responses among the 250 control experiments.

C7. **The serum level of a hormone (the Y factor) was measured to be 93 ± 1.5 units/ml (mean $\pm$ SEM) in 100 nonpregnant women and 110 ± 2.3 units/ml in 100 women in the first trimester of pregnancy. The difference between the means is $110 - 93 = 17$, with a 95% CI extending from 11.6 to 22.4 units/ml.**
   a. **Interpret that CI.**
   b. **Estimate the P value.**
   c. **What assumptions are you making to answer a and b?**
   d. **Would this be a useful test for diagnosing pregnancy?**

   a. If we assume that both samples are representative of the overall populations, we can be 95% sure that the mean level of the Y factor in pregnant women is between 11.6 and 22.4 units/ml higher than it is in nonpregnant women.
   b. The null hypothesis is that the two means are the same, so the difference between means is zero. Because the 95% CI does not include zero, the P value must be less than 0.05. Because the 95% CI doesn't come near zero, the P value must be a lot less than 0.05.
   c. We assume that the two samples are representative of the overall populations of pregnant and nonpregnant women of the appropriate age range and that the distribution of Y values is approximately Gaussian. Because the samples are so large, this assumption is not very important so long as the distributions aren't really weird. We do have to assume that the two populations have similar SDs and that the subjects have each been selected independently.
   d. Don't get trapped by thinking only about the P value and the CI. Those results convince us that the difference between the mean values is extremely unlikely to be a coincidence. It certainly is not surprising to find that the mean value of the hormone is elevated in pregnant women (it might be surprising, however, to find a hormone whose level does not change in pregnancy!).

But would the test be useful for diagnosing pregnancy? To answer this question, we need to think beyond means and consider the actual distribution of the values. The problem doesn't present the raw data, but it does present the SEM and the sample size. It is easy to compute the SDs. This is done by multiplying the SEM by the square root of the sample size. Therefore, the SD of the nonpregnant women is 15; the SD of the pregnant women is 23. You expect about 95% of the values to lie within 2 SDs on either side of the mean. For the nonpregnant women, this interval extends from 63 to 123. For pregnant women, it extends from 64 to 156. The two distributions overlap considerably, so the test would be completely useless as a diagnostic test.

Don't confuse the CI of the mean with the range of the data.

C8. **Cohen et al. (1993) investigated the use of active cardiopulmonary resuscitation (CPR). In standard CPR, the resuscitator compresses the victim's chest to force the heart to pump blood to the brain (and elsewhere) and then lets go to let the chest expand. Active CPR is done with a suction device. This enables the resuscitator to both pull up on the chest to expand it and press down to compress it. These investigators randomly assigned cardiac arrest patients to receive either standard or active CPR. Eighteen of 29 patients treated with active CPR were resuscitated. In contrast, 10 of 33 patients treated with standard CPR were resuscitated.**

   a. **Using Fisher's test, the two-sided P value is 0.0207. Explain what this means in plain language.**
   b. **What would a Type I error mean in this context?**
   c. **What would a Type II error mean in this context?**

   a. The null hypothesis is that the resuscitation rate is the same for active and standard CPR. If that hypothesis were true, the chance of observing a difference as large as or larger than that observed in a study this size is 2.07%.
   b. If you set $\alpha = 0.05$, you would conclude that the resuscitation rates are significantly different. You'd be making a Type I error if the resuscitation rates in the overall populations were really identical. You would conclude that there is a real difference, when in fact (but unknown to you) the difference you observed is just a coincidence of sampling.
   c. If you set $\alpha = 0.01$, you would conclude that the resuscitation rates are not significantly different. You'd be making a Type II error if the rates in the overall populations were really different.

C9. **Which of the following are inconsistent?**
   a. **Mean difference = 10. The 95% CI is –20 to 40. P = 0.45.**
   b. **Mean difference = 10. The 95% CI is –5 to 15. P = 0.02.**
   c. **Relative risk = 1.56. The 95% CI is 1.23 to 1.89. P = 0.013.**
   d. **Relative risk = 1.56. The 95% CI is 0.81 to 2.12. P = 0.04.**
   e. **Relative risk = 2.03. The 95% CI is 1.01 to 3.13. P < 0.001.**

a. Consistent P > 0.05 so the 95% CI must include 0.0.
b. Inconsistent. The 95% CI includes zero, so the P value must be greater than 0.05.
c. Consistent P < 0.05 so the 95% CI can not include 1.0.
d. Inconsistent. The P value is less than 0.05, so the 95% CI for the relative risk cannot include 1.0.
e. Inconsistent. The P value is way less than 0.05, so the 95% CI must not include 1.0 and shouldn't even come close. But the 95% CI starts at a value just barely greater than 1.00.

C10. **In response to many case reports of connective tissue diseases after breast implants, the FDA called for a moratorium on breast implants in 1992. Gabriel and colleagues (1994) performed a prospective study to determine whether there really was an association between breast implants and connective tissue (and other) diseases. They studied 749 women who had received a breast implant and twice that many control subjects. They analyzed their data using survival analysis to account for different times between implant and disease and to correct for differences in age. You can analyze the key findings more simply. They found that 5 of the cases and 10 of the controls developed connective tissue disease.**
a. **What null hypothesis would a P value test?**
b. **What is that P value (hint: no calculations or computer needed)?**

a. The null hypothesis is that there is no association between breast implants and connective tissue diseases. If the null hypothesis is true, the overall incidence of connective tissue diseases among cases ought to equal the incidence among controls.
b. The investigators used twice as many controls as patients with breast implants, and exactly twice as many controls as cases developed connective tissue disease. This is exactly what you expect under the null hypothesis. The relative risk is exactly one. The data provide no evidence whatsoever of any association between breast implants and connective tissue disease.

The P value (two-tailed) must be 1.0. The P value answers the question, If the null hypothesis is true, what is the chance of randomly sampling subjects and obtaining such strong (or stronger) evidence that there is an association? Because there was absolutely no evidence of association in this study, 100% of all studies would produce evidence this strong or stronger.

C11. **Figure 47.2 shows data from 461 meta-analyses that all showed a statistically significant effect (P < 0.05). A meta-analysis pools data from several (or many) studies to obtain one P value and CI that take into account all the data. Each symbol in the graph represents one such study. The graph shows that studies with larger samples tend to show a smaller effect size than do smaller (presumably less accurate) studies. Ioannidis (2008) presents that as evidence for the winner's curse explained at the end of Chapter 44.**

Figure 47.2 shows Spearman's correlation coefficient and the corresponding P value.

a. **Interpret the meaning of the P value.**
b. **The graph is plotted using a logarithmic scale for both axes, but it isn't clear whether the correlation was computed using the raw data or the log-transformed data. Can you tell from the figure? Does it matter? Would the two P values be different or identical?**

a. Spearman's nonparametric correlation coefficient quantifies the correlation between the ranks of the X values (sample size) and the ranks of the Y values (odds ratio). The null hypothesis is that there is no correlation between the two sets of ranks. If that null hypothesis were true, there would be less than a 0.01% chance that random chance would lead to a correlation coefficient this far away (or even farther) from zero.

It only makes sense to interpret the P value if these data are sampled from a larger population. Indeed, it makes sense to think of the meta-analyses summarized here to be representative of other meta-analyses not included here or yet to be published.

b. The nonparametric correlation only analyzes the ranks, not the raw data. The ranks of sample size and odds ratio are the same before and after transforming to logarithms. Thus, the value of r_s will be identical for the raw data and the log-transformed data, and so will the corresponding P value.

C12. **Two gentleman, John and Frank, decided to fly in a hot air balloon across the Atlantic Ocean, from North Carolina to Europe. After a frightening five-week journey, the pair finally spotted land. Frank thought that they were now over Ireland. They saw a man in a tweed suit walking along a path. John shouted down to the man, "Where are we?" The man stopped walking, looked up, thought for five seconds, and replied, "You're in a balloon." Frank then said to John, "That man must be a statistician." John asked, "How did you know?" What was Frank's answer?**

Frank answered, "He must be a statistician because his answer was 100% correct, but not the answer to the question that I had in mind." I'm not sure of the origin of this joke, but I think it does a good job of illustrating some of the miscommunication and occasional tension between biomedical scientists and statisticians.

D. PROBLEMS ABOUT SAMPLE SIZE AND POWER

D1. **Which will require more subjects, detecting a difference between two proportions when p1 = 0.10 and p2 = 0.20 or detecting a difference between two proportions when p1 = 0.20 and p2 = 0.40?**

Assume you are looking for the same difference with the same power and are using samples of equal size.

Equation 26.2 shows that the needed sample size is proportional to $p_{av}(1-p_{av})$, where p_{av} is the average of the two anticipated proportions. For the first scenario, this product is $0.15 \times 0.85 = 0.13$. For the second scenario, the product is $0.30 \times 0.70 = 0.21$. So the second scenario will need more subjects (about 50% more).

D2. **You read the following in a journal: "Before starting the study, we calculated that with a power of 80% and a significance level of 5%, 130 patients would be required in each group to demonstrate a 15-percentage-point reduction in mortality (from the expected rate of 33 down to 18%)." Explain in plain language what this means.**

If the null hypothesis were really true (the two treatments are identical in terms of mortality), then there is a 5% chance of obtaining a significant difference by a coincidence of random sampling. (The statement that the "significance level is 5%" means that $\alpha = 0.05$.) If the alternative hypothesis is true (that the mortality drops from 33 to 18%), then there is an 80% chance that a study with 130 patients in each group will end up with a statistically significant difference and a 20% chance of ending up with a conclusion that the difference is not statistically significant.

D3. **Lymphocytes contain β-adrenergic receptors. Epinephrine binds to these receptors and modulates immune responses. It is possible to count the average number of receptors on human lymphocytes using a small blood sample. You wish to test the hypothesis that people with asthma have fewer receptors. By reading various papers, you learn that there are about 1,000 receptors per cell and that the CV in a normal population is about 25%.**

   a. **How many asthmatic subjects do you need to determine the receptor number to plus or minus 100 receptors per cell with 95% confidence?**

   b. **You want to compare a group of normal subjects with a group of asthmatics. How many subjects do you need in each group to have 80% power to detect a mean change of 10% of the receptors using $\alpha = 0.05$ (two tailed)?**

   c. **How many subjects do you need in each group to have 95% power to detect a mean difference in receptor number of 5% with $\alpha = 0.01$?**

   a. Equation 26.3 computes approximate sample size to determine a mean with desired precision. To use this equation, you need to know s, the expected SD of the values. The problem states that the coefficient of variation is 25%, so s equals 25% of the mean (1,000 receptors/cell), which is 250 receptors/cell. The desired margin of error, w, is stated to be 100 receptors/cell. To use Equation 26.3, we need a value for C, which combines power and α (or confidence level) as defined in Table 26.1. When computing sample size needed to obtain a confidence interval with a specified margin of error, it is customary to use 50% power, so that it is equally likely that the confidence interval will be wider or narrower than desired. From Table 26.1, to have 95% confidence ($\alpha = 0.05$) and 50% power, C = 3.8. Now we can use Equation 26.3:

$$n = C\left(\frac{s}{w}\right)^2 = \frac{3.8 \cdot 250^2}{100^2} = 24$$

What does this mean? Assume the values follow a Gaussian distribution, that the SD is 250 sites/cell, and that you collect many samples of n = 24 and compute the 95% CI of each. You would expect the margin of error of a 95% CI to equal 100 receptors/cell on average. Half of the samples would have a CI narrower than 100 receptors/cell and half the samples would have a CI wider than 100 receptors/cell.

b. Equation 26.1 presents the equation to compute sample size for the difference between two means. The problem states that the coefficient of variation is 25%, so s equals 25% of the mean (1,000 receptors/cell), or 250 receptors/cell. The problem states that the margin of error, w, is 10% of the mean receptor number, so $0.10 \times 1,000$, or 100. Look up the value of C in Table 26.1. Given 80% power and $\alpha = 0.05$, C = 7.8. Now use Equation 26.1:

$$n = 2C\left(\frac{s}{w}\right)^2 = 2 \cdot 7.8\left(\frac{250}{100}\right)^2 = 97.5 \approx 98$$

Assume the values follow a Gaussian distribution, that the SD is 250 sites/cell, and that the two population means differ by 100 receptors per cell. You now perform many experiments with n = 98 in each group and calculate an unpaired t test in each. You would expect to find a statistically significant difference ($P < 0.05$) in 80% (the power) of those experiments and a difference that is not statistically significant in the other 20%.

c. This problem has changed three things from the previous one. The study is looking for a difference that is half as large (5% vs. 10%). This alone would increase the sample size by a factor of four. You also want more power (95% vs. 80%), which also increases the needed sample size. Furthermore, now the definition of statistical significance is stricter, with $P < 0.01$ instead of $P < 0.05$. Each of these changes increases the required sample size. All three changes will increase the sample size a lot. How much? Table 26.1 tells you that C = 17.8 (with 95% power and $\alpha = 0.01$). Use Equation 26.1:

$$n = 2C\left(\frac{s}{w}\right)^2 = 2 \cdot 17.8\left(\frac{250}{50}\right)^2 = 890$$

You need 890 subjects in each group.

D4. **You are preparing a grant for a study that will test whether a new drug treatment lowers blood pressure substantially. From previous experience, you know that 15 rats in each group is a large enough sample size and that the SD of measurements in different rats is about 10 mmHg. Prepare a convincing power analysis to justify that sample size.**

Start with the equation for computing n for comparing two means, plug in the numbers we know, and solve for w. Setting C to 7.8 (standard for 80% power and $\alpha = 0.05$):

$$n = 2 \cdot 7.8 \left(\frac{10}{w} \right)^2 = 15$$

$$w^2 = \frac{2 \cdot 7.8 \cdot 10^2}{15} = 104$$

$$w = \sqrt{104} = 10.2 \approx 10$$

Now you could write in a grant: "From previous work, we expected the SD of blood pressure to be 10 mmHg. Setting the significance level to 0.05, we calculated that a study with 15 animals in each group would have 80% power to detect a mean difference in blood pressures of 10 mmHg."

As this example shows, sample size calculations do not always proceed from the assumptions to the sample size. Sometimes, the sample size is chosen first and then justified. This is OK as long as the assumptions are reasonable.

D5. **The overall incidence of a disease is 1 in 10,000. You think that a risk factor increases the risk. How many subjects do you need if you want to have 80% power to detect a relative risk as small as 1.1 in a prospective study?**

From the problem, p_1 equals 0.00010 and p_2 equals 1.1 times p_1, or 0.00011, so $p_{av} = 0.000105$. The margin of error, w, equals the difference between the two proportions, 0.00001. You want 80% power. The problem doesn't specify a significance level, so let's use the standard value of 0.05. Thus C from Table 26.1 equals 7.8. Plug these values into Equation 26.2:

$$n = \frac{2C \cdot p_{av} \left(1 - p_{av} \right)}{w^2} = 16,483,269$$

The required sample size is huge—more than 16 million per group! Why such a large sample size? First, the disease is rare, with an incidence of only 1 in 10,000. So it will take a large number of subjects to get a reasonable number with the disease. Because you are looking for a small change in incidence, you will need even larger samples. Detecting small changes in rare diseases is virtually impossible in prospective studies. The required sample size is enormous. That is why case-control studies are so useful.

D6. **You are planning a clinical study. You know that the mean heart rate for the placebo-treated control group will be about 65 beats per minute with a standard deviation of about 8 beats per minute. You don't expect the heart rate to change with treatment (a LDL cholesterol-lowering agent). The study will set the Type I error rate, α, to the standard value of 0.05. Someone reviewing the protocol asks you what the power of the study is to detect a tiny change in pulse (1 beat per minute). Estimate that power without doing any calculations.**

Because the reviewer is asking about a change that is tiny compared to the expected standard deviation, your first thought might be that the study has zero power. But, in fact, the power can never go as low as zero. How low can it go?

If there is no difference between mean pulse rates, there is a 5% chance of obtaining a statistically significant difference by chance. That's the definition of alpha. So what is the chance (the power) of obtaining a statistically significant difference if the true difference is tiny? The answer will be greater than 5%, but not much greater.

Look back at Figures 20.1 and 26.1. Note that the power with a tiny difference is 5%, not 0%.

E. PROBLEMS ABOUT CORRELATION AND REGRESSION

E1. **Figure 47.3 shows the relationship between the population of countries (or territories) and their area. Each dot is a separate country. All 237 countries are shown. The left graph shows linear axes. The right graph shows both axes using a logarithmic axis.**
   a. Interpret the P value.
   b. Were the two variables properly assigned to the X-axis and Y-axis?
   c. Would the correlation coefficient or P value be different if the definitions of X and Y were swapped?
   d. What are the advantages of using logarithmic axes?

   a. P values help you generalize from sample data to make conclusions about the population. They help you extrapolate from your data to a more general situation. These data represent every single country there is. No generalization makes sense, so the P value is entirely without meaning.
   b. When cause and effect are clear, the horizontal axis is used to plot the variable that causes the variable plotted in the vertical axis to change. Here, cause and effect are not clear. Do countries with large areas tend to develop large populations? Or do countries with large populations tend to go to war with neighboring countries to increase their area? I suspect the former is more common, so I would plot the data the other way, with population on the vertical axis and area on the horizontal axis. But it is not a clear decision.

c. The calculation of the P value and the correlation coefficient would be identical if the axes were swapped.

d. The graph on the left uses linear axes. All but a dozen countries are in the lower-left corner of the graph, where they cannot be differentiated. A logarithmic scale stretches out lower values and scrunches up larger values. With these data, it is much easier to see the trend.

E2. **Figure 47.4 shows the same data as Figure 47.3 but is fit with a linear regression program.**

a. **What are the units of the slope? What does it mean?**

b. **Interpret the P value.**

c. **Imagine that the graph of data remained the same but the definitions of the independent (X) and dependent (Y) variables were switched. Now imagine that linear regression is used to find the line that best predicts population (horizontal axis) from area (vertical axis). Would the best-fit line be the same or different?**

d. **On logarithmic axes, why is the "line" curved?**

e. **On the logarithmic graph, why does the "line" stray so far from many points?**

f. **Figure 47.5 was also fit by linear regression. Why does it show a different line and different R_2 than those shown in Figure 47.4?**

a. The slope quantifies how much Y changes for each unit change in X. So the slope is expressed in units of the Y-axis divided by units of the X-axis, square kilometers per person. Here the slope is 0.007128. So for every increase in population (by one person), the area (on average) increases by 0.007128 square kilometers.

b. Normally the P value is easy to interpret from linear regression. The null hypothesis is that the true slope is horizontal (i.e., slope = 0.0). The P value answers the question, If this null hypothesis is true, what is the chance that randomly sampled data would have a slope as far from 0.0 as the actual data? But that only makes sense when the data are sampled from a larger population. This graph shows data for every single country that exists. No extrapolation or generalization is needed, so the P value is meaningless.

c. Linear regression minimizes the sum of the square of the discrepancies between actual and predicted Y values. Which variable is defined to be X and which is defined to be Y makes a difference. If you swapped the definitions of the two variables, a different best-fit line would result. If you kept the axes the same, then linear regression with X and Y defined differently would minimize the sum of the square of the horizontal distances of the points from the line. This would be a different line (in almost all cases) than the line that minimizes the sum of the square of the vertical distances of the points from the line. Although the line would be different, R^2 and the P value would be the same.

d. The curve shown on the graph was indeed drawn by linear regression after fitting the model: area = intercept + slope × population. When drawn on a graph with linear axes, this model appears as a straight line. On a graph with logarithmic axes, the fit of that model appears curved. A linear regression model only appears as a straight line when plotted on ordinary (not logarithmic) axes.

e. The curve on the graph is actually a linear regression line that minimizes the sum of the square of the difference between the actual area and the predicted area. On a linear graph, this method minimizes the sum of squares of the vertical distances between the dots and the line. But the logarithmic scale distorts the scale of the data, so the visible distances of the points from the curve are misleading.

f. Here the data were transformed to logarithms. The linear regression program was given the log(population) values as X and the log(area) values as Y. The line on the graph fits the model:

log(area) = intercept + slope × log(population).

This is a different model than was fit before, so it yields a different best-fit line on the graph and a different R^2 value.

E3a. **Figure 47.6 (left) shows a log(dose)–response curve, fit with nonlinear regression. The shaded region shows the 95% CI for the EC50 (the dose that provokes a response halfway between the lowest and highest possible responses). Why is this range so wide?**

The results next to the graph show that the nonlinear regression program was asked to find the best-fit values of four parameters. It fit not only the log(EC50) and the Hill slope (a measure of how steep the curve is), but also the top and bottom plateaus of the curve. These data simply do not plateau, so the data really give no hint about where the curve will level off. The best-fit value of the top parameter is quite uncertain, with a 95% CI ranging from 72 to 209. This top plateau defines 100%. If you can't define 100% (or 0%) very precisely, then you can't define 50% very precisely, and so the EC50 will be uncertain.

E3b. **Figure 47.6 (right) shows the same data, also with a log(dose)–response curve fit by nonlinear regression. Why is the shaded area denoting the 95% CI of the logEC50 so much narrower?**

The Y-axis is labeled "percent response," suggesting that the data have been normalized to some controls. The results shown on the graph on the right side of the figure were done with constraints so that the nonlinear regression program forces the curve to run from 0 to 100, without fitting the top and bottom plateaus. You can see that this is the case because the program does not provide best-fit values, or CI, for the parameters Top and Bottom. With this fit, there is no ambiguity about the top and bottom plateaus of the curve, so there is no ambiguity about the best-fit value of the log(EC50). Accordingly, the CI for the EC50 is much narrower.

E3c. **One of the points shows a response greater than 100%. Should it be removed because that is impossible?**

No! Data points should only be removed when they are clearly in error (and even then some would disagree). The 100% value is the average response with maximum drug. You expect half of the measurements at high drug concentrations to be less than 100% and half to be greater than 100%, so you may have an average greater than 100%. Values greater than 100% are not impossible!

E4a. **Figure 47.7 shows data and a line fit by linear regression. Interpret the R2 and P value shown on the graph.**

The R^2 is 0.895. This means that the linear regression model can explain 89.5% of the variation among the Y values, leaving 10.5% of the variation that is scatter from the regression line. The P value is tiny, less than 0.0001. That means that random sampling of points would lead to such a large R^2 (or a slope so far from horizontal) in less than 0.01% of experiments. The large R^2 and tiny P convince you that a linear regression model fits the data much better than a horizontal line. But it should not convince you that linear regression is the best model. A glance at the graph tells you that the linear regression model is incomplete. The data are curved.

E4b. **Figure 47.7 also shows results from a runs test. This is not a test covered in this book. Read about it in other books or on the Web and interpret what the P value means.**

When you look at Figure 47.7, it seems as though the data points form a curve, not a straight line. If you had an alternative model in mind, you could fit both models and compare as explained in Chapter 35. The runs test is simpler. It just asks whether the data deviate systematically from the fit of a line or curve. A run is one or more sequential points on the same side of the line or curve. These data only have three runs: 5 points on the left side below the line, 14 points in the middle above the line, and 6 points on the right side below the line. The null hypothesis for the runs test is that whether each point ends up above or below the line (or curve) is determined randomly and independently for each point. The P value answers the question, What is the chance that random sampling would result in as few or fewer runs than observed? Here, the P value is tiny. From this, you can conclude that the curvature of the data is statistically significant and that the straight-line model is probably not the best model for these data.

E5. **Arad and colleagues (2000) performed a large study to see whether measuring calcium in coronary arteries can help predict heart attacks. They studied 1,177 people who were free of cardiac symptoms. They measured calcium in the coronary arteries using a noninvasive method called electron beam–computed tomography. The results are reported as a coronary artery calcium score. They followed all these people for three to four years and recorded cardiac** events (3 people died, 15 had a nonfatal myocardial infarction, and 21 required a revascularization procedure to relieve chest pain). The question is whether the calcium score predicts the occurrence of these cardiac events better than the standard cardiac risk factors (age, high cholesterol, high blood pressure, diabetes). Complete data were available for 787 of the subjects. The data were analyzed by logistic regression, and the results are shown in Table 47.2.

a. Odds ratios compare the risk in one group with the risk in another. When the logistic regression model computes odds ratios, what group is the comparison group?

b. Which of the five variables make a statistically significant ($P < 0.05$) contribution to the model? Which has the highest P value?

c. Write out the logistic regression model.

d. What is the predicted odds ratio for someone who is older than 55 and has a coronary artery calcium score greater than 160?

e. Did this study have enough subjects?

f. Most of the cardiac events (21 of 39) were people requiring angioplasty or coronary artery surgery to restore blood flow in the coronary arteries. In some cases, the decision to perform these procedures might have been a difficult one. How would the results be affected if the decision to perform the procedures was partially influenced by knowing that the coronary calcium score was high?

g. Explain in plain language what the odds ratio of 14.3 means.

a. The comparison group is the one defined by setting all independent variables to zero. With this model, the comparison group is people who have a coronary artery calcium score less than 160; are less than 55 years old; and do not have elevated cholesterol, hypertension, or diabetes.

b. None of the five CIs includes an odds ratio of 1.0 (the null hypothesis), so all five P values must be less than 0.05. The odds ratio for hypertension is closest to 1.0, so it will have the highest P value.

c. The model is

$$\ln(OR) = X_1 \ln(14.3) + X_2 \ln(3.3) + X_3 \ln(4.0) + X_4 \ln(2.6) + X_5 \ln(4.8)$$

The five X variables are defined in Table 48.4.

d. Using the definitions of Table 48.4, X_1 and X_2 are equal to 1.0 and the rest of the X variables are equal to 0. So,

$$\ln(OR) = \ln(14.3) + \ln(3.3) = 2.66 + 1.19 = 3.85$$
$$OR = e^{3.85} = 47.2$$

If you assume that coronary events are rare, this can be interpreted as a relative risk. Compared with people younger than 55 with low coronary artery calcium scores, people older than 55 with high coronary

IF . . .	THEN SET . . .	OTHERWISE . . .
Coronary calcium score > 160	$X_1 = 1$	$X_1 = 0$
Age > 55 years old	$X_2 = 1$	$X_2 = 0$
Elevated cholesterol	$X_3 = 1$	$X_3 = 0$
Hypertension	$X_4 = 1$	$X_4 = 0$
Diabetes	$X_5 = 1$	$X_5 = 0$

Table 48.4. Definitions of the independent variables for multiple regression.

artery calcium scores have about 47 times the risk of dying from a cardiac event, having a myocardial infarction, or requiring coronary artery surgery or angioplasty.

e. Chapter 38 discussed sample size for studies analyzed with multiple logistic regression. There are no firm rules, but Chapter 38 gives several guidelines. The number of subjects needed is based on the number of independent variables included in the regression model (m) and the number of events (n). In this study, there were 39 events (3 people died, 15 had a nonfatal myocardial infarction, and 21 required a revascularization procedure to relieve chest pain), so n = 39. The model has five independent variables, so m = 5 (actually, they included and then rejected a few other variables, but we won't discuss those here). So there were about eight events per variable (n = 7.8 m). The sample size was a bit lower than the lowest recommended. Of course, when planning a study, the investigators rarely can choose the number of events. Instead, they choose the number of subjects and must predict the number of events.

f. The whole point of the study is to ask whether the high coronary artery calcium score is associated with more cardiac events. More than half of the events tracked by this study were procedures (angioplasty or coronary artery surgery) to increase flow through blocked coronary arteries. If the decision to perform these procedures was based on the high coronary calcium levels, then the study would be meaningless. It would show that high calcium levels in coronary arteries are associated with procedures performed because the calcium level is high. Testing that kind of tautology is not helpful.

The authors recognized this problem and tried to avoid it. They did not count procedures performed on patients without acute symptoms and did not count procedures performed because of the high calcium levels. But it seems that the people deciding whether to perform the angioplasty or coronary artery bypass knew the calcium score and could have used that score as part of their decision-making process. If this happened, the results would be impossible to interpret.

g. It is easiest to think of an odds ratio as an approximation of a relative risk. The odds ratio reported by logistic regression for having a coronary artery calcium score greater than 160 equals 14.3. Logistic regression is used when the outcome is binary. Here, the outcome is having a coronary event (death, heart attack, or need for angioplasty or coronary artery surgery) within three to four years.

Someone with a calcium score greater than 160 has 14.3 times the risk of having a coronary event than someone who has a calcium score less than 160 but is in the same age group (greater or less than 55) and is matched for the other three variables (presence or absence of elevated cholesterol, elevated blood pressure, or diabetes).

Of course, this odds ratio is meaningful only within the population these investigators studied. They decided not to include anyone who already had a heart attack, so this study cannot compute the odds ratio for having a second heart attack. The study excluded anyone who had chest pains, so the results cannot be used to predict future coronary events in someone with chest pains.

E6. **The "Oscar curse" is a phenomenon in which the career of an actor or actress often plummets just after he or she wins an Academy Award (Oscar) for best or best supporting actor or actress. Some examples of actors and actresses who are said to have been hexed by the Oscar curse include Halle Berry, Mira Sorvino, Cuba Gooding Jr, Marisa Tomei, Roberto Benigni, F. Murray Abraham, Adrien Brody, and Marlee Matlin. From Chapter 1, which statistical/probability concept is exemplified by the "Oscar curse"?**

The "Oscar curse" is an example of regression to the mean. This assumes that winning a major Academy Award requires not just talent, but also luck. An actor or actress usually can't give an Oscar-winning performance unless he or she is lucky enough to get a role in a great movie, with a great director and a great cast. Many Oscar-winning actors and actresses won't be so lucky again. Some notable exceptions of actors and actresses who seem to never have been hexed by the Oscar curse include Meryl Streep, Daniel Day Lewis, Jack Nicholson, and Katharine Hepburn.

PART J

Appendices

APPENDIX A

❖❖❖

Statistics with GraphPad

WHAT IS GRAPHPAD PRISM?

All the figures in this book were created with GraphPad Prism, and most of the analyses mentioned in this book can be performed with Prism (although not multiple and logistic regression).

GraphPad Prism, available for both Windows and Macintosh computers, combines scientific graphing, curve fitting, and basic biostatistics. It differs from other statistics programs in many ways:

- Statistical guidance. Prism helps you make the right choices and make sense of the results. If you like the style of this book, then you'll appreciate the help built into Prism.
- Analysis checklists. After completing any analysis in Prism, click the clipboard icon to see an analysis checklist. Prism shows you a list of questions to ask yourself to make sure you've picked an appropriate analysis.
- Nonlinear regression. Some statistics programs can't fit curves with nonlinear regression and others provide only the basics. Nonlinear regression is one of Prism's strengths, and it provides many options (e.g., remove outliers, compare models, compare curves, interpolate standard curves, etc.).
- Automatic updating. Prism remembers the links among data, analysis choices, results, and graphs. When you edit or replace the data, Prism automatically updates the results and graphs.
- Analysis choices can be reviewed and changed at any time.
- Error bars. You don't have to decide in advance. Enter raw data and then choose whether to graph each value or the mean with SD, SEM, or CI. Try different ways to present the data.

WHAT YOU NEED TO KNOW BEFORE
LEARNING GRAPHPAD PRISM

Prism is designed so you can just plunge in and use it without reading instructions. But avoid the temptation to sail past the Welcome dialog so you can go straight to the analyses. Instead, spend a few moments understanding the choices on the Welcome dialog (see Figure A1). This is where you start a new data table and graph, open a Prism file, or clone a graph from one you've made before.

The key to using Prism effectively is to choose the right kind of table for your data, because Prism's data tables are arranged differently than those used by most statistics programs. For example, if you want to compare three means with one-way ANOVA, you enter the three sets of data into three different columns in a data table formatted for entry of column data. With some other programs, you'd enter all the data into one column and also enter a grouping variable into another column to define the group to which each value belongs. Prism does not use grouping variables.

To understand how the various tables are organized, start by choosing some of the sample data sets built into Prism. These sample data sets come with explanations about how the data are organized and how to perform the analysis you want. After experimenting with several sample data sets, you'll be ready to analyze your own data.

You can download a free demo from www.graphpad.com. The demo is fully functional for 30 days.

Figure A1. The Welcome dialogue of GraphPad Prism.

Before entering data, you must choose which kind of data table is appropriate for your data. You can also choose sample data sets, accompanied by instructions.

ABOUT GRAPHPAD SOFTWARE

History

I wrote the first GraphPad program in 1984 and founded GraphPad Software, Inc., in 1989. The statistics programs available then were wonderful tools for statisticians but were overkill for scientists and researchers with limited statistical backgrounds. I created GraphPad Software to provide data analysis software designed for students and scientists, with plenty of statistical guidance built in. Today, a dozen full-time programmers develop GraphPad programs. More than 100,000 scientists and students around the world use Prism, and thousands use our free Web QuickCalcs every day.

GraphPad StatMate

GraphPad StatMate helps you decide how many data points you'll need for an experiment (see Figure A2). The concepts of power and sample size were explained in Chapter 26.

Some programs ask how much statistical power you desire and how large an effect you are seeking and then tell you what sample size you should use. The problem with this approach is that most people find those questions impossible to answer. You want to design a study with very high power to detect very small effects and with a very strict definition of statistical significance. But doing so requires a huge number of subjects, more than you can afford.

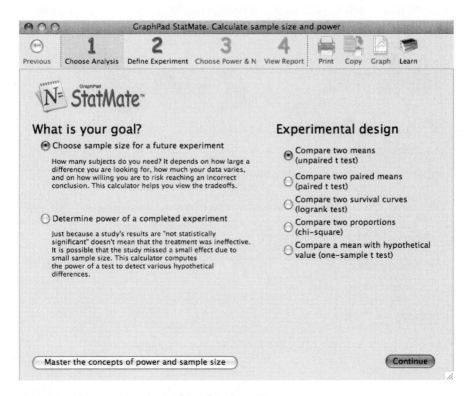

Figure A2. The opening screen of GraphPad StatMate.

StatMate uses a unique approach. First, you define the experimental system. If you are comparing two means, you must estimate the SD (from previous data). StatMate then shows you a table of the size of the effect (difference between means) that can be detected as a function of power (various values shown) and sample size. This table shows you the trade-offs among power, sample size, and detectable effect.

StatMate also helps you interpret experiments in which the effect is not statistically significant. This doesn't prove that a treatment had zero effect (see Chapter 19), only that the observed effect is not greater than you'd often see by chance alone. StatMate shows you the power of your completed experiment to detect various hypothetical differences.

GraphPad InStat

GraphPad InStat makes it really simple to do basic statistical calculations. Unlike Prism, it doesn't make customizable graphs, doesn't fit curves with nonlinear regression, doesn't analyze survival data, and cannot do two-way ANOVA. The advantage of InStat is that it guides you through every step, so there is no learning curve at all.

Free GraphPad Web QuickCalcs

QuickCalcs are free calculators on www.graphpad.com. We call them calculators because you enter data and instantly get back results, without saving or opening files.

The QuickCalcs system offers a diverse array of calculations (see Figure A3). The most popular calculators include t test, outlier detection, chi-square, CI of a proportion or count, and calculating a P value from a statistical ratio (i.e., t or F). All of the calculators are either self-explanatory or link to longer explanations.

Figure A3. The main screen of the free QuickCalcs at www.graphpad.com.

APPENDIX B

❖❖❖

Statistics with Excel

THE PROS AND CONS OF USING EXCEL
FOR STATISTICAL CALCULATIONS

Microsoft Excel is widely used and is a great program for managing and wrangling data sets. Excel has some statistical capabilities, and many also use it to do some statistical calculations.

The use of Excel for statistics is somewhat controversial, and some recommend that Excel not be used for statistics. One problem is that Excel is far from a complete statistics program. It lacks nonparametric tests, multiple comparisons tests following ANOVA, and many other tests. Another problem is that Excel reports statistical results without including all the supporting details other programs provide.

More seriously, Excel uses some poor algorithms for computing statistics, which can lead to incorrect results (Knusel, 2005; McCullough & Wilson, 2005). Microsoft has responded to these criticisms and fixed many issues in Excel 2003 (Microsoft, 2006). Don't use earlier versions of Excel for statistical work.

Unfortunately, some errors remain in Excel 2007 for Windows and Excel 2008 for Mac. McCullough and Hellser (2008) have pointed out many erroneous results produced by Excel 2007 (especially its Solver) and have concluded, "Microsoft has repeatedly proved itself incapable of providing reliable statistical functionality." Yalta (2008) reached a similar conclusion, stating that "the accuracy of various statistical functions in Excel 2007 range from unacceptably bad to acceptable but inferior." In contrast, Pace (2008) concluded that Microsoft has fixed the important bugs, leaving only statistical bugs that are trivial or obscure. He contended that Excel 2007 is a reasonable choice for analyzing the kinds of data most academics and professionals collect.

Given these problems, you should use another program to check important calculations, especially if your data seem unusual or include missing values.

WHAT YOU NEED TO KNOW BEFORE USING
EXCEL FOR STATISTICS

- If you want to compute the mean of a range of numbers, there is no MEAN() function. Use the AVERAGE() function.
- Excel has no function to compute the SEM. You'll need to compute from the SD and n using this equation:

= STDEV (range)/SQRT (COUNT (range))

- To compute most CIs, you must know the critical value of t (called t* in this book) for specified number of df and confidence (C, in percent, usually 95). Do not use Excel's CONFIDENCE() function, which is based on the z (normal) distribution, rather than the t distribution, and so has limited utility. Use this syntax:

=TINV (1–0.01 * C, df)

- The Help within Excel does not always provide useful information about statistical functions. Google for other sources of information.
- Excel provides an Analysis ToolPak, which can perform some statistical tests. You need to use the Add-in Manager to install it before you can use it. Unlike Excel equations, the results of the ToolPak are not linked to the data. If you edit the data, the results will remain fixed until you run the analysis again.
- Beware of Excel's RANK() function. Nonparametric tests assign all tied values the average of the ranks for which they tie. Excel's RANK() function assigns all tied values the lowest of the ranks for which they tie, and the other ranks are not used.
- Beware of the NORMDIST(z) function (Gaussian distribution). It looks like it ought to be similar to the TDIST(t) function (Student t distribution), but the two work very differently. Experiment with sample data before using this function.
- The excellent book by Pace (2008) gives many details about using Excel to do statistical calculations. It can be purchased as either a printed book or as a pdf download.
- Starting with Excel 2010 (Windows) and 2011 (Mac), Excel includes a new set of statistical functions. When you have a choice, use the newer ones, which all are named using several words separated by periods. For example, the old function NORMINV (which still exists) has been replaced with a function called NORM.INV (which is more accurate).

APPENDIX C

❖❖❖

Statistics with R

Some things in R are incredibly easy to do; other tasks are bizarrely difficult. Such an environment can be strange and perplexing at first.

MATT BRIGGS (2008c)

WHAT IS R?

R is a popular software environment (for Windows, Macintosh, and Unix computers) that is used for manipulating, analyzing, and graphing data. Short introductions are provided by Burns (2005) and Briggs (2008c) and a longer one by Robinson (2008). Learn about R and download it from www.r-project.org.

WHAT YOU NEED TO KNOW BEFORE LEARNING R

- R is free and widely available.
- R is a programming environment. You write commands and get back answers. This gives it great power but also makes it hard to get started.
- You don't interact with R via dialogs, menus, and buttons like most Windows and Mac software. Instead, using R feels like having an instant message session with the computer.
- Almost every statistical and bioinformatics analysis has been implemented in R. It is comprehensive.
- Running some analyses requires that you first install add-on software (called a package), as well as R.
- When you run an analysis, the results are stored in an object and R presents only a few key results. You can then enter new commands to query that object to get more details.
- The R language is case sensitive. If you accidentally enter an upper-case letter ("A") when R expects a lowercase letter ("a"), the command won't work.
- You can save R commands in files and then run those R programs. This means you can share R files.
- R is related to S. If you read about a statistical method implemented in S (or S-plus), it will almost certainly also work in R.

- R comes with only a rudimentary spreadsheet-like data editor. You can type values into its session or ask it to read text files. Extensions provide table editors and let it access databases or files created by other statistics programs.
- R includes lots of online help to explain its syntax, but this online help generally assumes you already understand the tests you are using and the results they produce.

APPENDIX D

❖❖❖

Values of the *t* Distribution Needed to Compute Confidence Intervals

The margin of error of many CIs equals the standard error times a critical value of the t distribution tabulated in Table D1. This value depends on the desired confidence level and the number of df, which equals n minus the number of parameters estimated. For example, when computing a confidence interval of a mean,

df	DESIRED CONFIDENCE LEVEL				df	DESIRED CONFIDENCE LEVEL			
	80%	90%	95%	99%		80%	90%	95%	99%
1	3.0777	6.3138	12.7062	63.6567	27	1.3137	1.7033	2.0518	2.7707
2	1.8856	2.9200	4.3027	9.9248	28	1.3125	1.7011	2.0484	2.7633
3	1.6377	2.3534	3.1824	5.8409	29	1.3114	1.6991	2.0452	2.7564
4	1.5332	2.1318	2.7764	4.6041	30	1.3104	1.6973	2.0423	2.7500
5	1.4759	2.0150	2.5706	4.0321	35	1.3062	1.6896	2.0301	2.7238
6	1.4398	1.9432	2.4469	3.7074	40	1.3031	1.6839	2.0211	2.7045
7	1.4149	1.8946	2.3646	3.4995	45	1.3006	1.6794	2.0141	2.6896
8	1.3968	1.8595	2.3060	3.3554	50	1.2987	1.6759	2.0086	2.6778
9	1.3830	1.8331	2.2622	3.2498	55	1.2971	1.6730	2.0040	2.6682
10	1.3722	1.8125	2.2281	3.1693	60	1.2958	1.6706	2.0003	2.6603
11	1.3634	1.7959	2.2010	3.1058	65	1.2947	1.6686	1.9971	2.6536
12	1.3562	1.7823	2.1788	3.0545	70	1.2938	1.6669	1.9944	2.6479
13	1.3502	1.7709	2.1604	3.0123	75	1.2929	1.6654	1.9921	2.6430
14	1.3450	1.7613	2.1448	2.9768	80	1.2922	1.6641	1.9901	2.6387
15	1.3406	1.7531	2.1314	2.9467	85	1.2916	1.6630	1.9883	2.6349
16	1.3368	1.7459	2.1199	2.9208	90	1.2910	1.6620	1.9867	2.6316
17	1.3334	1.7396	2.1098	2.8982	95	1.2905	1.6611	1.9853	2.6286
18	1.3304	1.7341	2.1009	2.8784	100	1.2901	1.6602	1.9840	2.6259
19	1.3277	1.7291	2.0930	2.8609	150	1.2872	1.6551	1.9759	2.6090
20	1.3253	1.7247	2.0860	2.8453	200	1.2858	1.6525	1.9719	2.6006
21	1.3232	1.7207	2.0796	2.8314	250	1.2849	1.6510	1.9695	2.5956
22	1.3212	1.7171	2.0739	2.8188	300	1.2844	1.6499	1.9679	2.5923
23	1.3195	1.7139	2.0687	2.8073	350	1.2840	1.6492	1.9668	2.5899
24	1.3178	1.7109	2.0639	2.7969	400	1.2837	1.6487	1.9659	2.5882
25	1.3163	1.7081	2.0595	2.7874	450	1.2834	1.6482	1.9652	2.5868
26	1.3150	1.7056	2.0555	2.7787	500	1.2832	1.6479	1.9647	2.5857

Table D1

$df = (n - 1)$, because you are only estimating one parameter (the mean); when computing a confidence interval for a slope obtained by linear regression, $df = (n - 2)$, because you estimated two parameters (the slope and the intercept).

These values were computed using the following formula from Excel 2011 (C is the desired confidence level in percent):

$$=T.INV.2T (1-0.01*C, df)$$

If you have an older version of Excel (prior to 2010) use the TINV function instead of T.INV. Don't use Excel's CONFIDENCE() function, which is based on the z (normal) distribution rather than the t distribution and so has limited utility.

APPENDIX E

❖❖❖

A Review of Logarithms

COMMON (BASE 10) LOGARITHMS

The best way to understand logarithms is through an example. If you take 10 to the third power ($10\times10\times10$), the result is 1,000. The logarithm is the inverse of that power function. The logarithm (base 10) of 1,000 is the power of 10 that gives the answer 1,000. So the logarithm of 1,000 is 3. If you multiply 10 by itself 3 times, you get 1,000.

You can take 10 to a negative power. For example, taking 10 to the –3 power is the same as taking the reciprocal of 10^3. So 10^{-3} equals $1/10^3$, or 0.001. The logarithm of 0.001 is the power of 10 that equals 0.001, which is –3.

You can take 10 to a fractional power. Ten to the ½ power equals the square root of 10, which is 3.162. So the logarithm of 3.162 is 0.5.

Ten to the zero power equals 1, so the logarithm of 1.0 is 0.0.

You can take the logarithm of any positive number. The logarithms of values between zero and one are negative; the logarithms of values greater than one are positive. The logarithms of zero and all negative numbers are undefined, because there is no power of 10 that results in a negative number or zero.

OTHER BASES

The logarithms shown in the previous section are called base 10 logarithms, because the computations take 10 to some power. These are also called *common logarithms*.

You can compute logarithms for any power. Mathematicians prefer natural logarithms, using base e (2.7183 . . .). By convention, natural logarithms are used in logistic and proportional hazards regression (see Chapter 37).

Biologists sometimes use base 2 logarithms, often without realizing it. The base 2 logarithm is the number of doublings it takes to reach a value. So the log base 2 of 16 is 4, because if you start with 1 and double it four times (2, 4, 8, and 16), the result is 16. Immunologists often serially dilute antibodies by factors of 2, so they often graph data on a log2 scale. Cell biologists use base 2 logarithms to convert cell counts to number of doublings.

Logarithms using different bases are proportional to each other. Consequently, converting from natural logs to common logs is sort of like changing units. Divide a natural logarithm by 2.303 to compute the common log of the same value. Multiply a common log by 2.303 to obtain the corresponding natural log.

NOTATION

Unfortunately, the notation is used inconsistently.

The notation "log(x)" usually means the common (base 10) logarithm, but some computer languages use it to mean the natural logarithm.

The notation "ln(x)" always means natural logarithm.

The notation "$\log_{10}(x)$" clearly shows that the logarithm uses base 10.

LOGARITHMS CONVERT MULTIPLICATION INTO ADDITION

Logarithms are popular because of this equation:

$$\log(A \cdot B) = \log(A) + \log(B)$$

Similarly, as shown in Chapter 11, logarithms transform a lognormal distribution into a Gaussian distribution.

ANTILOGARITHMS

The antilogarithm (also called an antilog) is the inverse of the logarithm transform. Because the logarithm (base 10) of 1,000 equals 3, the antilogarithm of 3 is 1,000. To compute the antilogarithm of a base 10 logarithm, take 10 to that power.

To compute the antilogarithm of a natural logarithm, take e to that power. The natural logarithm of 1,000 is 6.908. So the antilogarithm of 6.908 is $e^{6.908}$, which is 1,000. Spreadsheets and computer languages use the notation exp(6.908).

❖❖❖

Choosing a Statistical Test

If you are not sure which statistical test to use, the tables in this appendix may help you decide. Reviewing these tables is also a good way to review your understanding of statistics. Of course, this appendix can't include every possible statistical test (and some kinds of data require developing new statistical tests).

OUTCOME: CONTINUOUS DATA FROM GAUSSIAN DISTRIBUTION

Examples	• Cholesterol plasma level (mg/dL)
	• Change in systolic blood pressure (mmHg)
Test Gaussian assumption	• Normality tests
	• Outlier tests
Describe one sample	• Frequency distribution
	• Sample mean
	• Minimum and maximum value and range
	• 25th and 75th percentiles
	• Sample SD
Make inferences about one population	• One-sample t test
Compare two unmatched (unpaired) groups	• Unpaired t test
Compare two matched (paired) groups	• Paired t test
Compare three or more unmatched (unpaired) groups	• One-way ANOVA, followed by multiple comparisons tests
Compare three or more matched (paired) groups	• Repeated-measures ANOVA followed by multiple comparisons tests
Quantify association between two variables	• Pearson's correlation
Explain/predict one variable from another	• Simple linear regression
	• Simple nonlinear regression (*simple* means you are predicting the outcome from a single independent variable)
Explain/predict one variable from several others	• Multiple linear regression (*multiple* means you are predicting the outcome from more than one independent variable)
	• Multiple nonlinear regression

513

OUTCOME: CONTINUOUS DATA FROM NON-GAUSSIAN DISTRIBUTION (OR RANKED DATA)

Examples	• 17β-estradiol levels (pg/mL) • IIEF-5 scoring system (5–25) • Gleason score for prostate cancer (2–10) • Confidence (5 = very high, 4 = high, 3 = moderate, 2 = low, 1 = very low) • Number of headaches in a week
Describe one sample	• Frequency distribution • Sample median • Minimum and maximum value and range • 25th and 75th percentiles
Make inferences about one population	• Wilcoxon's rank-sum test
Compare two unmatched (unpaired) groups	• Mann–Whitney test
Compare two matched (paired) groups	• Wilcoxon's matched pairs test
Compare three or more unmatched (unpaired) groups	• Kruskal–Wallis test • Dunn's posttest
Compare three or more matched (paired) groups	• Friedman's test • Dunn's posttest
Quantify association between two variables	• Spearman's correlation

OUTCOME: SURVIVAL TIMES (OR TIME TO AN EVENT)

Examples	• Months until death for prostate cancer patient • Number of days until cold symptoms go away • Minutes until REM sleep begins
Describe one sample	• Kaplan–Meier survival curve • Median survival time • Five-year survival percentage
Make inferences about one population	• Confidence bands around survival curve • CI of median survival
Compare two unmatched (unpaired) groups	• Log-rank test • Gehan–Breslow test • CI of ratio of median survival times • CI of hazard ratio
Compare two matched (paired) groups	• Conditional proportional hazards regression
Compare three or more unmatched (unpaired) groups	• Log-rank test • Gehan–Breslow test
Compare three or more matched (paired) groups	• Conditional proportional hazards regression
Explain/predict one variable from one or several others.	• Cox's proportional hazards regression

OUTCOME: BINOMIAL

Examples	• Cure (yes/no) of acute myeloid leukemia within a specific time period • Success (yes/no) of preventing motion sickness • Recurrence of infection (yes/no)
Describe one sample	• Proportion
Make inferences about one population	• CI of proportion • Binomial test to compare observed distribution with a theoretical (expected) distribution
Compare two unmatched (unpaired) groups	• Fisher's exact test
Compare two matched (paired) groups	• McNemar's test
Compare three or more unmatched (unpaired) groups	• Chi-square test • Chi-square test for trend
Compare three or more matched (paired) groups	• Cochran's Q
Explain/predict one variable from one or several others.	• Logistic regression

REFERENCES

❖

Agnelli, G., Buller, H. R., Cohen, A., Curto, M., Gallus, A. S., Johnson, M., Porcari, et al. (2012). Apixaban for extended treatment of venous thromboembolism. *New England Journal of Medicine, 368,* 699–708. doi:10.1056/NEJMoa1207541.

Agresti, A., & Coull, B. A. (1998). Approximate is better than exact for interval estimation of binomial proportions. *American Journal of Statistics, 52,* 119–126.

Allen, M. C., Donohue, P. K., & Dusman, A. E. (1993). The limit of viability—Neonatal outcome of infants born at 22 to 25 weeks' gestation. *New England Journal of Medicine, 329,* 1597–1601.

Altman, D. G. (1990). *Practical statistics for medical research.* London: Chapman & Hall/CRC.

Altman, D. G., & Bland, J. M. (1995). Absence of evidence is not evidence of absence. *BMJ (Clinical Research Ed.), 311,* 485.

_____. (1998). Time to event (survival) data. *BMJ (Clinical Research Ed.), 317,* 468–469.

Anscombe, F. J. (1973). Graphs in statistical analysis. *American Statistician, 27,* 17–21.

Arad, Y., Spadaro, L. A., Goodman, K., Newstein, D., & Guerci, A. D. (2000). Prediction of coronary events with electron beam computed tomography. *Journal of the American College of Cardiology, 36,* 1253–1260.

Arden, R., Gottfredson, L. S., Miller, G., & Pierce, A. (2008). Intelligence and semen quality are positively correlated. *Intelligence, 37,* 277–282.

Austin, P. C., & Goldwasser, M. A. (2008). Pisces did not have increased heart failure: Data-driven comparisons of binary proportions between levels of a categorical variable can result in incorrect statistical significance levels. *Journal of Clinical Epidemiology, 61,* 295–300.

Austin, P. C., Mamdani, M. M., Juurlink, D. N., & Hux, J. E. (2006). Testing multiple statistical hypotheses resulted in spurious associations: A study of astrological signs and health. *Journal of Clinical Epidemiology, 59,* 964–969.

Babyak, M. A. (2004). What you see may not be what you get: A brief, nontechnical introduction to overfitting in regression-type models. *Psychosomatic Medicine, 66,* 411–421.

Bailar, J.C. (1997). The promise and problems of meta-analysis. *New England Journal of Medicine, 337,* 559–561.

Bakhshi, E., Eshraghian, M. R., Mohammad, K., & Seifi, B. (2008). A comparison of two methods for estimating odds ratios: Results from the National Health Survey. *BMC Medical Research Methodology, 8,* 78.

Barter, P. J., Caulfield, M., Eriksson, M., Grundy, S. M., Kastelein, J. J., Komajda, M., Lopez-Sendon, J., Mosca, L., et al. (2007). Effects of torcetrapib in patients at high risk for coronary events. *New England Journal of Medicine, 357,* 2109–2122.

Bausell, R. B. (2007). *Snake oil science: The truth about complementary and alternative medicine.* Oxford, UK: Oxford University Press. ISBN=0195313682.

Begley, C. G. C., & Ellis, L. M. L. (2012). Drug development: Raise standards for preclinical cancer research. *Nature, 483,* 531–533.

Benjamini, Y., & Hochberg, Y. (1995). Controlling the false discovery rate: A practical and powerful approach to multiple testing. *Journal of Royal Statistical Society, B, 57,* 290–300.

Bennett, C. M., Baird, A. A., Miller, M. B., & Wolford, G. L. (2011). Neural correlates of interspecies perspective taking in the post-mortem Atlantic salmon: An argument for proper multiple comparisons correction. *Journal of Serendipitous and Unexpected Results, 1,* 1–5.

Bernstein, C. N., Nugent, Z., Longobardi, T., & Blanchard, J. F. (2009). Isotretinoin is not associated with inflammatory bowel disease: A population-based case–control study. *American Journal of Gastroenterology, 104,* 2774–2778.

Berry, D. A. (2007). The difficult and ubiquitous problems of multiplicities. *Pharmaceutical Statistics, 6,* 155–160.

Bickel, P. J., Hammel, E. A., & O'Connell, J. W. (1975). Sex bias in graduate admissions: Data from Berkeley. *Science, 187,* 398–404.

Bishop, D. (2013) Interpreting unexpected significant results. Accessed July 2, 2013 from deevybee.blogspot.co.uk/2013/06/interpreting-unexpected-significant.html

Bland, J. M. J., & Altman, D. G. D. (2011). Comparisons against baseline within randomised groups are often used and can be highly misleading. *Trials, 12,* 264–264.

Blumberg, M. S. (2004). *Body heat: Temperature and life on earth.* Cambridge, MA: Harvard University Press. ISBN=0674013697.

Bleyer, A. & Welch, H. G. (2012). Effect of Three Decades of Screening Mammography on Breast-Cancer Incidence. N Engl J Med 367, 1998–2005.

Boos, D. D., & Stefanski, L. A. (2011). P-value precision and reproducibility. *American Statistician, 65,* 213–221.

Borkman, M., Storlien, L. H., Pan, D. A., Jenkins, A. B., Chisholm, D. J., & Campbell, L. V. (1993). The relation between insulin sensitivity and the fatty-acid composition of skeletal-muscle phospholipids. *New England Journal of Medicine, 328,* 238–244.

Briggs, W. M. (2008a). On the difference between mathematical ability between boys and girls. *William M. Briggs* (blog), July 25. Accessed June 21, 2009, from wmbriggs.com/blog/?p=163/.

_____. (2008b). Do not calculate correlations after smoothing data. *William M. Briggs* (blog), February 14. Accessed June 21, 2009 25, 2008, from wmbriggs.com/blog/?p=86/.

_____. (2008c). *R.* wmbriggs.com/public/briggs_chap05.pdf

Brown, L. D., Cai, T. T., & DasGupta, A. (2001). Interval estimation for a binomial proportion. *Statistical Science, 16,* 101–133.

Burnham, K., & Anderson, D. (2003). *Model selection and multi-model inference,* 2nd ed. New York: Springer. ISBN=0387953647.

Burns, P. (2005). *A guide for the unwilling S user.* Accessed January 26, 2009, at www. burns-stat.com/pages/Tutor/unwilling_S.pdf.

Bryant, F. B., & Brockway, J. H. (1997). Hindsight bias in reaction to the verdict in the O. J. Simpson criminal trial. *Basic and Applied Social Psychology, 19,* 225–241.

Campbell, M. J. (2006). *Statistics at square two,* 2nd ed. London: Blackwell. ISBN=1-4051-3409-9.

Cantor, W. J., Fitchett, D. L., Borgundyagg, B., Ducas, J., Heffernam, M., Cohen, E. A., Morrison, L. J., et al. (2009). Routine early angioplasty after fibrinolysis for acute myocardial infarction. *New England Journal of Medicine, 360,* 2705–2718.

Cardiac Arrhythmia Suppression Trial (CAST) Investigators. (1989). Preliminary report: Effect of encainide and lecainide on mortality in a randomized trial of arrhythmia suppression after myocardial infarction. *New England Journal of Medicine, 3212,* 406–412.

Carroll, L. (1871). *Through the looking glass.* Accessed December 19, 2012, at www. gutenberg.org/ebooks/12.

Central Intelligence Agency (CIA). (2012). *The world factbook.* Accessed August 26, 2012, at www.cia.gov/library/publications/the-world-factbook/fields/2018.html.

Chan, A. W., Hrobjartsson, A., Haahr, M. T., Gotzsche, P. C., & Altman, D. G. (2004). Empirical evidence for selective reporting of outcomes in randomized trials: Comparison of protocols to published articles. *Journal of the American Medical Association, 291,* 2457–2465.

Clopper, C. J., & Pearson, E. S. (1934). The use of confidence or fiducial limits illustrated in the case of the binomial. *Biometrika, 26,* 404–413.

Cohen, J. (1988). *Statistical power analysis for the behavioral sciences,* 2nd ed. Hillsdale, NJ: Erlbaum. ISBN=0805802835.

Cohen, T. J., Goldner, B. G., Maccaro, P. C., Ardito, A. P., Trazzera, S., Cohen, M. B., & Dibs, S.R. (1993). A comparison of active compression–decompression cardiopulmonary resuscitation with standard cardiopulmonary resuscitation for cardiac arrests occurring in the hospital. *New England Journal of Medicine, 329,* 1918–1921.

Colquhoun, D. D. (2003). Challenging the tyranny of impact factors. *Nature, 423,* 479–480.

Cramer, H. (1999). *Mathematical methods of statistics.* Princeton, NJ: Princeton University Press. ISBN=0691005478.

Crichton, M. (2003). Environmentalism as religion. Accessed November 9, 2008, at www. michaelcrichton.net/speech-environmentalismaseligion.html.

_____. (2005). The case for skepticism on global warming. Accessed November 9, 2008, at www.michaelcrichton.net/speech-ourenvironmentalfuture.html.

Cumming, G. (2008). Replication and p intervals: P values predict the future only vaguely, but confidence intervals do much better. *Perspectives on Psychological Science, 3,* 286–300.

_____. (2011). *Understanding The New Statistics: Effect Sizes, Confidence Intervals, and Meta-Analysis.* New York: Routledge, ISBN=978-0415879682.

Cumming, G., Fidler, F., & Vaux, D. L. (2007). Error bars in experimental biology. *Journal of Cell Biology, 177,* 7–11.

Dallal, G. E. (2012). *The Little Handbook of Statistical Practice eBook.* Kindle Direct Publishing. ASIN: B00847SM6A.

Darwin, C. (1876). *The effects of cross and self fertilisation in the vegetable kingdom.* London: Murray.

Davenas, E., Beauvais, F., Amara, J., Oberbaum, M., Robinzon, B., Miadonna, A., Tedeschi, A., Pomeranz, B., Fortner, P., & Belon, P. (1988). Human basophil degranulation triggered by very dilute antiserum against IgE. *Nature, 333,* 816–818.

Denes-Raj, V., & Epstein, S. (1994). Conflict between intuitive and rational processing: When people behave against their better judgment. *Journal of Personality and Social Psychology, 66,* 819–829.

Environmental Protection Agency (2011). Exposure Factors Handbook. Accessed July 7, 2013 at www.epa.gov/ncea/efh/pdfs/efh-chapter08.pdf

Evans, S. (2007). When and how can endpoints be changed after initiation of a randomized clinical trial. *PLoS Hub for Clinical Trials, 2,* e18.

Ewigman, B. G., Crane, J. P., Frigoletto, F. D., LeFevre, M. L., Bain, R. P., & McNellis, D. (1993). Effect of prenatal ultrasound screening on perinatal outcome. RADIUS Study Group. *New England Journal of Medicine, 329,* 821–827.

Eyding, D. D., Lelgemann, M., Grouven, U., Härter, M., Kromp, M., Kaiser, T., Kerekes, M. F., Gerken, M., & Wieseler, B. (2010). Reboxetine for acute treatment of major depression: Systematic review and meta-analysis of published and unpublished placebo and selective serotonin reuptake inhibitor controlled trials. *British Medical Journal, 341,* c4737.

Federal Election Commission. (2012). Official 2012 presidential general election results. Accessed June 2, 2013, at www.fec.gov/pubrec/fe2012/2012presgeresults.pdf.

Feinstein, A. R., Sosin, D. M., & Wells, C. K. (1985). The Will Rogers phenomenon. Stage migration and new diagnostic techniques as a source of misleading statistics for survival in cancer. *New England Journal of Medicine, 312,* 1604–1608.

Fisher, R. A. (1935). *The design of experiments.* New York: Hafner Pub.Co. ISBN=0198522290.

_____. (1936). Has Mendel's work been rediscovered? *Annals of Science, 1,* 115–137.

Fisher, L. D, & Van Belle, G. (1993). *Biostatistics. A methodology for the health sciences.* New York: Wiley Interscience. ISBN=0-471-58465-7.

Fleming, T. R. (2006). Standard versus adaptive monitoring procedures: A commentary. *Statistics in Medicine, 25,* 3305–3512; discussion 3313–3314, 3326–3347.

Flom, P. L., & Cassell, D. L. (2007). Stopping stepwise: Why stepwise and similar selection methods are bad, and what you should use. NorthEast SAS Users Group. Accessed September 17, 2012, at www.nesug.org/proceedings/nesug07/sa/sa07.pdf.

Frazier, E. P., Schneider, T., & Michel, M. C. (2006). Effects of gender, age and hypertension on beta-adrenergic receptor function in rat urinary bladder. *Naunyn-Schmiedeberg's Archives of Pharmacology, 373,* 300–309.

Freedman, D. (1983). A note on screening regression equations. *American Statistician, 37,* 152–155.

_____. (2007). *Statistics,* 4th ed. New York: Norton. ISBN=978-0393929720.

Freese, J. (2008). The problem of predictive promiscuity in deductive applications of evolutionary reasoning to intergenerational transfers: Three cautionary tales. In *Intergenerational caregiving,* ed. A. Booth, A. C. Crouter, S. M. Bianchi, & J. A. Seltzer. Urban Institute Press. ISBN=978-0877667476.

Fung, K. (2011). The statistics of anti-doping. Accessed December 9, 2012, at junkcharts. typepad.com/numbersruleyourworld/2011/03/the-statistics-of-anti-doping.html.

Gabriel, S. E., O'Fallon, W. M., Kurland, L. T., Beard, C. M., Woods, J. E., & Melton, L. J. (1994). Risk of connective-tissue diseases and other disorders after breast implantation. *New England Journal of Medicine, 330,* 1697–1702.

Gallup. (2012). Romney has support among lowest income voters. September 12. www.gallup.com/poll/157508/romney-support-among-lowest-income-voters. aspx.

García-Berthou, E., & Alcaraz, C. (2004). Incongruence between test statistics and P values in medical papers. *BMC Medical Research Methodology,4;13-18* .

Gelman, A. (2009). How does statistical analysis differ when analyzing the entire population rather than a sample? *Statistical Modeling, Causal Inference, and Social Science* (blog), July 3. Accessed December 11, 2012, at andrewgelman.com/2009/07/how_does_statis/.

_____. (2012). Overfitting. *Statistical Modeling, Causal Inference, and Social Science* (blog), July 27. Accessed November 15, 2012, at andrewgelman.com/2012/07/27/15864/.

_____. (2013). Don't let your standard errors drive your research agenda. *Statistical Modeling, Causal Inference, and Social Science* (blog), February 1. Accessed February 8, 2013, at andrewgelman.com/2013/02/dont-let-your-standard-errors-drive-your-research-agenda/.

Gelman, A., & Feller, A. (2012) Red versus blue in a new light. *New York Times,* September 12. Accessed January 2013 at campaignstops.blogs.nytimes.com/2012/11/12/red-versus-blue-in-a-new-light/.

Gelman, A., & Stern, H. (2006). The difference between "significant" and "not significant" is not itself statistically significant. *American Statistician, 60,* 328–331.

Gelman, A., & Tuerlinckx, F. (2000). Type S error rates for classical and Bayesian single and multiple comparison procedures. *Computational Statististics, 15,* 373–390.

Gelman, A., & Weakliem, D. (2009). Of beauty, sex and power: Statistical challenges in estimating small effects. *American Scientist, 97,* 310–311.

Gilovich, T. (1985). The hot hand in basketball: On the misperception of random sequences. *Cognitive Psychology*, 17, 295–314.

Gigerenzer, G. (2002). *Calculated risks.* New York: Simon and Schuster. ISBN=0-7432-0556-1.

Glantz, S. A., & Slinker, B. K. (2000). *Primer of applied regression and analysis of variance,* 2nd ed. New York: McGraw–Hill. ISBN=0071360867.

Goddard, S. (2008). Is the earth getting warmer, or cooler? *The Register*, May 2. Accessed June 13, 2008, at www.theregister.co.uk/2008/05/02/a_tale_of_two_thermometers/.

Goldacre, B. (2013) Bad Pharma.. London: Faber & Faber. ISBN=978-0865478008.

Goldstein, D. (2006). The difference between significant and not significant is not statistically significant. *Decision Science News.* Accessed February 8, 2013, at www.decisionsciencenews.com/2006/12/06/the-difference-between-significant-and-not-significant-is-not-statistically-significant/.

Good, P. I., & Hardin, J. W. (2006). *Common errors in statistics (and how to avoid them).* Hoboken, NJ: Wiley. ISBN=0471794317. 1.

Goodman, S. (2008). A dirty dozen: Twelve p-value misconceptions. *Seminars in Hematology, 45,* 135–140.

Goodwin, P. (2010). Why hindsight can damage foresight. *Foresight, 17,* 5–7.

Gotzsche, P. C. (2006). Believability of relative risks and odds ratios in abstracts: Cross sectional study. *BMJ (Clinical Research Ed.), 333,* 231–234.

Gould, S. J. (1997). *Full house: The spread of excellence from Plato to Darwin.* New York: Three Rivers Press. ISBN=0609801406.

Hankins, M. *Still not significant.* Psychologically Flawed April 21, 2013. Accessed June 28, 2013 at mchankins.wordpress.com/2013/04/21/still-not-significant-2/

Hanley, J. A., & Lippman-Hand, A. (1983). If nothing goes wrong, is everything alright? *Journal of the American Medical Association, 259,* 1743–1745.

Harter, H. L. (1984). Another look at plotting positions. *Communications in Statistics—Theory and Methods, 13,* 1613–1633.

Hartung, J. (2005). Statistics: When to suspect a false negative inference. In *American Society of Anesthesiology 56th annual meeting refresher course lectures,* Lecture 377, 1–7. Philadelphia: Lippincott.

Heal, C. F., Buettner, P. G., Cruickshank, R., & Graham, D. (2009). Does single application of topical chloramphenicol to high risk sutured wounds reduce incidence of wound infection after minor surgery? Prospective randomised placebo controlled double blind trial. *British Medical Journal, 338,* 211–214.

Henderson, B. (2005). Open letter to Kansas School Board. *Church of the Flying Spaghetti Monster.* Accessed December 8, 2012, at www.venganza.org/about/open-letter/.

Hetland, M. L., Haarbo, J., Christiansen, C., & Larsen, T. (1993). Running induces menstrual disturbances but bone mass is unaffected, except in amenorrheic women. *American Journal of Medicine, 95,* 53–60.

Hoenig, J. M., & Heisey, D. M. (2001). The abuse of power: The pervasive fallacy of power. Calculations for data analysis. *American Statistician, 55,* 1–6.

Hollis, S., & Campbell, F. (1999). What is meant by intention to treat analysis? Survey of published randomized controlled trials. *BMJ (Clinical Research Ed.), 319,* 670–674.

Hsu, J. (1996). *Multiple comparisons: Theory and methods.* Boca Raton, FL: Chapman & Hall/CRC. ISBN=0412982811.

Huber, P. J. (2003). *Robust statistics.* Hoboken, NJ: Wiley–Interscience. ISBN=0471650722.

Hunter, D. J., Manson, J. E., Colditz, G. A., Stampfer, M. J., Rosner, B., Hennekens, C., Speizer, F. E., & Willet, W. C. (1993). A prospective study of the intake of vitamins C, E, and A and the risk of breast cancer. *New England Journal of Medicine, 329,* 234–240.

Hyde, J. S., Lindberg, S. M., Linn, M. C., Ellis, A. B., & Williams, C. C. (2008). Diversity. Gender similarities characterize math performance. *Science, 321,* 494–495.

Ioannidis, J. P. (2005). Why most published research findings are false. *PLoS Medicine, 2,* e124.

———. (2008). Why most discovered true associations are inflated. *Epidemiology, 19,* 640–648.

Julious, S. A. (2005). Sample size of 12 per group rule of thumb for a pilot study. *Pharmaceutical Statistics, 4,* 287–291.

Kane, J. M., & Mertz, J. E. (2012). Debunking myths about gender and mathematics performance. *Notices of the AMS, 59,* 10–21.

Katz, M. H. (2006). *Multivariable analysis: A practical guide for clinicians.* Cambridge: Cambridge University Press. ISBN=052154985x.

Kaul, A., & Diamond, G. (2006). Good enough: A primer on the analysis and interpretation of noninferiority trials. *Annals of Internal Medicine, 145,* 62–69.

Kelley, K., & Maxwell, S. E. (2008). Sample size planning with applications to multiple regression: Power and accuracy for omnibus and targeted effects. Pages 166–192 in *The SAGE handbook of social research methods,* ed. P. Alasuutari, L. Bickman, & J. Brannen. Thousand Oaks CA: SAGE Publications.

Kerr, N. L. (1998). HARKing: Hypothesizing after the results are known. *Personality and Social Psychology Review, 2,* 196–217.

Kirk, A. P., Jain, S., Pocock, S., Thomas, H. C., & Sherlock, S. (1980). Late results of the Royal Free Hospital prospective controlled trial of prednisolone therapy in hepatitis B surface antigen negative chronic active hepatitis. *Gut, 21,* 78–83.

Kline, R. B. (2004). *Beyond significance testing: Reforming data analysis methods in behavioral research.* New York: American Psychological Association. IBSN:1591471184.

Knol, M. J., Pestman, W. R., & Grobbee, D. E. (2011). The (mis)use of overlap of confidence intervals to assess effect modification. *European Journal of Epidemiology, 26,* 253–254.

Knottnerus, J. A. J., & Bouter, L. M. L. (2001). The ethics of sample size: Two-sided testing and one-sided thinking. *Journal of Clinical Epidemiology, 54,* 109–110.

Knusel, L. (2005). On the accuracy of statistical distributions in Microsoft Excel 2003. *Computational Statistics and Data Analysis, 48,* 445–449.

Kriegeskorte, N., Simmons, W. K., Bellgowan, P. S. F., & Baker, C. I. (2009). Circular analysis in systems neuroscience: The dangers of double dipping. *Nature Neuroscience, 12,* 535–540.

Kruschke, J.K. (2011), *Doing Bayesian Data Analysis: A Tutorial with R and BUGS* New York: Academic Press / Elsevier. ISBN=9780123814852.

Kuehn, B. (2006). Industry, FDA warm to "adaptive" trials. *Journal of the American Medical Association, 296,* 1955–1957.

Lamb, E. L. (2012). 5 sigma—What's that? *Scientific American Blog Network,* July 17. Accessed November 16, 2012, at blogs.scientificamerican.com/observations/2012/07/17/five-sigmawhats-that/.

Lanzante, J. R. (2005). A cautionary note on the use of error bars. *Journal of Climate, 18,* 3699–3703.

Lazic, S. E. (2010). The problem of pseudoreplication in neuroscientific studies: Is it affecting your analysis? *BMC Neuroscience, 11,* 5.

Laupacis, A., Sackett, D. L., & Roberts, R. S. (1988). An assessment of clinically useful measures of the consequences of treatment. *New England Journal of Medicine, 318,* 1728–1733.

Lee, K. L., McNeer, J. F., Starmer, C. F., Harris, P. J., & Rosati, R. A. (1980). Clinical judgment and statistics. Lessons from a simulated randomized trial in coronary artery disease. *Circulation, 61,* 508–515.

Lehman, E. (2007). *Nonparametrics: Statistical methods based on ranks.* New York: Springer. ISBN=0387352120.

Lenth, R. V. (2001). Some practical guidelines for effective sample size determination. *American Statistician, 55,* 187–193.

Levine, M., & Ensom, M. H. (2001). Post hoc power analysis: An idea whose time has passed? *Pharmacotherapy, 21,* 405–409.

Levins, R. (1966). The strategy of model building in population biology. *American Scientist, 54,* 421–431.

Lewin, T. (2008). Math scores show no gap for girls, study finds. *New York Times,* July 25. Accessed July 26, 2008, at www.nytimes.com/2008/07/25/education/25math.html.

Limpert, E., Stahel, W. A., & Abbt, M. (2001). Log-normal distributions across the sciences: Keys and clues. *Biosciences, 51,* 341–352.

Lucas, M. E. S., Deen, J. L., von Seidlein, L., Wang, X., Ampuero, J., Puri, M., Mohammad Ali, et al. (2005). Effectiveness of mass oral cholera vaccination in Beira, Mozambique. *New England Journal of Medicine, 352,* 757–767.

Ludbrook, L., & Lew, M. J. (2009). Estimating the risk of rare complications: Is the "rule of three" good enough? *Australian and New Zealand Journal of Surgery, 79,* 565–570.

Mackowiak, P. A., Wasserman, S. S., & Levine, M. M. (1992). A critical appraisal of 98.6 degrees F, the upper limit of the normal body temperature, and other legacies of Carl Reinhold August Wunderlich. *Journal of the American Medical Association, 268,* 1578–1580.

Manly, B. F. J. (2006). *Randomization, bootstrap and Monte Carlo methods in biology,* 3rd ed. London: Chapman & Hall/CRC. ISBN=1584885416.

Marsupial, D. (2013). Comment on: Interpretation of p-value in hypothesis testing. *Cross Validated,* January 3. Accessed January 5, 2013, at stats.stackexchange.com/questions/46856/interpretation-of-p-value-in-hypothesis-testing.

Masicampo, E. J., & Lalande, D. R. (2012). A peculiar prevalence of p values just below .05. *Quarterly Journal of Experimental Psychology,* August 2. doi:10.1080/1747021 8.2012.711335.

McCullough, B. D., & Hellser, D. A. (2008). On the accuracy of statistical procedures in Microsoft Excel 2007. *Computational Statistics and Data Analysis, 52,* 4570–4578.

McCullough, B. D., & Wilson, B. (2005). On the accuracy of statistical procedures in Microsoft Excel 2003. *Computational Statistics and Data Analysis, 49,* 1244.

Mendel, G. J. (1866) Versuche über Pflanzen-Hybriden. Vorgelegt in den Sitzungen vom 8. Februar und 8. März 1865. Verhandlungen des naturforschenden Vereines in Brünn, Band IV, Heft 1 (1865): 3-47. Brünn: Verlag des Vereines.

Messerili, F. H. (2012). Chocolate consumption, cognitive function, and Nobel laureates. *New England Journal of Medicine, 367,* 1562–1564.

Metcalfe, C. (2011) Holding onto power: Why confidence intervals are not (usually) the best basis for sample size calculations. *Trials, 12,* A101.

Meyers, M. A. (2007). *Happy accidents: Serendipity in modern medical breakthroughs.* New York: Arcade. ISBN=1559708190.

Micceri, T. (1989). The unicorn, the normal curve, and other improbable creatures. *Psychological Bulletin, 105,* 156–166.

Microsoft. (2006). Description of improvements in the statistical functions in Excel 2003 and in Excel 2004 for Mac. Accessed December 16, 2008, at support.microsoft.com/kb/828888.

Mills, J. L. (1993). Data torturing. *New England Journal of Medicine, 329,* 1196.

Montori, V. M., & Guyatt, G. H. (2001). Intention-to-treat principle. *Canadian Medical Association Journal, 165,* 1339–1341.

Montori, V. M., Kleinbart, J., Newman, T. B., Keitz, S., Wyer, P. C., Moyer, V., & Guyatt, G. (2004). Tips for learners of evidence-based medicine: 2. Measures of precision (confidence intervals). *Canadian Medical Association Journal, 171,* 611–615.

Moser, B. K., & Stevens, G. R. (1992). Homogeneity of variance in the two-sample means test. *American Statistician, 46,* 19–21.

Motulsky, H. J. (2012). GraphPad curve fitting guide. Accessed January 2013 at www.graphpad.com/guides/prism/6/curve-fitting/.

Motulsky, H. J., & Brown, R. E. (2006). Detecting outliers when fitting data with nonlinear regression—A new method based on robust nonlinear regression and the false discovery rate. *BMC Bioinformatics, 7,* 123.

Motulsky, H., & Christopoulos, A. (2004). *Fitting models to biological data using linear and nonlinear regression: A practical guide to curve fitting.* New York: Oxford University Press. ISBN=978-0195171808.

Motulsky, H. J., O'Connor, D. T., & Insel, P. A. (1983). Platelet alpha 2-adrenergic receptors in treated and untreated essential hypertension. *Clinical Science, 64,* 265–272.

Moyé, L. A., & Tita, A. T. N. (2002). Defending the rationale for the two-tailed test in clinical research. *Circulation, 105,* 3062–3065.

Munger, K. L, Levin, L. I., Massa, J., Horst, R., Orban, T., and Ascherio, A. (2013). Preclinical serum 25-hydroxyvitamin D levels and risk of Type 1 diabetes in a cohort of US military personnel. *American Journal of Epidemiology,* first published online February 3, 2013. doi:10.1093/aje/kws243.

New Scientist. (2007). NASA blows millions on flawed airline safety survey. November 10. Accessed May 25, 2008, at www.newscientist.com/channel/opinion/mg19626293. 900-nasa-blows-millions-on-flawed-airline-safety-survey.html.

Nieuwenhuis, S., Forstmann, B. U., & Wagenmakers, E.-J. (2011). Erroneous analyses of interactions in neuroscience: A problem of significance. *Nature Neuroscience, 14,* 1105–1107.

Oremus, W., The Wedding Industry's Pricey Little Secret, Slate,. Accessed June 13, 2013 at slate.me/1a4KLuH

Pace, L. A. (2008). *The Excel 2007 data and statistics cookbook,* 2nd ed. Anderson, SC: TwoPaces. ISBN=978-0-9799775-2-7.

Parker, R. A., & Berman, N. G. (2003). Sample size: More than calculations. *American Statistician, 57,* 166–170.

Paulos, J. A. (2008). *Irreligion: A mathematician explains why the arguments for God just don't add up.* New York: Hill and Wang. ISBN=0809059193.

Payton, M. E., Greenstone, M. H., & Schenker, N. (2003). Overlapping confidence intervals or standard error intervals: What do they mean in terms of statistical significance? *Journal of Insect Science, 3,* 34–40.

Pielke, R. (2008). Prometheus: Forecast verification for climate science, part 3. January 9. Accessed April 20, 2008, at sciencepolicy.colorado.edu/prometheus/archives/climate_ change/001315forecast_verificatio.html.

Prinz, F. F., Schlange, T. T., & Asadullah, K. K. (2011). Believe it or not: How much can we rely on published data on potential drug targets? *Nature Reviews Drug Discovery, 10,* 712.

Price, A. L., Zaitlen, N. A., Reich, D., & Patterson, N. (2010). New approaches to population stratification in genome-wide association studies. *Nature Reviews Genetics, 11,* 459–463.

Pullan, R. D., Rhodes, J., Gatesh, S., Mani, V., Morris, J. S., Williams, G. T., Newcombe, R. G., et al. (1994). Transdermal nicotine for active ulcerative colitis. *New England Journal of Medicine, 330,* 811–815.

Ridker, P. M., Danielson, E., Fonseca, F. A. H., Genest, J., Gotto, A. M., Jr., Kastelein, J. J. P., Koening, W., et al. (2008). Rosuvastatin to prevent vascular events in men and women with elevated C-reactive protein. *New England Journal of Medicine, 359,* 2195–2207.

Roberts, S. (2004). Self-experimentation as a source of new ideas: Ten examples about sleep, mood, health, and weight. *Behavioral and Brain Sciences, 27,* 227–262; discussion 262–287.

Robinson, A. (2008) *icebreakeR.* Accessed January 26, 2009, from cran.r-project.org/doc/contrib/Robinson-icebreaker.pdf.

Rosman, N. P., Colton, T., Labazzo, J., Gilbert, P. L., Gardella, N. B., Kaye, E. M.,Van Bennekom, C., & Winter, M. R. (1993). A controlled trial of diazepam administered during febrile illnesses to prevent recurrence of febrile seizures. *New England Journal of Medicine, 329,* 79–84.

Rothman, K. J. (1990). No adjustments are needed for multiple comparisons. *Epidemiology, 1,* 43–46.

Russo, J. E., & Schoemaker, P. J. H. (1989). *Decision traps. The ten barriers to brilliant decision-asking and how to overcome them.* New York: Simon & Schuster. ISBN=0671726099.

Samaniego, F. J. A (2008). Conversation with Myles Hollander. *Statistical Science, 23,* 420–438.

Sawilowsky, S. S. (2005). Misconceptions leading to choosing the t test over the Wilcoxon Mann–Whitney test for shift in location parameter. *Journal of Modern Applied Statistical Methods, 4,* 598–600.

Schoemaker, A. L. (1996). What's normal?—Temperature, gender, and heart rate. *Journal of Statistics Education, 4,* 2. Accessed May 5, 2007, at www.amstat.org/publications/jse/v4n2/datasets.shoemaker.html.

Seaman, M. A., Levin, J. R., & Serlin, R. C. (1991). New developments in pairwise multiple comparisons: Some powerful and practicable procedures. *Psychological Bulletin, 110,* 577–586.

Sheskin, D. J. (2007). *Handbook of parametric and nonparametric statistical procedures,* 4th ed. New York: Chapman & Hall/CRC. ISBN=1584888148.

Shettles, L. B. (1996). *How to choose the sex of your baby. Monroe, WA:* Main Street Books. ISBN=978-0385485623.

Silver, N. (2012). Uncertainty clouds polling, but Obama remains electoral college favorite. *Five Thirty Eight* (blog), fivethirtyeight.blogs.nytimes.com/2012/10/22/oct-21-uncertainty-clouds-polling-but-obama-remains-electoral-college-favorite/

Simmons, J. P., Nelson, L. D., & Simonsohn, U. (2011). False-positive psychology: Undisclosed flexibility in data collection and analysis allows presenting anything as significant. *Psychological Science, 22,* 1359–1366.

_____. (2013). P-Curve: A Key to the File Drawer. J. Experimental Psychology: General. In press.

Simon, S. (2005). Stats: Standard deviation versus standard error. May 16. Accessed December 20, 2012, at www.pmean.com/05/StandardError.html.

Snapinn, S. M. (2000). Noninferiority trials. *Current Control Trials in Cardiovascular Medicine, 1,* 19–21.

Sparling, B. (2001). Ozone history. NASA Advanced Supercomputing Division. Accessed June 13, 2008, at www.nas.nasa.gov/About/Education/Ozone/history.html.

Spector, R., & Vesell, E. S. (2006a). The heart of drug discovery and development: Rational target selection. *Pharmacology, 77,* 85–92.

_____. (2006b). Pharmacology and statistics: Recommendations to strengthen a productive partnership. *Pharmacology, 78,* 113–122.

Squire, P. (1988). Why the 1936 *Literary Digest* poll failed. *Public Opinion Quarterly, 52,* 125–133.

Staessen, J. A., Lauwerys, R. R., Buchet, J. P., Bulpitt, C. J., Rondia, D., Vanrenterghem, Y., Amery, A., & the Cadmibel Study Group. (1992). Impairment of renal function with increasing blood lead concentrations in the general population. The Cadmibel Study Group. *New England Journal of Medicine, 327,* 151–156.

Statwing. (2012). The ecological fallacy. *Statwing* (blog), December 20. Accessed February 8, 2013, at blog.statwing.com/the-ecological-fallacy/.

Svensson, S., Menkes, D. B. & Lexchin, J. Surrogate outcomes in clinical trials: a cautionary tale. (2013). *JAMA Intern Med., 173,* 611–612.

Taubes, G. (1995). Epidemiology faces its limits. *Science, 269,* 164–169.

Thun, M. J., & Sinks, T. (2004). Understanding cancer clusters. *CA: A Cancer Journal for Clinicians, 54,* 273–280.

Tierney, J. (2008). A spot check of global warming. *New York Times,* January 10. Accessed April 20, 2008, at tierneylab.blogs.nytimes.com/2008/01/10/a-spot-check-of-global-warming/.

Turner, E. H., Matthews, A. M., Linardatos, E., Tell, R. A., & Rosenthal, R. (2008). Selective publication of antidepressant trials and its influence on apparent efficacy. *New England Journal of Medicine, 358,* 252–260.

Twain, M. (1883) Life on the Misissippi.

U.S. Census (2011). Accessed January 30, 2012, at www.census.gov/hhes/www/income/data/historical/household/2011/H08_2011.xls.

U.S. Food and Drug Administration (FDA). (2012). Preface. *Approved drug products with therapeutic equivalence evaluations.* Accessed December 23, 2012, at www.fda.gov/Drugs/DevelopmentApprovalProcess/ucm079068.htm.

Van Belle, G. (2008). *Statistical rules of thumb,* 2nd ed. New York: Wiley–Interscience. ISBN=0470144483.

Vaux, D. L., Fidler, F., & Cumming, G. (2012). Replicates and repeats—What is the difference and is it significant? *EMBO Reports, 13,* 291–296.

Vedula, S. S., Bero, L., Scherer, R. W., & Dickersin, K. (2009). Outcome reporting in industry-sponsored trials of gabapentin for off-label use. *New England Journal of Medicine, 361,* 1963–1971.

Velleman, P. F., & Wilkinson, L. (1993). Nominal, ordinal, interval, and ratio typologies are misleading. *American Statistician, 47,* 65–72.

Vera-Badillo, F. E., Shapiro, R., Ocana, A., Amir, E., & Tannock, I. F. (2013). Bias in reporting of end points of efficacy and toxicity in randomized, clinical trials for women with breast cancer. *Annals of Oncology.* doi:10.1093/annonc/mds636.

Vickers, A. J. (2006). Shoot first and ask questions later: How to approach statistics like a real clinician. *Medscape Business of Medicine, 7,* 2. Accessed June 19, 2009, at www.medscape.com/viewarticle/540898.

———. (2010). *What is a P-value anyway?* Boston: Addison-Wesley. ISBN=0-321-629930-2.

Vittinghoff, E., Glidden, D. V., Shiboski, S. C., & McCulloch, C. E. (2007). *Regression methods in biostatistics: Linear, logistic, survival, and repeated measures models (statistics for biology and health).* New York: Springer. ISBN=0387202757.

von Hippel, P.T, *Mean, Median, and Skew: Correcting a Textbook Rule.* Journal of Statistics Education, 13(2).

Vos Savant, M. (1997). *The power of logical thinking: Easy lessons in the art of reasoning . . . and hard facts about its absence in our lives.* New York: St. Martin's Griffin. ISBN=0312156278.

Wasserman, L. (2012). P values gone wild and multiscale madness. *Normal Deviate,* August 16. Accessed December 10, 2012, at normaldeviate.wordpress.com/2012/08/16/p-values-gone-wild-and-multiscale-madness/.

Wellek, S. (2002). *Testing statistical hypotheses of equivalence.* Boca Raton, FL: Chapman & Hall/CRC. ISBN=1584881607.

Westfall, P., Tobias, R., Rom, D., Wolfinger, R., & Hochberg,Y. (1999) *Multiple comparisons and multiple tests using the SAS system.* Cary, NC: SAS Publishing. ISBN=1580253970.

Wilcox, R. R. (2001). *Fundamentals of modern statistical methods: Substantially improving power and accuracy.* New York: Springer-Verlag. ISBN=0387951571.

———. (2010). *Fundamentals of modern statistical methods: Substantially improving power and accuracy,* 2nd ed. New York: Springer. ISBN=978-1441955241.

Wilcox, A. J., C. R. Weinberg, & Baird, D. D. (1995). Timing of sexual intercourse in relation to ovulation. Effects on the probability of conception, survival of the pregnancy, and sex of the baby. *New England Journal of Medicine, 333,* 1517–1521.

Winston, W. (2004) Introduction to optimization with the Excel Solver tool. Accessed March 6, 2013, at office.microsoft.com/en-us/excel-help/introduction-to-optimization-with-the-excel-solver-tool-HA001124595.aspx.

Wolff, A. (2002, Jan. 21). That old black magic. *SI.com,* August 27. Accessed May 25, 2008, at sportsillustrated.cnn.com/2003/magazine/08/27/jinx/.

Wormuth, D. W. (1999) Actuarial and Kaplan-Meier survival analysis: There is a difference. *Journal of Thoracic and Cardiovascular Surgery, 118,* 973.

Wright, S. P. (1992). Adjusted P-values for simultaneous inference. *Biometrics, 48,* 1005–1013.

Xu, F., & Garcia, V. (2008). Intuitive statistics by 8-month-old infants. *Proceedings of the National Academy of Sciences of the United States of America, 105,* 5012–5015.

Yalta, A. T. (2008). The accuracy of statistical distributions in Microsoft Excel 2007. *Computational Statistics and Data Analysis, 52,* 4579–4586.

Zhang, J. H., Chung, T. D., & Oldenburg, K. R. (1999). A simple statistical parameter for use in evaluation and validation of high throughput screening assays. *Journal of Biomolecular Screening, 4,* 67–73.

Zhang, J., & Yu, K. F. (1998). What's the relative risk? A method of correcting the odds ratio in cohort studies of common outcomes. *Journal of the American Medical Association, 280,* 1690–1691

Ziliak, S., & McCloskey, D. N. (2008). *The cult of statistical significance: How the standard error costs us jobs, justice, and lives.* Ann Arbor: University of Michigan Press. ISBN=0472050079.

Zollner, S., & Pritchard, J. K. (2007). Overcoming the winner's curse: Estimating penetrance parameters from case–control data. *American Journal of Human Genetics, 80,* 605–615.

INDEX